Quantum Nonlinear Optics

E. Hanamura Y. Kawabe A. Yamanaka

Quantum Nonlinear Optics

With 123 Figures

Springer

Professor Dr. Eiichi Hanamura
Professor Dr. Yutaka Kawabe
Professor Dr. Akio Yamanaka

Chitose Institute of Science and Technology
758-65 Bibi, Chitose-shi
Hokkaido 066-8655, Japan
e-mail: hanamura@photon.chitose.ac.jp
y-kawabe@photon.chitose.ac.jp
a-yamana@photon.chitose.ac.jp

Translation from the original Japanese edition of
Ryoshi Kogaku (Quantum Optics) by Eiichi Hanamura

ISBN 978-3-642-07610-7 e-ISBN 978-3-540-68484-8

Springer is a part of Springer Science+Business Media.

springer.com

Softcover reprint of the hardcover 1st edition 2007

Cover concept: eStudio Calamar Steinen
Cover design: WMX Design GmbH, Heidelberg

Preface

It was more than ten years ago that an original version of this monograph was published with the title *Quantum Optics* in Japanese from Iwanami Shoten in Tokyo. Therefore, making the best use of this chance to translate the book into an English version, we have tried to include the exciting developments of the relevant subjects in these ten years, especially novel nonlinear optical responses of materials. The first example of these nonlinear optical phenomena is laser cooling and subsequent observation of Bose–Einstein and Fermi condensation of neutral atoms. Second, it is now possible to generate femtosecond laser pulses. Then higher-harmonics in the extreme ultraviolet and soft X-ray regions and higher-order Raman scattering can be generated by irradiating these ultrashort laser pulses on atomic and molecular gases and crystals. These multistep signals are applied to the generation of attosecond laser pulses. Third, interference effects of the second harmonics are used to observe the ferroelectric and antiferromagnetic domain structures of crystals with a strongly correlated electronic system.

These novel nonlinear optical phenomena could not be treated without the quantized radiation field. We already have classical textbooks treating, individually, the quantum theory of the radiation field and nonlinear optics. Taking account of these situations, we have described these exciting nonlinear optical responses as well as laser oscillation and supperradiance, based upon the quantum theory of the radiation field. At the same time, we have changed the title of this monograph to *Quantum Nonlinear Optics*.

We start Chap. 1 with standard quantization of the radiation field and then treat several states of the radiation field, such as the coherent state, the quadrature squeezed state, and the photon-number squeezed state. After obtaining the Hamiltonian describing the interaction between the radiation field and electrons in Chap. 2, we discuss the suppression and enhancement of spontaneous emission, and the laser cooling and subsequent condensation of neutral atoms which have been achieved by using these interactions effectively. The statistical characteristics of the radiation field are classified by introducing the correlation function of the radiation field in Chap. 3. Here a degree of coherence is defined and some examples are also discussed. In terms of these statistics, some properties of lasers are characterized in Chap. 4. Here the mechanism of laser oscillator is mathematically formulated and some examples of laser are introduced. Many interesting games are played by using these lasers as demonstrated in Chap. 5 on "Dynamics of Light." In the first half of this chapter we will discuss Q-switching, mode locking and pulse compression, and soliton formation and chirping of the laser pulse depending upon whether we have anomalous or normal dispersion. The experimental and theoretical aspects of superradiance will be discussed in the second half of Chap. 5. Chapters 6 and 7 are devoted to nonlinear optical response. The electron–radiation interaction can usually be treated by perturbation methods in conventional nonlinear optical responses shown in Chap. 6. These examples are second-harmonic generation, sum-frequency generation and parametric amplification and oscillation. The third-order optical response contains colorful phenomena such as coherent anti-Stokes Raman scattering, optical bistability, the Kerr effect and third-harmonic generation. Second-harmonic generation is used to determine the ferroelectric and antiferromagnetic domain structures in crystals with strongly correlated electrons. These subjects will be discussed in Chap. 6. The technological development of ultrashort laser pulses has made it possible to produce novel nonlinear optical responses which should be treated beyond the perturbation method of the electron–radiation interaction. These topics will be treated in Chap. 7. Here, for example, high harmonics beyond the 100th order can be generated from atoms irradiated by femtosecond laser pulses. When a vibrational mode of molecules or crystals is resonantly excited by two incident laser beams, higher-order Raman lines are observed. A series of these spectral lines is mode locked so that sometimes attosecond laser pulses become available as predicted from Fourier transformation of these spectral lines. These will be discussed in Chap. 7.

It took five years to complete this monograph as the CREST research project was running under the sponsorship of the Japan Science and Technology agency (JST). Some of these results are shown in this monograph. We thank Professor Takuo Sugano and other members of JST for warm support of this project, and Miss Saika Kanai, a member of JST project, for helping us to complete this monograph as well as to run our research smoothly for these five years. Finally we are very grateful to Dr. Claus E. Ascheron for his patience and encouragement over these years.

March 2006

Eiichi Hanamura
Yutaka Kawabe
Akio Yamanaka

Contents

1

Quantization of the Radiation Field

Quantum theory was born on December 14th, 1900, when Max Planck proposed the quantum hypothesis. In order to explain the frequency distribution of electromagnetic radiation from a cavity surrounded by walls of temperature T, Planck postulated that the exchange of energy between the walls and the field takes place as the absorption and emission of electromagnetic energy with a discrete quantity $h\nu$ or its multiples. The distribution of radiation energy calculated under this hypothesis reproduced the spectrum observed in a smelting furnace, which is an example of a cavity surrounded by high-temperature walls. In 1905, Einstein extended the ideas of Planck, and explained the photoemission effect by assuming that optical radiation is equivalent to the flux of optical quanta. After the development of quantum theory from electromagnetic wave phenomena, which had been described by Maxwell's equations in classical physics, it was extended by de Broglie's concept of matter wave in 1923 which proposed that particles also have wave properties like the photon. In 1926, Schrödinger introduced the wave equation for a particle wave, and completed the construction of quantum mechanics. The duality of particle and wave properties in an electromagnetic field and matter was understood as complementary descriptions of particle and wave features mediated by Heisenberg's uncertainty principle.

In this chapter, the quantization of the radiation field will be formulated first, and then Heisenberg's uncertainty relation will be derived, respectively, in Sects. 1.1 and 1.2 [1–7]. Next, the generation method and physical properties of coherent states will be discussed in Sect. 1.3. A coherent state is one of the minimum uncertainty states where the fluctuations of two components, in quadrature, of the electric field have equivalent magnitude. In addition to the standard topics in quantum optics, we introduce recent progress regarding nonclassical light in the last section. In principle, the coherent state is not a unique solution satisfying the minimum uncertainty conditions. It is possible to reduce the noise of one component without limit, while sacrificing the uncertainty of the other quadrature component. Likewise, it is also possible to reduce the fluctuation of the photon number in a mode with the sacrifice of the phase fluctuation, because the photon number and phase are conjugated observables. These two states are called the quadrature-phase squeezed state and photon-number squeezed state. The details of these states will be discussed in Sect. 1.4. The uncertainty principle plays an important role also in measurement processes as well as generation processes. The product of the uncertainty of a measured value and its reaction to the conjugated observable must be larger than a certain constant. Therefore, if we can make a system in which the reaction of the measurement does not affect the measured value, it is possible to enhance the accuracy of the measurement to any degree. This is realized as the quantum nondemolition measurement of light. The squeezed states and the quantum nondemolition measurement are expected to be utilized in the detection of gravitational waves and also to enhance the performance of optical communication systems.

1.1 Maxwell's Equations and Hamiltonian Formalism

Maxwell's equations for the electromagnetic fields $\boldsymbol{E}$ and $\boldsymbol{H}$ in free space take the form

$$\mathrm{rot}\boldsymbol{E} + \frac{\partial}{\partial t}\boldsymbol{B} = 0, \tag{1.1}$$

$$\mathrm{rot}\boldsymbol{H} - \frac{\partial}{\partial t}\boldsymbol{D} = 0, \tag{1.2}$$

$$\mathrm{div}\boldsymbol{B} = 0, \tag{1.3}$$

$$\mathrm{div}\boldsymbol{D} = 0. \tag{1.4}$$

The electric displacement $\boldsymbol{D}$ and magnetic flux density $\boldsymbol{B}$ are expressed in terms of the electric permittivity ϵ_0 and magnetic permeability μ_0 of free space as

$$\boldsymbol{D} = \epsilon_0\boldsymbol{E}, \qquad \boldsymbol{B} = \mu_0\boldsymbol{H}. \tag{1.5}$$

Because $\boldsymbol{B}$ is derived from a vector potential $\boldsymbol{A}$ according to

$$\boldsymbol{B} = \mathrm{rot}\boldsymbol{A}, \tag{1.6}$$

equation (1.3) is automatically satisfied. Reducing the magnetic flux density $\boldsymbol{B}$ by substitution of (1.6) into (1.1), we can obtain a relation as follows:

$$\mathrm{rot}\left[\boldsymbol{E} + \frac{\partial}{\partial t}\boldsymbol{A}\right] = 0.$$

Therefore, we can define a scalar potential ϕ as

$$\boldsymbol{E} + \frac{\partial}{\partial t}\boldsymbol{A} = -\mathrm{grad}\phi, \tag{1.7}$$

because the relation rot(gradϕ) = 0 is always satisfied. It follows from (1.6) and (1.7) that $\boldsymbol{E}$ and $\boldsymbol{B}$ are invariant under the following gauge transformation:

$$\boldsymbol{A}(\boldsymbol{r}t) \rightarrow \boldsymbol{A}'(\boldsymbol{r}t) = \boldsymbol{A}(\boldsymbol{r}t) + \nabla F(\boldsymbol{r}t),$$
$$\phi(\boldsymbol{r}t) \rightarrow \phi'(\boldsymbol{r}t) = \phi(\boldsymbol{r}t) - \frac{\partial}{\partial t} F(\boldsymbol{r}t),$$

where $F(\boldsymbol{r}t)$ is an arbitrary function of space $\boldsymbol{r}$ and time t.

In order to eliminate the arbitrariness of the vector and scalar potentials, we should choose a gauge condition. In this book, the Coulomb gauge is employed which is defined by

$$\mathrm{div}\boldsymbol{A} = 0, \tag{1.8}$$

and grad$\phi = 0$ is chosen for simplicity. Under the Coulomb gauge, the electric field can be given by the vector potential as

$$\boldsymbol{E} = -\frac{\partial}{\partial t}\boldsymbol{A}. \tag{1.9}$$

Substituting (1.5), (1.6) and (1.9) into (1.2), the equation for the vector potential $\boldsymbol{A}$ can be derived as

$$\mathrm{rot}(\mathrm{rot}\boldsymbol{A}) + \epsilon_0\mu_0\frac{\partial^2}{\partial t^2}\boldsymbol{A} = 0. \tag{1.10}$$

By using the formula rot(rot) = grad(div)$-\Delta$ and the relation $c^2 = 1/\epsilon_0\mu_0$, (1.10) is transformed into the wave equation

$$\Delta\boldsymbol{A} - \frac{1}{c^2}\frac{\partial^2}{\partial t^2}\boldsymbol{A} = 0. \tag{1.11}$$

Equation (1.4) is satisfied automatically because of the relation div$\boldsymbol{A} = 0$ for the Coulomb gauge.

Considering a cubic volume with dimension $L(0 \leq x, y, z \leq L)$ and imposing a periodic boundary condition for simplicity, a solution of (1.11) is given as:

$$\boldsymbol{A}(\boldsymbol{r}t) = \boldsymbol{A}_0 e^{i(\boldsymbol{k}\cdot\boldsymbol{r}-\omega t)}, \tag{1.12}$$
$$\boldsymbol{k} \equiv (k_x, k_y, k_z) = \frac{2\pi}{L}(n_x, n_y, n_z), \tag{1.13}$$

where n_x, n_y, n_z are arbitrary integers. The angular frequency ω is related to the wavenumber vector $\boldsymbol{k}$ as $\omega = c|\boldsymbol{k}| = ck$. Because the relation $\boldsymbol{A}_0 \cdot \boldsymbol{k} = 0$ can be obtained from the Coulomb gauge condition (1.8), the vector potential $\boldsymbol{A}$ is found to be a transverse wave. By defining the unit vector parallel to polarization direction as $\boldsymbol{A} = A_0\boldsymbol{e}$, the electric and magnetic fields $\boldsymbol{E}$ and $\boldsymbol{B}$ can be expressed as

$$\boldsymbol{E}(\boldsymbol{r}t) = i\omega \boldsymbol{e} A_0 \exp[i(\boldsymbol{k}\cdot\boldsymbol{r} - \omega t)], \tag{1.14}$$
$$\boldsymbol{B}(\boldsymbol{r}t) = i[\boldsymbol{k}\times\boldsymbol{e}] A_0 \exp[i(\boldsymbol{k}\cdot\boldsymbol{r} - \omega t)], \tag{1.15}$$

by using (1.6) and (1.9).

In the general case, the vector potential $\boldsymbol{A}$ can be given by a superposition of single-mode solution (1.12) as

$$\boldsymbol{A}(\boldsymbol{r}t) = \frac{1}{\sqrt{V}} \sum_{\boldsymbol{k}} \sum_{\gamma=1}^{2} \boldsymbol{e}_{\boldsymbol{k}\gamma} \{ q_{\boldsymbol{k}\gamma}(t) e^{i\boldsymbol{k}\cdot\boldsymbol{r}} + q^*_{\boldsymbol{k}\gamma}(t) e^{-i\boldsymbol{k}\cdot\boldsymbol{r}} \}, \tag{1.16}$$

where γ represents the two independent polarization directions. Complex conjugate terms are added in order to make the observables $\boldsymbol{E}$ and $\boldsymbol{B}$ real numbers. The time dependence is included in the amplitude $q_{\boldsymbol{k}\gamma}(t)$. From now on, the letter λ will be used as the parameter of the electromagnetic mode instead of $\boldsymbol{k}\gamma$. The energy density of the electromagnetic field is given by the following equation:

$$U(\boldsymbol{r}t) = \frac{1}{2}(\boldsymbol{D}\cdot\boldsymbol{E} + \boldsymbol{B}\cdot\boldsymbol{H}). \tag{1.17}$$

Therefore, the energy of the electromagnetic field in a volume of $V = L^3$ is denoted as

$$\begin{aligned} \mathcal{H} &= \int U(\boldsymbol{r}t) d^3\boldsymbol{r} \\ &= \epsilon_0 \sum_{\lambda} \omega_\lambda^2 (q^*_\lambda q_\lambda + q_\lambda q^*_\lambda). \end{aligned} \tag{1.18}$$

Using real variables

$$Q_\lambda(t) = q_\lambda(t) + q^*_\lambda(t), \tag{1.19}$$
$$\dot{Q}_\lambda(t) = -i\omega_\lambda(q_\lambda - q^*_\lambda), \tag{1.20}$$

instead of the Fourier expansion coefficients q_λ, q^*_λ, we can transform (1.18) into the more familiar canonical form:

$$\mathcal{H} = \frac{\epsilon_0}{2} \sum_{\lambda} (\dot{Q}_\lambda^2 + \omega_\lambda^2 Q_\lambda^2). \tag{1.21}$$

When we introduce the generalized momentum $P_\lambda = \epsilon_0 \dot{Q}_\lambda$ which is conjugate to the coordinate Q_λ, we can replace the momentum P_λ with the differential operator with respect to the canonical conjugate coordinate Q_λ:

$$P_\lambda \to -i\hbar \frac{\partial}{\partial Q_\lambda}. \tag{1.22}$$

Then we find that the Hamiltonian has the following form

$$\mathcal{H} = \sum_{\lambda} \left(-\frac{\hbar^2}{2\epsilon_0} \frac{\partial^2}{\partial Q_\lambda^2} + \frac{\epsilon_0}{2} \omega_\lambda^2 Q_\lambda^2 \right). \tag{1.23}$$

Because this form is the same as the Hamiltonian of an ensemble of harmonic oscillators, the eigenenergy can be given by

$$E_{\{n_\lambda\}} = \sum_{\lambda} \hbar\omega_\lambda \left(n_\lambda + \frac{1}{2} \right), \tag{1.24}$$

where $n_\lambda = 0, 1, 2, \ldots$, and $\{n_\lambda\}$ represents the photon number distribution over all modes. Now, an annihilation operator $\hat{a}_\lambda$ and a creation operator $\hat{a}_\lambda^\dagger$ of the photon in a mode $\lambda \equiv (\boldsymbol{k}, \gamma)$ can be introduced by:

$$\hat{a}_\lambda = \sqrt{\frac{\epsilon_0\omega_\lambda}{2\hbar}} \left(Q_\lambda + \frac{i}{\epsilon_0\omega_\lambda} P_\lambda \right), \tag{1.25}$$

$$\hat{a}_\lambda^\dagger = \sqrt{\frac{\epsilon_0\omega_\lambda}{2\hbar}} \left(Q_\lambda - \frac{i}{\epsilon_0\omega_\lambda} P_\lambda \right). \tag{1.26}$$

As is well known from the theory of the harmonic oscillator, application of these operators on an eigenstate $|n_1, n_2, n_3, \ldots, n_\lambda, \ldots\rangle$ of the Hamiltonian (1.23) transforms it into other states as shown below:

$$\hat{a}_\lambda|\cdots, n_\lambda, \ldots\rangle = \sqrt{n_\lambda}|\cdots, n_\lambda - 1, \ldots\rangle,$$
$$\hat{a}_\lambda^\dagger|\cdots, n_\lambda, \ldots\rangle = \sqrt{n_\lambda + 1}|\cdots, n_\lambda + 1, \ldots\rangle.$$

The Hamiltonian can be expressed with creation and annihilation operators $\hat{a}_\lambda^\dagger$ and $\hat{a}_\lambda$:

$$\mathcal{H} = \sum_{\lambda} \hbar\omega_\lambda \left(\hat{a}_\lambda^\dagger \hat{a}_\lambda + \frac{1}{2} \right). \tag{1.27}$$

1.2 Quantization of the Radiation Field and Heisenberg's Uncertainty Principle

As shown in the previous section, the most essential aspect of the field quantization is to render the canonical variables Q_λ and P_λ into noncommutative operators satisfying the following relationship:

$$[P_\lambda, Q_\lambda] \equiv P_\lambda Q_\lambda - Q_\lambda P_\lambda = -i\hbar. \tag{1.28}$$

This relation can be confirmed by replacing P_λ by the differential operator indicated by (1.22). The canonical variables of different modes are commutative, i.e.,

$$[P_\lambda, Q_\mu] = 0 \qquad (\lambda \neq \mu),$$

and the commutator among the same types of operators also vanishes:

$$[P_\lambda, P_\mu] = [Q_\lambda, Q_\mu] = 0.$$

The commutation relation between the creation and annihilation operators is derived from the substitution of (1.25) and (1.26) into (1.28):

$$[\hat{a}_\lambda, \hat{a}_\lambda^\dagger] = \delta_{\lambda\mu}. \tag{1.29}$$

Products of the uncertainties of noncommutative quantities must be larger than a certain value; that is, they must satisfy Heisenberg's uncertainty relation. For example, the position and momentum of a particle cannot be determined simultaneously to arbitrary precision. In this case, the product of the uncertainties of the position Δq and that of the momentum Δp cannot be smaller than $\hbar/2$. The uncertainty principle reflects the probabilistic feature of the wavefunction in quantum mechanics. If an ensemble of identical particles in an identical state is separated into two groups, and their positions are measured for the first group and the momenta for the other group, the dispersion of measured quantities Δq, and Δp must obey the uncertainty relationship ($\Delta p \cdot \Delta q \geq \hbar/2$). This uncertainty relation can be generalized to the case of any physical quantities which satisfy the following commutation relation:

$$[A, B] = AB - BA = iC. \tag{1.30}$$

In this case, the uncertainties of ΔA, and ΔB will satisfy the following inequality:

$$\Delta A \cdot \Delta B \geq \frac{|\langle C \rangle|}{2}. \tag{1.31}$$

Let us prove inequality (1.31) by using Schwartz's inequality in the following way. The square of the uncertainty and expectation value are defined, respectively, by the following expressions:

$$(\Delta A)^2 = \langle \psi, (A - \langle A \rangle)^2 \psi \rangle, \tag{1.32}$$

$$\langle C \rangle = \langle \psi, C\psi \rangle. \tag{1.33}$$

Here $\langle A \rangle$ means the expectation value of the observable A for the normalized wavefunction ψ. Then, we can show that

$$\begin{aligned}(\Delta A)^2 \cdot (\Delta B)^2 &\equiv \langle \psi, (A - \langle A \rangle)^2 \psi \rangle \langle \psi, (B - \langle B \rangle)^2 \psi \rangle \\ &\geq |\langle (A - \langle A \rangle)\psi, (B - \langle B \rangle)\psi \rangle|^2 \\ &= |\langle \psi, (A - \langle A \rangle)(B - \langle B \rangle)\psi \rangle|^2. \end{aligned} \tag{1.34}$$

The product of the operators $\Delta\tilde{A} \equiv A - \langle A \rangle$ and $\Delta\tilde{B} \equiv B - \langle B \rangle$ can be rewritten as

$$\Delta\tilde{A} \cdot \Delta\tilde{B} = \frac{1}{2}[\Delta\tilde{A}, \Delta\tilde{B}] + \frac{1}{2}\{\Delta\tilde{A}, \Delta\tilde{B}\}, \tag{1.35}$$

where $\{a, b\}$ means the anticommutation relation defined as $ab + ba$. By using the relation $[\Delta\tilde{A}, \Delta\tilde{B}] = [A, B] = iC$, the following inequality can be obtained:

$$\begin{aligned}|\langle\psi, \Delta\tilde{A} \cdot \Delta\tilde{B}\psi\rangle|^2 &= \frac{1}{4}|\langle\psi, [\Delta\tilde{A}, \Delta\tilde{B}]\psi\rangle + \langle\psi, \{\Delta\tilde{A}, \Delta\tilde{B}\}\psi\rangle|^2 \\ &\geq \frac{1}{4}|\langle\psi, [\Delta\tilde{A}, \Delta\tilde{B}]\psi\rangle|^2 = \frac{1}{4}|\langle C\rangle|^2. \end{aligned} \tag{1.36}$$

The last inequality is derived from the facts that (1) $i[\Delta\tilde{A}, \Delta\tilde{B}]$ and $\{\Delta\tilde{A}, \Delta\tilde{B}\}$ are both Hermitian operators, and (2) the first term in the absolute value in the upper formula is purely imaginary and the second term is real. Finally, the relation (1.31) can be proven by combining (1.34) and (1.36).

Fourier expansion of the vector potential (1.16) gives the coefficients $q_\lambda(t)$ and $q_\lambda^*(t)$, and these are rewritten by the creation and annihilation operators using (1.19), (1.20), (1.25) and (1.26):

$$q_\lambda(t) = \sqrt{\frac{\hbar}{2\epsilon_0\omega_\lambda}}\hat{a}_\lambda, \qquad q_\lambda^*(t) = \sqrt{\frac{\hbar}{2\epsilon_0\omega_\lambda}}\hat{a}_\lambda^\dagger. \tag{1.37}$$

The electric field $\boldsymbol{E}$ can be transformed into an expansion form by its definition (1.9) and the substitution of (1.37) into (1.16):

$$\boldsymbol{E} = i\sqrt{\frac{1}{2V}}\sum_\lambda e_\lambda\sqrt{\frac{\hbar\omega_\lambda}{\epsilon_0}}\{\hat{a}_\lambda e^{i\boldsymbol{k}\cdot\boldsymbol{r}} - \hat{a}_\lambda^\dagger e^{-i\boldsymbol{k}\cdot\boldsymbol{r}}\}. \tag{1.38}$$

The electric field $\boldsymbol{E}$ and the photon number operator n_λ are noncommutative:

$$[\hat{n}_\lambda, \boldsymbol{E}] \neq 0. \tag{1.39}$$

Therefore, when n_λ has a certain value, that is, the radiation field has a constant energy, the electric field cannot have a certain value but greatly fluctuates around an averaged value.

The eigenstate of the Hamiltonian (1.23) or (1.27), with the energy given by (1.24), can be expressed as

$$\Psi_{\{n\}} \equiv |n_1, n_2, \ldots, n_\lambda, \ldots\rangle = \prod_\lambda \frac{(\hat{a}_\lambda^\dagger)^{n_\lambda}}{\sqrt{n_\lambda!}}|0\rangle, \tag{1.40}$$

where $|0\rangle$ indicates the vacuum state of the photon.

The electric field $\boldsymbol{E}$ can be expressed with the photon-number operator $\hat{n}_\lambda$ and the phase operator $\hat{\phi}_\lambda$ by using the following relations:

$$\hat{a}_\lambda = e^{i\hat{\phi}_\lambda}\sqrt{\hat{n}_\lambda}, \qquad \hat{a}_\lambda^\dagger = \sqrt{\hat{n}_\lambda}e^{-i\hat{\phi}_\lambda}. \tag{1.41}$$

From the commutation relation of the creation and annihilation operators given by (1.29), the relation

$$e^{i\hat{\phi}_\lambda}\hat{n}_\lambda - \hat{n}_\lambda e^{i\hat{\phi}_\lambda} = e^{i\hat{\phi}_\lambda} \tag{1.42}$$

can be obtained if $\hat{\phi}_\lambda$ and $\hat{n}_\lambda$ satisfy the commutation relation

$$\hat{\phi}_\lambda \hat{n}_\lambda - \hat{n}_\lambda \hat{\phi}_\lambda = -i. \tag{1.43}$$

Applying (1.43) to the Heisenberg's uncertainty relationship (1.31), the product of the fluctuation of the photon number and that of the phase can be given by

$$(\Delta n_\lambda) \cdot (\Delta \phi_\lambda) \geq \frac{1}{2}. \tag{1.44}$$

If the photon number of a mode λ is known, it is impossible to know the accurate phase of the mode. On the other hand, the photon number is unknown when the phase can be determined. When there are two waves and only the phase difference of them is known, it is possible to determine the total photon number, but there is no method to determine to which waves a photon belongs.

1.3 Coherent State of Light

The coherent state is the quantum state of light most proximate to the classical radiation field. For the coherent state, the product of the fluctuations of two noncommutative physical quantities takes the minimum value. The electric field of the coherent state can be approximately expressed by a well-defined amplitude and phase as with classical waves. The coherent state is an eigenstate of the non-Hermitian annihilation operator $\hat{a}_\lambda$, and it can be expressed as a superposition of eigenstates $|\{n_\lambda\}\rangle$ of the radiation field. In this section, the generation and physical characteristics of the state will be discussed.

Because an eigenstate of the total radiation field can be expressed as the direct product of eigenstates of each mode, one can start the discussion from the formulation of the coherent state $|\alpha\rangle$ for a single mode λ. Therefore, the subscript λ for the mode will be eliminated from now on in this chapter. The coherent state is defined as an eigenstate of the annihilation operator $\hat{a}$ as

$$\hat{a}|\alpha\rangle = \alpha|\alpha\rangle. \tag{1.45}$$

In order to construct the coherent state, it is convenient to use a normalized single-mode photon-number state $|n\rangle$ given by

$$|n\rangle = \frac{1}{\sqrt{n!}}(\hat{a}^\dagger)^n|0\rangle. \tag{1.46}$$

Here, $|0\rangle$ means the vacuum state of the photon mode λ. This definition can be naturally understood from the multimode formula (1.40). The coherent state can be given as a linear combination of the number states:

$$|\alpha\rangle = \exp\left(-\frac{1}{2}|\alpha|^2\right)\sum_n \frac{\alpha^n}{\sqrt{n!}}|n\rangle. \tag{1.47}$$

The normalized expression of the coherent state (1.47) can be derived from the definition of the state (1.45). The Hermitian conjugate of the relation $\hat{a}^\dagger|n\rangle = \sqrt{n+1}|n+1\rangle$ can be expressed as

$$\langle n|\hat{a} = \sqrt{n+1}\langle n+1|.$$

By applying $|\alpha\rangle$ on the right on both sides, and using the relation $\langle n|\hat{a}|\alpha\rangle = \alpha\langle n|\alpha\rangle$ – easily obtained from (1.45) – the following equation can be derived:

$$\sqrt{n+1}\langle n+1|\alpha\rangle = \alpha\langle n|\alpha\rangle. \tag{1.48}$$

This equation and the following chain of equations:

$$\begin{aligned} &\sqrt{n}\langle n|\alpha\rangle = \alpha\langle n-1|\alpha\rangle, \qquad \sqrt{n-1}\langle n-1|\alpha\rangle = \alpha\langle n-2|\alpha\rangle, \\ &\sqrt{n-2}\langle n-2|\alpha\rangle = \alpha\langle n-3|\alpha\rangle, \quad \cdots, \quad \langle 1|\alpha\rangle = \alpha\langle 0|\alpha\rangle \end{aligned}$$

finally give the relation:

$$\langle n|\alpha\rangle = \frac{\alpha^n}{\sqrt{n!}}\langle 0|\alpha\rangle. \tag{1.49}$$

As the photon-number states constitute a complete system as

$$\sum_{n=0}^{\infty}|n\rangle\langle n| = 1, \tag{1.50}$$

the coherent state $|\alpha\rangle$ can be expanded with these number states $|n\rangle$ as

$$|\alpha\rangle = \sum_{n=0}^{\infty}|n\rangle\langle n|\alpha\rangle = \langle 0|\alpha\rangle\sum_{n=0}^{\infty}\frac{\alpha^n}{\sqrt{n!}}|n\rangle. \tag{1.51}$$

The coefficient $\langle 0|\alpha\rangle$ can be determined from the normalization condition

$$\langle\alpha|\alpha\rangle = |\langle 0|\alpha\rangle|^2\sum_{n=0}^{\infty}\frac{|\alpha|^{2n}}{n!} = |\langle 0|\alpha\rangle|^2\exp(|\alpha|^2) = 1.$$

Then, the expression for the coherent state (1.47) can be obtained from this relation and (1.51).

The parameter α in the coherent state is an arbitrary complex number. The averaged photon number is given by $|\alpha|^2$, and the expectation value of the photon number n obeys a Poisson distribution as

$$|\langle n|\alpha\rangle|^2 = \frac{(|\alpha|^2)^n}{n!}\exp(-|\alpha|^2). \tag{1.52}$$

The coherent state with $\alpha = 0$ is identical to the photon-number state with $n = 0$ (that is, $|0\rangle$).

The coherent state can also be obtained from the vacuum state by a unitary transformation $D(\alpha)$ as

$$|\alpha\rangle = D(\alpha)|0\rangle = \exp(\alpha\hat{a}^\dagger - \alpha^*\hat{a})|0\rangle. \tag{1.53}$$

This is an important representation in order to study the generation of a coherent state described in the last half of this section. Before the derivation of (1.53), let us prove that the unitary operator $D(\alpha)$ is equivalent to the displacement operator $D(\beta)$ as follows:

$$D^{-1}(\beta)\hat{a}D(\beta) = \hat{a} + \beta, \tag{1.54}$$
$$D^{-1}(\beta)\hat{a}^\dagger D(\beta) = \hat{a}^\dagger + \beta^*. \tag{1.55}$$

After operating with $D^{-1}(\beta)$ on right on both sides of (1.54), apply it to the state $|\alpha\rangle$. Then we get the following relation:

$$D^{-1}(\beta)\hat{a}|\alpha\rangle = \alpha D^{-1}(\beta)|\alpha\rangle = (\hat{a} + \beta)D^{-1}(\beta)|\alpha\rangle. \tag{1.56}$$

Hence, if $\alpha = \beta$,

$$\hat{a}D^{-1}(\alpha)|\alpha\rangle = 0. \tag{1.57}$$

Because this equation means that

$$D^{-1}(\alpha)|\alpha\rangle = |0\rangle, \tag{1.58}$$

one can obtain

$$|\alpha\rangle = D(\alpha)|0\rangle. \tag{1.59}$$

From the preceding discussion, it is known that the coherent state can be generated by applying the displacement operator $D(\alpha)$ to the vacuum state of the photon. Next, let us determine the form of $D(\alpha)$ in the following discussion. If $\alpha = 0$, we obtain

$$D(0) = 1 \tag{1.60}$$

from (1.59). Operating with $D(\beta)$ on the left of both (1.54) and (1.55), and replacing β with the infinitesimal quantity $d\alpha$ give the following:

$$\hat{a}D(d\alpha) = D(d\alpha)(\hat{a} + d\alpha), \tag{1.61}$$
$$\hat{a}^\dagger D(d\alpha) = D(d\alpha)\left(\hat{a}^\dagger + (d\alpha)^*\right). \tag{1.62}$$

The solution of these simultaneous equations can be expressed by a linear combination of $\hat{a}$ and $\hat{a}^\dagger$ as

$$D(d\alpha) = 1 + A\hat{a}^\dagger + B\hat{a}, \tag{1.63}$$

because of the boundary condition $D(0) = 1$ and the fact that $d\alpha$ and $d\alpha^*$ are infinitesimal. Next, $\hat{a}$, and $\hat{a}^\dagger$ are applied on the left of (1.62) and (1.61), respectively, and then subtraction of each side leads to

$$\begin{aligned} \hat{a} \times (1.62) - \hat{a}^\dagger \times (1.61) &= (\hat{a}\hat{a}^\dagger - \hat{a}^\dagger\hat{a})D(d\alpha) = D(d\alpha) \\ &= \hat{a}D(d\alpha)\hat{a}^\dagger + \hat{a}D(d\alpha)(d\alpha)^* - \hat{a}^\dagger D(d\alpha)\hat{a} - \hat{a}^\dagger D(d\alpha)d\alpha. \end{aligned} \tag{1.64}$$

Comparison of this equation with (1.63) gives $A = d\alpha$ and $B = -(d\alpha)^*$, and the final expression is

$$D(d\alpha) = 1 + \hat{a}^\dagger d\alpha - \hat{a}(d\alpha)^*. \tag{1.65}$$

Now, we introduce a real parameter λ as $d\alpha = \alpha d\lambda$ and assume the relation

$$D[\alpha(\lambda + d\lambda)] = D(\alpha d\lambda)D(\alpha\lambda). \tag{1.66}$$

Then the differential equation of the operators is given as

$$\frac{d}{d\lambda}D(\alpha\lambda) = (\alpha\hat{a}^\dagger - \alpha^*\hat{a})D(\alpha\lambda). \tag{1.67}$$

Integration of the equation finally gives (1.53) by taking $\lambda = 1$.

Two kinds of expressions $|\alpha\rangle$ were given in the preceding discussion: one was given by the displacement operator (1.53), the other given by a linear combination of photon-number states $|n\rangle$ in (1.47). The two expressions can be proven to be equivalent. In general, if the operators A and B satisfy the relations $[[A, B], A] = [[A, B], B] = 0$, then it can be shown that

$$\exp(A + B) = \exp(A)\exp(B)\exp\left\{-\frac{1}{2}[A, B]\right\}. \tag{1.68}$$

In our case, substituting $A = \alpha\hat{a}^\dagger$, and $B = -\alpha^*\hat{a}$ into the formula, $D(\alpha)$ is given by

$$D(\alpha) = \exp(\alpha\hat{a}^\dagger - \alpha^*\hat{a}) = \exp\left(-\frac{1}{2}|\alpha|^2\right)\exp(\alpha\hat{a}^\dagger)\exp(-\alpha^*\hat{a}). \tag{1.69}$$

Application of $\exp(-\alpha^*\hat{a})$ to the photon vacuum state $|0\rangle$ does not change the state, as can be seen from the following expansion:

$$\exp(-\alpha^*\hat{a})|0\rangle = \left\{1 - \alpha^*\hat{a} + \frac{(\alpha^*\hat{a})^2}{2!} + \cdots\right\}|0\rangle = |0\rangle. \tag{1.70}$$

Hence, application of the displacement operator on $|0\rangle$ leads to

$$\begin{aligned} D(\alpha)|0\rangle &= \exp\left(-\frac{1}{2}|\alpha|^2\right)\exp(\alpha\hat{a}^\dagger)|0\rangle = \exp\left(-\frac{1}{2}|\alpha|^2\right)\sum_{n=0}^{\infty}\frac{1}{n!}(\alpha\hat{a}^\dagger)^n|0\rangle \\ &= \exp\left(-\frac{1}{2}|\alpha|^2\right)\sum_{n=0}^{\infty}\frac{1}{\sqrt{n!}}\alpha^n|n\rangle, \end{aligned} \tag{1.71}$$

which corresponds to the expression (1.47).

The coherent state $|\alpha\rangle$ is proven to be a minimum uncertainty state which is the most proximate to a classical electromagnetic wave. From (1.25) and (1.26), the generalized canonical coordinate Q and its conjugate momentum P are expressed by creation and annihilation operators $\hat{a}^\dagger$ and $\hat{a}$ as

$$Q = \sqrt{\frac{\hbar}{2\epsilon_0\omega}}(\hat{a} + \hat{a}^\dagger), \qquad P = i\sqrt{\frac{\epsilon_0\hbar\omega}{2}}(\hat{a}^\dagger - \hat{a}). \tag{1.72}$$

The expectation values of the position and momentum in the coherent state $|\alpha\rangle$ are given as

$$\begin{aligned}\langle\alpha|Q|\alpha\rangle &= \sqrt{\frac{\hbar}{2\epsilon_0\omega}}(\alpha + \alpha^*),\\ \langle\alpha|P|\alpha\rangle &= i\sqrt{\frac{\epsilon_0\hbar\omega}{2}}(\alpha^* - \alpha).\end{aligned} \tag{1.73}$$

Also the expectation values of P^2 and Q^2 can be calculated to be

$$\begin{aligned}\langle\alpha|Q^2|\alpha\rangle &= \frac{\hbar}{2\epsilon_0\omega}\{(\alpha + \alpha^*)^2 + 1\},\\ \langle\alpha|P^2|\alpha\rangle &= \frac{\epsilon_0\hbar\omega}{2}\{1 - (\alpha^* - \alpha)^2\}.\end{aligned} \tag{1.74}$$

Therefore, the variances ΔP and ΔQ for electromagnetic field are given by:

$$\begin{aligned}(\Delta Q)^2 &\equiv \langle\alpha|Q^2|\alpha\rangle - (\langle\alpha|Q|\alpha\rangle)^2 = \frac{\hbar}{2\epsilon_0\omega},\\ (\Delta P)^2 &\equiv \langle\alpha|P^2|\alpha\rangle - (\langle\alpha|Q|\alpha\rangle)^2 = \frac{\epsilon_0\hbar\omega}{2}.\end{aligned} \tag{1.75}$$

This result shows that the minimum uncertainty relation $\Delta Q \cdot \Delta P = \hbar/2$ is satisfied in the coherent state.

The coherent states constitute a complete system in a given mode, but states with different α value are not orthogonal. These mathematical characteristics are discussed in the following paragraphs.

(1) Closure

If the complex number α is expressed with amplitude $|\alpha| \equiv r$ and phase ϕ as $\alpha = re^{i\phi}$, then

$$\begin{aligned}\int d^2\alpha|\alpha\rangle\langle\alpha| &= \int_0^\infty rdr \int_0^{2\pi} d\phi e^{-r^2} \sum_m \sum_n \frac{r^m}{\sqrt{m!}} e^{im\phi} \frac{r^n}{\sqrt{n!}} e^{-in\phi} |m\rangle\langle n|\\ &= 2\pi \sum_n \frac{1}{n!} \int_0^\infty rdr r^{2n} e^{-r^2} |n\rangle\langle n| = \pi.\end{aligned} \tag{1.76}$$

Here, we take into account the completeness of photon number states $|n\rangle$, which is given as the relation $\sum_n |n\rangle\langle n| = 1$, and the following orthogonality relation:

$$\int_0^{2\pi} d\phi e^{i(m-n)\phi} = 2\pi\delta_{mn}. \tag{1.77}$$

From (1.76), the following closure property is proven:

$$\frac{1}{\pi}\int d^2\alpha|\alpha\rangle\langle\alpha| = 1. \tag{1.78}$$

(2) Nonorthogonality

The overlap of two coherent states $|\alpha\rangle$ and $|\beta\rangle$ can be calculated as

$$\begin{aligned}\langle\beta|\alpha\rangle &= \sum_m\sum_n \frac{(\beta^*)^m}{\sqrt{m!}}\frac{\alpha^n}{\sqrt{n!}}\exp\left\{-\frac{1}{2}(|\alpha|^2+|\beta|^2)\right\}\langle m|n\rangle \\ &= \sum_n \frac{(\alpha\beta^*)^n}{n!}\exp\left\{-\frac{1}{2}(|\alpha|^2+|\beta|^2)\right\} \\ &= \exp\left\{-\frac{1}{2}(|\alpha|^2+|\beta|^2)+\beta^*\alpha\right\}.\end{aligned} \tag{1.79}$$

Finally, we obtain $|\langle\beta|\alpha\rangle|^2 = \exp\{-|\alpha-\beta|^2\}$, the result of which shows that two states are approximately orthogonal only when the distance between two complex numbers α and β is large enough.

(3) Over-closure

From these two features, it is known that a coherent state $|\alpha\rangle$ can be expressed by a superposition of other coherent states as

$$\begin{aligned}|\alpha\rangle &= \frac{1}{\pi}\int d^2\beta|\beta\rangle\langle\beta|\alpha\rangle \\ &= \frac{1}{\pi}\int d^2\beta|\beta\rangle\exp\left\{-\frac{1}{2}(|\alpha|^2+|\beta|^2)+\beta^*\alpha\right\}.\end{aligned} \tag{1.80}$$

This is an expression for over-closure.

Lasers, operating under population inversion much greater than that at the threshold condition, generate light in the coherent state. As shown in Chap. 5, when a laser oscillates, the phases of the transition dipole moments of all contributing atoms (molecules) are synchronized. Such electronic motion behaves as a classical current. It can be shown that the radiation from a classical current gives coherent light in the following. The coherent state of a multimode can be given as the product of multiple coherent modes:

$$|\{\alpha_\lambda\}\rangle = \prod_\lambda |\alpha_\lambda\rangle. \tag{1.81}$$

The vector potential $\boldsymbol{A}(\boldsymbol{r}t)$ can be expanded into a linear combination of creation $\hat{a}_\lambda^\dagger$ and annihilation operators $\hat{a}_\lambda$ for all modes as

$$\boldsymbol{A}(\boldsymbol{r}t) = \frac{1}{\sqrt{V}} \sum_\lambda \sqrt{\frac{\hbar}{2\epsilon_0 \omega_\lambda}} \boldsymbol{e}_\lambda [\hat{a}_\lambda e^{i\boldsymbol{k}\cdot\boldsymbol{r}-i\omega t} + \hat{a}_\lambda^\dagger e^{-i\boldsymbol{k}\cdot\boldsymbol{r}+i\omega t}]. \tag{1.82}$$

The interaction Hamiltonian made by a classical current density $\boldsymbol{J}(\boldsymbol{r}t)$ by atoms (molecules) and an electromagnetic field $\boldsymbol{A}(\boldsymbol{r}t)$ in a laser system is written as

$$\mathcal{V}(t) = -\int \boldsymbol{J}(\boldsymbol{r}t) \cdot \boldsymbol{A}(\boldsymbol{r}t) d^3\boldsymbol{r}. \tag{1.83}$$

The details of interaction Hamiltonian will be derived in Chap. 2. When the Hamiltonian of an atomic system, an electromagnetic field, and the interaction between them are denoted by $\mathcal{H}_{\rm atom}$, $\mathcal{H}_{\rm rad}$, and $\mathcal{V}$, respectively, (1.83) can be given in the interaction representation as

$$\mathcal{V}(t) = \exp\left(\frac{i\mathcal{H}_0 t}{\hbar}\right) \mathcal{V} \exp\left(\frac{-i\mathcal{H}_0 t}{\hbar}\right),$$

where $\mathcal{H}_0 = \mathcal{H}_{\rm atom} + \mathcal{H}_{\rm rad}$. The wavefunction of the total system $\phi(t)$ at time t is related to the wavefunction in the interaction representation $|t\rangle$ as

$$\phi(t) = \exp\left(\frac{i\mathcal{H}_0 t}{\hbar}\right) |t\rangle,$$

and $|t\rangle$ obeys the following time development equation:

$$i\hbar \frac{\partial}{\partial t} |t\rangle = \mathcal{V}(t)|t\rangle. \tag{1.84}$$

The temporal evolution of the wavefunction $|t\rangle$ can also be given by the propagator $U(t, t_0)$ as

$$|t\rangle = U(t, t_0)|t_0\rangle,$$

and the propagator is subject to the following differential equation

$$\frac{d}{dt} U(t, t_0) = B(t) U(t, t_0), \qquad U(t_0, t_0) = 1, \tag{1.85}$$

where

$$B(t) = \frac{i}{\hbar} \int \boldsymbol{J}(\boldsymbol{r}t) \cdot \boldsymbol{A}(\boldsymbol{r}t) d^3\boldsymbol{r}. \tag{1.86}$$

Integration of (1.85) gives a formal solution of $U(t, t_0)$ in perturbation expansion form as

$$
\begin{aligned}
U(t,t_0) &= \exp\left\{\int_{t_0}^{t} B(t')dt' + \frac{1}{2}\int_{t_0}^{t} dt' \int_{t_0}^{t'} dt''[B(t'),B(t'')]\right\} \\
&= \exp\left[\sum_{\lambda}\{\alpha_\lambda \hat{a}_\lambda^\dagger - \alpha_\lambda^* \hat{a}_\lambda\} + i\phi\right]. \qquad (1.87)
\end{aligned}
$$

Here, we assume the limit $t_0 \to -\infty$, and α_λ is expressed as

$$
\alpha_\lambda = \frac{i}{\hbar\sqrt{V}}\int_{-\infty}^{t} dt' \int d^3\boldsymbol{r}\sqrt{\frac{\hbar}{2\epsilon_0\omega_\lambda}}e^{-i(\boldsymbol{k}\cdot\boldsymbol{r}-\omega t')}\boldsymbol{e}_\lambda \cdot \boldsymbol{J}(\boldsymbol{r},t'). \qquad (1.88)
$$

The term including $[B(t'),B(t'')]$ in (1.87) is a c-number as known from (1.82) and (1.86), and it gives the phase $i\phi$. If the system is in a steady state, where α_λ and ϕ do not include time t, then we can choose an arbitrary phase as $\phi = 0$, and the photon system is known to be in the coherent state as shown in the following equation:

$$
\begin{aligned}
|t\rangle &= \prod_{\lambda}\exp(\alpha_\lambda \hat{a}_\lambda^\dagger - \alpha_\lambda^* \hat{a}_\lambda)|0\rangle \\
&= \prod_{\lambda}\exp\left(-\frac{1}{2}|\alpha_\lambda|^2\right)\exp(\alpha_\lambda \hat{a}_\lambda^\dagger)\exp(-\alpha_\lambda^* \hat{a}_\lambda)|0\rangle \\
&= |\{\alpha_\lambda\}\rangle. \qquad (1.89)
\end{aligned}
$$

Here, we assume that the photon system is in a vacuum at $t \to -\infty$. The radiation from the classical current is proven to generate a coherent state of the photon. In Chap. 5, lasers are shown to give a coherent state of the photon when it is operated by pumping sufficiently higher than its threshold.

1.4 Squeezed State of Light

As described in the preceding section, a laser operated under strong pumping emits light in a coherent state. In this section, the magnitude of the fluctuation of two quadrature (sine and cosine) components of the electric field of laser light (coherent light) is shown to be equivalent. Next, the mathematical features of a quadrature-phase squeezed state are given, where the uncertainty of one quadrature component is forced to be smaller than that of the conventional minimum uncertainty; however, it is accompanied by an increase in the fluctuation of the other component. The methods of generation of the squeezed state are also discussed. Finally, the characteristics and the generation of photon-number squeezed states are given, where the fluctuation of the photon number is reduced, with the sacrifice of the phase fluctuation of a given mode.

1.4.1 Quadrature-Phase Squeezed State

The fluctuations of two quadrature components in a coherent state are calculated first. Considering one mode in an electric field $\boldsymbol{E}(t)$, it can be written in trigonometric form as

$$\begin{aligned}\boldsymbol{E}(t) &= i\boldsymbol{E}_0(\hat{a}e^{-i(\omega t-\boldsymbol{k}\cdot\boldsymbol{r})} - \hat{a}^\dagger e^{i(\omega t-\boldsymbol{k}\cdot\boldsymbol{r})}) \\ &= 2\boldsymbol{E}_0[\hat{q}\sin(\omega t - \boldsymbol{k}\cdot\boldsymbol{r}) + \hat{p}\cos(\omega t - \boldsymbol{k}\cdot\boldsymbol{r})], \end{aligned} \tag{1.90}$$

where we use $\boldsymbol{E}_0 = \boldsymbol{e}_\lambda\sqrt{\hbar\omega/2\epsilon_0 V}$ given in (1.38). In this case, $\hat{q}$ and $\hat{p}$ are given as

$$\hat{q} = \frac{1}{2}(\hat{a} + \hat{a}^\dagger), \qquad \hat{p} = \frac{i}{2}(\hat{a} - \hat{a}^\dagger), \tag{1.91}$$

and these operators obey the following commutation relation

$$[\hat{q}, \hat{p}] = -\frac{i}{2}. \tag{1.92}$$

Then, the fluctuations Δq and Δp satisfy the uncertainty relation:

$$\Delta q \cdot \Delta p \geq \frac{1}{4}. \tag{1.93}$$

This is obtained from (1.31) and (1.92). On the other hand, the expectation values of $\hat{q}$ and $\hat{p}$ for the coherent state are given as

$$\langle\alpha|\hat{q}|\alpha\rangle = \frac{1}{2}(\alpha + \alpha^*), \qquad \langle\alpha|\hat{p}|\alpha\rangle = \frac{i}{2}(\alpha - \alpha^*), \tag{1.94}$$

and the expectation values of $\hat{q}^2$ and $\hat{p}^2$ are given as:

$$\begin{aligned}\langle\alpha|\hat{q}^2|\alpha\rangle &= \frac{1}{4}\langle\alpha|\{\hat{a}^2 + \hat{a}\hat{a}^\dagger + \hat{a}^\dagger\hat{a} + (\hat{a}^\dagger)^2\}|\alpha\rangle \\ &= \frac{1}{4}(\alpha + \alpha^*)^2 + \frac{1}{4}, \end{aligned} \tag{1.95}$$

$$\langle\alpha|\hat{p}^2|\alpha\rangle = \frac{1}{4} - \frac{1}{4}(\alpha - \alpha^*)^2. \tag{1.96}$$

Finally, it is shown that the squares of the fluctuations are calculated to be

$$(\Delta q)^2 \equiv \langle\alpha|(\hat{q} - \langle\hat{q}\rangle)^2|\alpha\rangle = \frac{1}{4}, \tag{1.97}$$

$$(\Delta p)^2 \equiv \langle\alpha|(\hat{p} - \langle\hat{p}\rangle)^2|\alpha\rangle = \frac{1}{4}, \tag{1.98}$$

where $\langle\hat{p}\rangle = \langle\alpha|\hat{p}|\alpha\rangle$ and $\langle\hat{q}\rangle = \langle\alpha|\hat{q}|\alpha\rangle$. From these results, it is proven that the fluctuations of both quadrature components are equal in coherent states and also that the minimum uncertainty relationship is satisfied as known from (1.93), (1.97) and (1.98).

Next, we discuss the squeezing of light for which the distribution of quantum noise is changed, that is, the uncertainties of the amplitudes of each quadrature component p and q differ from one another. The coherent state is made by the linear interaction (1.83) between the electromagnetic field and the electron system. On the other hand, when some kinds of nonlinear interaction work effectively, the operators $\{\hat{a}, \hat{a}^\dagger\}$ can be treated as a photon pair with the same amplitude and the opposite phase or a photon pair with phase conjugate quantum correlation. The microscopic mechanism of the nonlinear optical interaction will be introduced in the later part of this section.

Based on the analogy to the Bogoliubov transformation in Bose–Einstein condensation, new operators b and $b^\dagger$ are introduced as a linear combination with complex coefficients μ and ν as

$$\hat{b} = \mu\hat{a} + \nu\hat{a}^\dagger, \qquad \hat{b}^\dagger = \mu^*\hat{a}^\dagger + \nu^*\hat{a}, \tag{1.99}$$

$$\hat{a} = \mu^*\hat{b} - \nu\hat{b}^\dagger, \qquad \hat{a}^\dagger = \mu\hat{b}^\dagger - \nu^*\hat{b}. \tag{1.100}$$

Here, (1.99) is a linear canonical transformation under the condition that $|\mu|^2 - |\nu|^2 = 1$. The physical background of this transformation will be discussed later. Therefore, the transformation must be represented by a unitary transformation U_L as

$$\hat{b} = U_L\hat{a}U_L^{-1} = \mu\hat{a} + \nu\hat{a}^\dagger. \tag{1.101}$$

Then, we can introduce a pseudo-photon number operator $\hat{b}^\dagger\hat{b}$ and a pseudo-photon number state $|m\rangle\rangle$ by using the unitary transformation U_L as

$$N \equiv \hat{b}^\dagger\hat{b} = U_L\hat{a}^\dagger\hat{a}U_L^{-1},$$
$$N|m\rangle\rangle = m|m\rangle\rangle \qquad (m = 0, 1, 2, \ldots). \tag{1.102}$$

The squeezed state $|\beta\rangle\rangle$ is defined as an eigenstate of the $\hat{b}$ operator:

$$\hat{b}|\beta\rangle\rangle = \beta|\beta\rangle\rangle, \qquad \langle\langle\beta|\hat{b}^\dagger = \beta^*\langle\langle\beta|, \tag{1.103}$$

$$|\beta\rangle\rangle = U_L|\beta\rangle = D(\beta)|0\rangle\rangle, \tag{1.104}$$

where $D(\beta)$ is defined as

$$D(\beta) \equiv e^{\beta\hat{b}^\dagger - \beta^*\hat{b}}. \tag{1.105}$$

Fluctuations of the electromagnetic field components $\hat{p}$ and $\hat{q}$ in the squeezed state will be calculated next. For one quadrature component $\hat{q}$,

$$(\Delta q)_s^2 \equiv \frac{1}{4}\{\langle\langle\beta|(\hat{a} + \hat{a}^\dagger)|\beta\rangle\rangle\}^2$$
$$= \frac{1}{4}|\mu - \nu|^2. \tag{1.106}$$

Here the subscript s on the left-hand side means an average in the squeezed state. Likewise, for the other component $\hat{p}$,

$$(\Delta p)_s^2 = \frac{1}{4}|\mu + \nu|^2. \tag{1.107}$$

In this calculation, the operators $\hat{a}$ and $\hat{a}^\dagger$ are rewritten as $\hat{b}$ and $\hat{b}^\dagger$ using (1.100), and also the definition of the squeezed state (1.103) is used. The expectation value of $\hat{a}$ is also defined as

$$\langle\langle \hat{a} \rangle\rangle \equiv \langle\langle \beta|\hat{a}|\beta\rangle\rangle = \mu^*\beta - \nu\beta^* \equiv \hat{\beta} \equiv \beta_q + i\beta_p. \tag{1.108}$$

The off-diagonal terms of the fluctuation do not vanish in the squeezed state as shown in the following calculation:

$$\begin{aligned}\{\Delta(qp)\}_s &\equiv \langle\langle \beta|(\hat{q} - \beta_q)(\hat{p} - \beta_p)|\beta\rangle\rangle \\ &= \frac{1}{4}i(\mu^*\nu - \nu^*\mu + 1), \end{aligned} \tag{1.109}$$

$$\begin{aligned}\{\Delta(pq)\}_s &\equiv \langle\langle \beta|(\hat{p} - \beta_p)(\hat{q} - \beta_q)|\beta\rangle\rangle \\ &= \frac{1}{4}i(\mu^*\nu - \nu^*\mu - 1). \end{aligned} \tag{1.110}$$

These fluctuations can be diagonalized with the rotation given by

$$\begin{aligned}\hat{a}' &\equiv \hat{a}e^{i\phi} = (\hat{q} + i\hat{p})(\cos\phi + i\sin\phi) \\ &\equiv \hat{q}' + i\hat{p}' = \hat{q}\cos\phi - \hat{p}\sin\phi + i(\hat{q}\sin\phi + \hat{p}\cos\phi),\end{aligned} \tag{1.111}$$

where the constant ϕ can be determined so that the sum of the off-diagonal parts of the fluctuation $\{\Delta(q'p') + \Delta(p'q')\}_s$ vanishes. Therefore, it can be expressed as

$$\tan 2\phi = \frac{i(\mu^*\nu - \nu^*\mu)}{\mu\nu^* + \nu\mu^*}. \tag{1.112}$$

From these results, the variances of the coordinate $\hat{q}'$ and momentum $\hat{p}'$ after the rotation in such a generalized coordinate are calculated to be

$$\begin{aligned}(\Delta q')^2 &= \frac{1}{4}(|\mu| - |\nu|)^2, \\ (\Delta p')^2 &= \frac{1}{4}(|\mu| + |\nu|)^2.\end{aligned} \tag{1.113}$$

The fluctuation of $\hat{q}'$ is then smaller than the value in the coherent state, but the squeezed state is known to be a minimum uncertainty state from the following relations:

$$(\Delta q')^2(\Delta p')^2 = \frac{1}{16}(|\mu|^2 - |\nu|^2)^2 = \frac{1}{16}, \tag{1.114}$$

$$(\Delta q')^2 = \frac{1}{4}(|\mu| - |\nu|)^2 = \frac{1}{4}\frac{1}{(|\mu| + |\nu|)^2} < \frac{1}{4}. \tag{1.115}$$

This state is called the quadrature-phase squeezed state. Figures 1.1 and 1.2 give an intuitive understanding of the physical meaning of the squeezed state

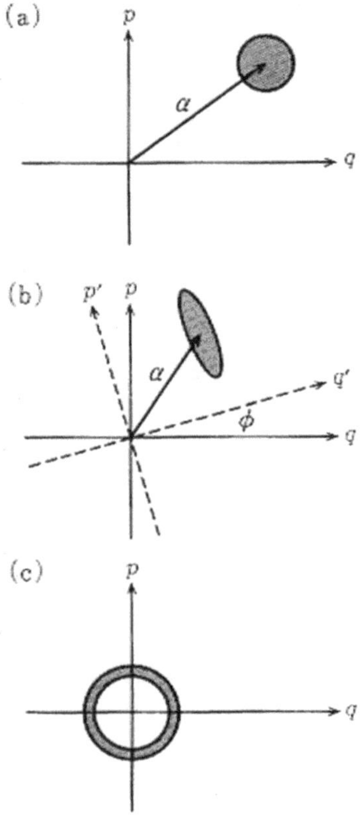

Fig. 1.1. Distribution of quantum noise plotted in phase space $(\hat{p}, \hat{q})$. $\hat{p}$ and $\hat{q}$ mean orthogonal quadrature components of electric field expressed as $\boldsymbol{E}(t) = 2\boldsymbol{E}_0(\hat{q}\cos\omega t + \hat{p}\sin\omega t)$. (**a**) Coherent state ($\Delta p = \Delta q$). (**b**) Quadrature phase squeezed state; this shows the situation for $\Delta q' < \Delta p'$ where $\hat{q}' = \hat{q}\cos\phi - \hat{p}\sin\phi$, $\hat{p}' = \hat{q}\sin\phi + \hat{p}\cos\phi$. (**c**) Photon-number squeezed state. See Sect. 1.4.2

and the coherent state. The uncertainties in the phase space $(\hat{q}, \hat{p})$ for the coherent state are indicated in Fig. 1.1(a), and those for the quadrature-phase squeezed state in Fig. 1.1(b). Figure 1.1(c) shows the photon-number state which will be discussed in the last half of this section. Figures 1.2(a)–(c) show the fluctuating behavior of the electric field for these three cases.

The next problem is to find the method to realize the unitary transformation (1.104) and (1.105) as a physical process. Second- and third-order nonlinear optical processes, which will be described in Chap. 6, play important roles: that is: (a) optical parametric amplification (second order); (b) second-harmonic generation (second order); and (c) degenerate four-wave mixing (third order).

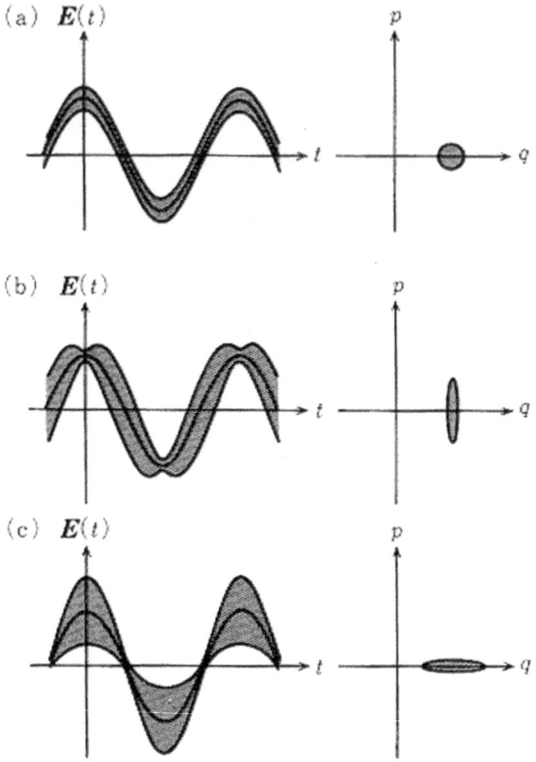

Fig. 1.2. (*Left*) Wave forms of the electric field and (*right*) the distribution of uncertainties in two conjugated variables in phase space $(\hat{q}, \hat{p})$. (**a**) Coherent state. (**b**) Quadrature phase squeezed state. Fluctuation in electric field amplitude is squeezed. (**c**) Fluctuation in phase is squeezed

In order to generate a squeezed state, the coherent state is used as a pump source. Considering the small nonlinear optical coefficients in general, the coherent pump light should be strong enough that it can be regarded as a classical electromagnetic wave. In this case, the annihilation and creation operators of the pump beam can be expressed as a c-number:

$$\hat{a}_p = c e^{-i\omega_p t}, \qquad \hat{a}_p^\dagger = c^* e^{i\omega_p t}. \tag{1.116}$$

The Hamiltonian describing the nonlinear interaction processes including photons to be squeezed $(\hat{a}, \hat{a}^\dagger)$ and pump light $(\hat{a}_p, \hat{a}_p^\dagger)$ are given as

$$\text{(a)} \qquad \mathcal{V}(t) = \hbar(\chi^{(2)} \hat{a}_p^\dagger \hat{a}\hat{a} + \chi^{(2)*} \hat{a}^\dagger \hat{a}^\dagger \hat{a}_p), \tag{1.117}$$

$$\text{(b)} \qquad \mathcal{V}(t) = \hbar(\chi^{(2)} \hat{a}^\dagger \hat{a}_p \hat{a}_p + \chi^{(2)*} \hat{a}_p^\dagger \hat{a}_p^\dagger \hat{a}), \tag{1.118}$$

$$\text{(c)} \qquad \mathcal{V}(t) = \hbar\chi^{(3)} [(\hat{a}_p^\dagger)^2 \hat{a}\hat{a} + \hat{a}^\dagger \hat{a}^\dagger (\hat{a}_p)^2]. \tag{1.119}$$

They correspond to three nonlinear optical processes (a), (b), and (c) mentioned in the preceding paragraph. Here, $\chi^{(2)}$ and $\chi^{(3)}$ represent the quantities proportional to the second- and the third-order nonlinear optical susceptibility, respectively. If the angular frequency of the photon to be squeezed is denoted as ω, the corresponding pump frequencies should be $\omega_p = 2\omega$ in (a) and $\omega_p = \omega$ in (c). For the case of the harmonic generation process (b), the pump frequency is determined by the relation $\omega = 2\omega_p$ and squeezing of the pumping light ω_p as well as the light ω can be expected.

When laser light of the coherent state $|\beta\rangle$ is incident in a nonlinear optical medium, it starts to interact with the atomic (molecular) system at $t = 0$ with the Hamiltonian $\mathcal{V}(t)$. If the propagator of the photon system is given by $U(t)$, the interaction process can be described by the equations:

$$|\beta\rangle\rangle = U_L|\beta\rangle = U(t)|\beta\rangle, \tag{1.120}$$

$$i\hbar\frac{\partial}{\partial t}U(t) = \mathcal{H}(t)U(t), \qquad U(0) = 1. \tag{1.121}$$

In order to discuss the evolution of the photon state, let us begin by calculating $U(t)$ first. In the nonlinear optical processes (a) and (c), the Hamiltonian $\mathcal{H}$ of the photon system can be written as:

$$\mathcal{H} = \hbar\omega\hat{a}^\dagger\hat{a} + \mathcal{V}(t) = \hbar\{\omega\hat{a}^\dagger\hat{a} + f_2^*\hat{a}^2 + f_2(\hat{a}^\dagger)^2\}. \tag{1.122}$$

By replacing $f_2(t)$ and $U(t)$ with $r(t)$ and $U'(t)$, respectively, in the following,

$$f_2(t) = r(t)e^{i\phi - 2i\omega t}, \tag{1.123}$$

$$U(t) = e^{-i\omega\hat{a}^\dagger\hat{a}t}U'(t), \tag{1.124}$$

the differential equation (1.121) is transformed into the form:

$$i\hbar\frac{\partial}{\partial t}U'(t) = \hbar r(t)\left\{e^{-i\phi}\hat{a}^2 + e^{i\phi}(\hat{a}^\dagger)^2\right\}U'(t). \tag{1.125}$$

Choosing $\phi = \pi/2$ and defining $R(t) = \int_0^t r(t')dt'$, the integration of (1.125) leads to the expression:

$$U'(t) = \exp[R(t)\{(\hat{a}^\dagger)^2 - \hat{a}^2\}]. \tag{1.126}$$

Finally, the unitary transformation determined by (1.124) and (1.126) gives the pseudo-photon operators $\hat{b}$ and $\hat{b}^\dagger$ from the definition (1.101):

$$\begin{aligned}\hat{b} &= U(t)\hat{a}U^{-1}(t)\\ &= e^{i\omega t}\cosh(2R)\hat{a} + e^{-i\omega t}\sinh(2R)\hat{a}^\dagger,\end{aligned} \tag{1.127}$$

$$\begin{aligned}\hat{b}^\dagger &= U(t)\hat{a}^\dagger U^{-1}(t)\\ &= e^{i\omega t}\sinh(2R)\hat{a} + e^{-i\omega t}\cosh(2R)\hat{a}^\dagger.\end{aligned} \tag{1.128}$$

This is a canonical transformation given by (1.99) and (1.100), as proven from the following calculation:

$$[\hat{b}, \hat{b}^\dagger] = \cosh^2(2R) - \sinh^2(2R) = 1. \tag{1.129}$$

These results give $|\mu| = \cosh(2R)$ and $|\nu| = \sinh(2R)$, and the dispersions of $\hat{p}'$ and $\hat{q}'$ given by (1.113) are calculated as

$$(\Delta q')^2 = \tfrac{1}{4}(|\mu| - |\nu|)^2 = \frac{1}{4}e^{-4R}, \tag{1.130}$$

$$(\Delta p')^2 = \tfrac{1}{4}(|\mu| + |\nu|)^2 = \frac{1}{4}e^{4R}. \tag{1.131}$$

In this expression, if the function $r(t)$ introduced in (1.123) is time independent, the value R is equal to $rt = rL/c^*$ where $t = L/c^*$ is a transit time of the electromagnetic wave in the nonlinear optical medium with length L, and c^* is the light speed in the medium. Equations (1.130) and (1.131) show that these three nonlinear optical processes will produce the squeezed state of light.

Experimental efforts to observe the squeezed states have been made since the middle of the 1980s. Here, we show an example done by Kimble's group [8], in which they reduced the noise of laser light to less than half by using a degenerate parametric transformation. A degenerate parametric transformation is a nonlinear optical process introduced in the preceding paragraph and (1.117) in which process the pump photons with frequency ω_p incident in a nonlinear optical medium are divided into two sets of photons with $\omega = \omega_\mathrm{p}/2$. They employed MgO:$LiNbO_3$ as a $\chi^{(2)}$ material, and inserted it into a high-Q Fabry–Pérot cavity as shown in Fig. 1.3.

The amplification gain of the optical parametric process is sensitive to the phase, and the amplification is enhanced due to the repetition of round-trip travelling of light in the cavity. They generated squeezed light of 1.06 μm wavelength by using second harmonics of a Nd^{3+}:YAG laser (0.53 μm wavelength) as a pump. Because one quadrature component of the light is squeezed as shown in Fig. 1.1(b), a phase-sensitive detection process is required in order to confirm the squeezing of the quadrature phase component. Therefore, for the measurement of squeezed noise, the electric field $E(t)$ must be detected at the instant when the fluctuation of $E(t)$ is the smallest, that is, when $E(t)$ has a minimum or maximum value, as indicated in Fig. 1.2(b). Measurement of the light intensity itself is inappropriate for the detection of quadrature-phase squeezed light, because we would measure the average of the noise from both quadrature components normal to each other. Homodyne detection is one of the methods of detecting the squeezed component. Homodyne detection was used for the superposition of the squeezed light and coherent light in the same radiation mode formed by a beamsplitter as shown in Fig. 1.3. In this case, 1.06 μm single-mode laser was employed as a light source for pumping of the second harmonics and a local oscillator. The beamsplitter divides both

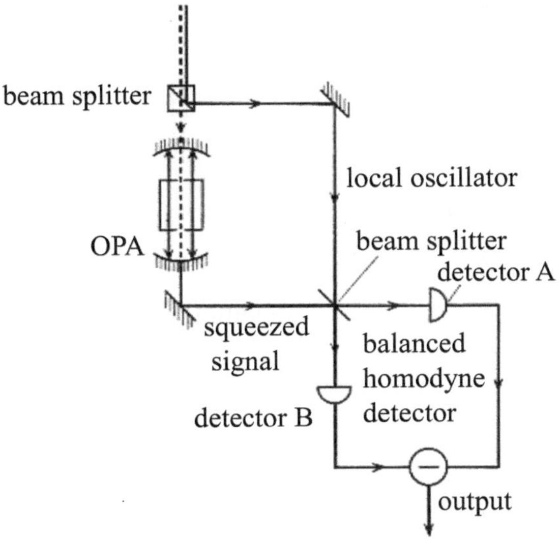

Fig. 1.3. Schematic diagram of the experiment of generation and measurement of squeezed light using a degenerate parametric process

the squeezed signal and the coherent local oscillator into 50:50, and it also mixes them simultaneously. The two light beams from the beamsplitter are received by detectors A and B, and the photocurrent from these two detectors is received by a balanced homodyne detector. The fluctuation from the local oscillator is suppressed by employing this detection process, and we can observe the squeezed quadrature phase component by choosing an adequate phase.

If the phase of the local oscillator is varied, for example by changing the length of the light path, we can pick up the fluctuation of one squeezed component $\hat{q}' \sin(\omega t)$ at a certain phase. Figure 1.4 shows that the noise level from the receiver becomes, in turn, larger and smaller than the shot noise level, by changing the phase of the local oscillator.

This reflects the difference of the uncertainty for two quadrature components. The shot noise level is observed by blocking the signal light. This is determined from the fluctuation in the vacuum, and is independent of the phase of the local oscillator. At an adequate phase position, we can observe the variance $(\Delta\hat{q}')^2$ of the squeezed state to be smaller than the shot noise, while at 90° phase difference, the observed dispersion, $(\Delta\hat{p}')^2$, of $\hat{p}'$ is larger than the vacuum fluctuation. Figure 1.5 plots the minimum fluctuation $(\Delta\hat{q}')^2$ and the maximum fluctuation $(\Delta\hat{p}')^2$ observed in Fig. 1.4 when the pump intensity is varied. The points show that the fluctuation satisfies the minimum uncertainty relation $(\Delta\hat{q}')^2(\Delta\hat{p}')^2 = 1/16$. This means that one component of the fluctuation is squeezed with the enhancement of the other fluctuation component. Squeezed states can also be generated by harmonic generation and degenerate four-wave mixing.

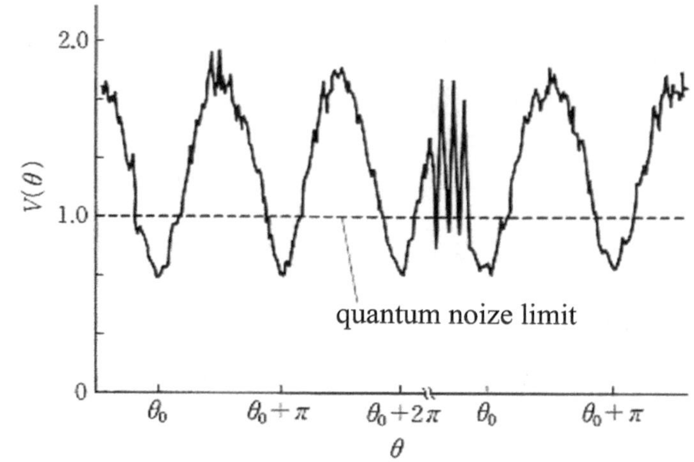

Fig. 1.4. Observation of quadrature phase squeezed state. The curve shows the output current from the balanced homodyne detector shown in Fig. 1.3 as a function of phase θ of the local oscillator. The signal undulates over and below the quantum noise limit shown by the dashed line, by changing the phase [8]

1.4.2 Antibunching Light and Photon-Number Squeezed State

One of the important results from quantization of the electromagnetic field is that all photon states must have unavoidable random fluctuations characterized for each state. Such a fluctuation appears as the variance of the photon number observed during a time interval T, and also as the ratio of successive counting of photons at two instances with temporal distant τ. Therefore, from photon-counting measurements we can get an insight into the intrinsic characteristics of photon states and their generation processes.

Before discussing the photon-number squeezed state, the photon statistics for the coherent state will be summarized briefly. From the fact that the coherent state is the most proximate to a classical electromagnetic wave, we can assume that photons do not have statistical correlation in the light beam and these arrive with random time intervals. In this case, the photon number observed in a time interval T obeys a Poisson distribution, as known from (1.52). In a Poisson distribution, the variance σ_n of the observed photon number is equal to the averaged photon number $\langle n \rangle$. The normalized ratio of simultaneous detection $g^{(2)}(\tau)$ (degree of second-order temporal coherence) is always 1 for any delay time τ, because photons arrive at the detector without any correlation with other photons in this case. The function $g^{(2)}(\tau)$ is defined from $G^{(2)}(\tau)$ which is the observed photon number per second at fixed temporal distance τ. It is normalized with the square of the photon number $\langle n \rangle$ per second:

$$g^{(2)}(\tau) = \frac{G^{(2)}(\tau)}{\langle n \rangle^2}. \tag{1.132}$$

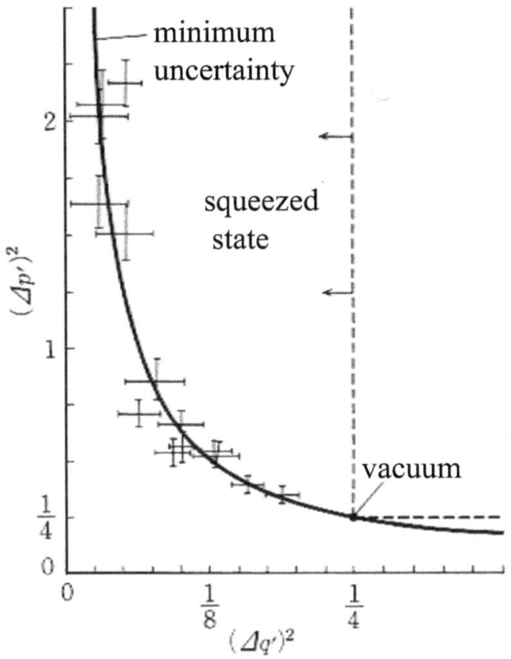

Fig. 1.5. Relation between the variances $(\Delta\hat{q}')^2$ and $(\Delta\hat{p}')^2$ of two variables $\hat{p}'$ and $\hat{q}'$ for the quadrature phase squeezed state generated by the degenerate parametric process [8]

If every photon arrives at the same temporal distance, the number of observed photons during the time interval T becomes constant, then the variance σ_n is zero. In this case, if the variance of the photon number σ_n is smaller than the average $\langle n \rangle$, the state is called sub-Poissonian. If the ratio of simultaneous detection $g^{(2)}(\tau)$ is smaller than 1 at $\tau = 0$, or when the slope of $g^{(2)}(\tau)$ is positive at $t = 0$, the light is known as antibunching light. The light in the photon-number squeezed state has such properties. Later, in Chap. 3, the light of thermal radiation determined by Planck's radiation law is shown to be bunching light where $g^{(2)}(\tau) > 1$ in the vicinity of $\tau = 0$. Because the photons are subjected to Bose statistics, they have a tendency to form photon pairs also in the temporal region. In the context of photon statistics, it is known that the coherent state forms the border between bunching and antibunching photon states. Single-mode light shows antibunching properties when $g^{(2)}(0) < g^{(2)}(\tau)$, and it gives photon-number squeezing or sub-Poissonian statistics when $g^{(2)}(0) < 1$ under the conditions which will be described in Chap. 3.

From the above discussion, it is known that the photon-number squeezed state can be generated by making anticorrelation among photons which exist at close temporal positions. If we let a photon be emitted without any control,

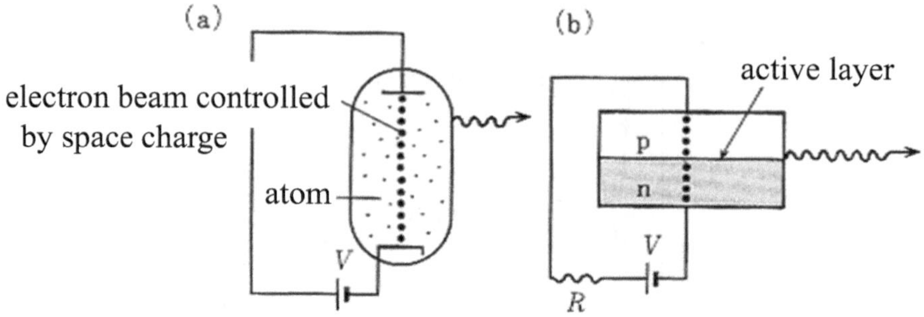

Fig. 1.6. (**a**) Schematic diagram of the experiment by Franck and Hertz. It shows fluorescence from atoms excited by an electron beam in a bipolar tube. (**b**) Solid-state version of the Franck–Hertz experiment. Light-emitting diode driven by a constant current source

the light will obey a Poisson distribution. If we can prevent light emission for a fixed period after the emission of the last photon, the temporal distribution of photons will be more regular.

As a first example, the experiment of Franck and Hertz is discussed, in which photons are emitted from atoms excited by inelastic collision with an electron beam. In their experiment, shown in Fig. 1.6(a), the current is stabilized under the balance between the Coulomb force to accelerate electrons due to the applied voltage between two electrodes in a bipolar tube and the Coulomb repulsion force among electrons travelling in the tube. The current is known as the space-charge limited current. There is usually shot noise in the anode current, because electron emission from the cathode is a random Poissonian process.

When a space-charge limited current is formed in a vacuum tube, however, the carrier distribution works as the potential for the emitted electrons, and the potential changes according to the emission rate of the electrons. Consequently, it gives negative feedback on the anode current, then the shot noise is suppressed. Therefore, this process controls the number of atoms excited by the elctron beam so as the photon is more regularly emitted from atoms. With such well controlled excitation of atoms, the fluctuation of the photon number in fluorescent radiation is squeezed.

It is also possible to reproduce a similar effect in a light-emitting diode operated by a constant current source under the space charge limit. It generates a photon-number squeezed state because one photon is emitted by one injected electron in the ideal case.

Yamamoto et al. [9, 10] also developed a semiconductor laser driven by a constant current source as shown in Fig. 1.6(b), and demonstrated the generation of a photon-number squeezed state. Yamamoto's experiment is actually the replacement of an atomic light source in the Franck–Hertz experiment with a solid state device operating under the same space-charge limited condition,

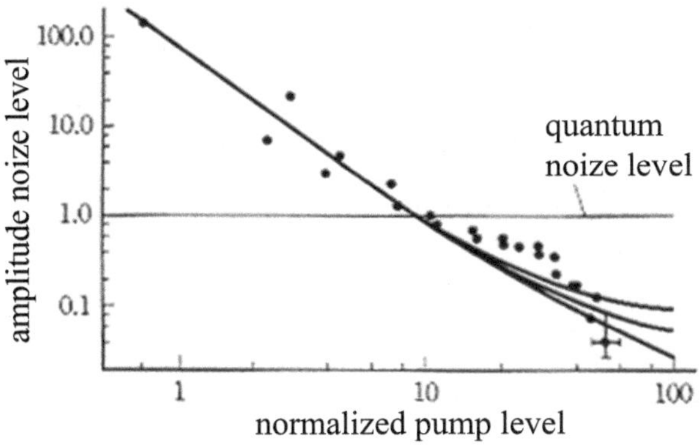

Fig. 1.7. Noise level of the photon-number squeezed state generated by a laser diode driven by a constant current source. Photon number squeezing (noise below the quantum limit) was observed under high pumping rate [9]

and the emission mechanism is replaced by the stimulated emission. Consequently, the magnitude of the noise, that is, the fluctuation of the photon number, was suppressed to 1/10 of the coherent state noise under the driving current which is sufficiently higher than that of the threshold, as shown in Fig. 1.7. This method has advantages such that it reduces the size of the equipment by using solid state devices; it is possible to generate a strong photon flux; and it is also possible to squeeze the photon-number fluctuation in a broad frequency range with high efficiency.

In order to describe the photon-number squeezed state in quantum optics, the photon number operator $\hat{n}$ and the phase operators $\hat{S}$ and $\hat{C}$ are introduced as the following:

$$\hat{n} = \hat{a}^\dagger \hat{a}, \tag{1.133}$$

$$\hat{S} = \frac{1}{2i}[(\hat{n}+1)^{-1/2}\hat{a} - \hat{a}^\dagger(\hat{n}+1)^{-1/2}], \tag{1.134}$$

$$\hat{C} = \frac{1}{2}[(\hat{n}+1)^{-1/2}\hat{a} + \hat{a}^\dagger(\hat{n}+1)^{-1/2}]. \tag{1.135}$$

It is impossible to define a Hermitian operator representing the phase as introduced by (1.41), but the operator $\hat{S}$ works as a phase operator when the photon number n in a mode is large enough. A photon-number state will be found to be a minimum-uncertainty state of $\hat{n}$ and $\hat{S}$ in the following. The commutation relation of these operators is expressed as

$$[\hat{n}, \hat{S}] = \frac{1}{2i}\left\{(\hat{n}+1)^{-1/2}[\hat{n}, \hat{a}] - [\hat{n}, \hat{a}^\dagger](\hat{n}+1)^{-1/2}\right\} = i\hat{C}. \tag{1.136}$$

From (1.31), the uncertainty relation

$$\langle(\Delta\hat{n})^2\rangle\langle(\Delta\hat{S})^2\rangle \geq \frac{1}{4}|\langle\hat{C}\rangle|^2 \tag{1.137}$$

is obtained. When the averaged photon number $n = \langle n \rangle$ is much larger than unity, the operators $\hat{a}$ and $\hat{a}^\dagger$ can be approximated by (1.41) as

$$\hat{a} \cong e^{i\hat{\phi}}\sqrt{\hat{n}}, \qquad \hat{a}^\dagger \cong \sqrt{\hat{n}}e^{-i\hat{\phi}}. \tag{1.138}$$

In this case, $\hat{C}$ and $\hat{S}$ are expressed as

$$\hat{C} \cong \frac{1}{2}\left[\frac{1}{\sqrt{\hat{n}+1}}e^{i\hat{\phi}}\sqrt{\hat{n}} + \sqrt{\hat{n}}e^{-i\hat{\phi}}\frac{1}{\sqrt{\hat{n}+1}}\right] \cong 1, \tag{1.139}$$

$$\hat{S} \cong \frac{1}{2i}\left[\frac{1}{\sqrt{\hat{n}+1}}e^{i\hat{\phi}}\sqrt{\hat{n}} - \sqrt{\hat{n}}e^{-i\hat{\phi}}\frac{1}{\sqrt{\hat{n}+1}}\right] \cong \hat{\phi}, \tag{1.140}$$

if the variation of phase ϕ is small enough. From (1.140), we find that $\langle(\Delta\hat{S})^2\rangle$ is nearly equal to $\langle(\Delta\phi)^2\rangle$. Then the uncertainty relation (1.137) can be rewritten as

$$\langle(\Delta\hat{n})^2\rangle\langle(\Delta\hat{\phi})^2\rangle \geq \frac{1}{4}. \tag{1.141}$$

If we apply this condition to the coherent state, where $n \equiv \langle\hat{n}\rangle = |\alpha|^2$ and $\langle(\Delta\hat{n})^2\rangle = |\alpha|^2$, the minimum phase fluctuation is obtained as

$$\langle(\Delta\hat{\phi})^2\rangle = \frac{1}{4n}. \tag{1.142}$$

Therefore, it is known that the phase fluctuation can be reduced as the averaged photon number n increases. In the photon-number squeezed state, the fluctuations of photon number $\hat{n}$ and phase $\hat{S}$ are different and are expressed as

$$\langle\langle(\Delta\hat{n})^2\rangle\rangle = \frac{1}{2}\langle\langle\hat{C}\rangle\rangle e^{-2R}, \tag{1.143}$$

$$\langle\langle(\Delta\hat{S})^2\rangle\rangle = \frac{1}{2}\langle\langle\hat{C}\rangle\rangle e^{2R}. \tag{1.144}$$

When the squeezing parameter R satisfies the condition $R > -(1/2)\ln(2n)$, the state has sub-Poissonian characteristics which are formulated as

$$\langle\langle(\Delta\hat{n})^2\rangle\rangle < n. \tag{1.145}$$

On the other hand, the fluctuation of the phase $\hat{S}$ is larger to compensate for the small fluctuation in n:

$$\langle\langle(\Delta\hat{S})^2\rangle\rangle \to \langle\langle(\Delta\hat{\phi})^2\rangle\rangle > \frac{1}{4n} \tag{1.146}$$

for the state with high mean photon number n.

The fluctuation of the photon-number squeezed state in phase space (q', p') is drawn in Fig. 1.1(c). The temporal variation of its electric field is given

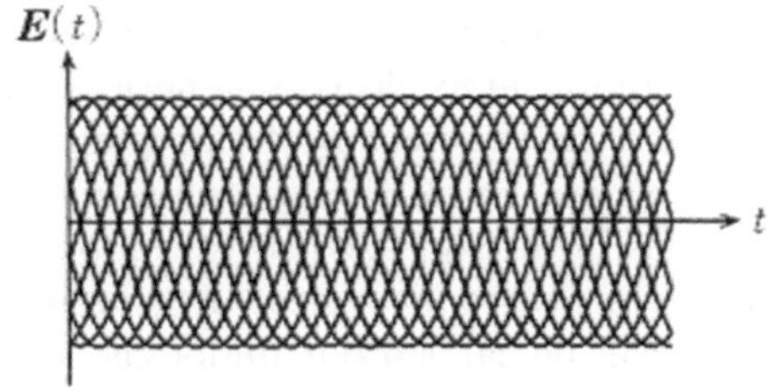

Fig. 1.8. Drawing of the temporal behavior of the electric field $\boldsymbol{E}(t)$ in the photon-number squeezed state. In a real case, sinusoidal curves must be drawn continuously

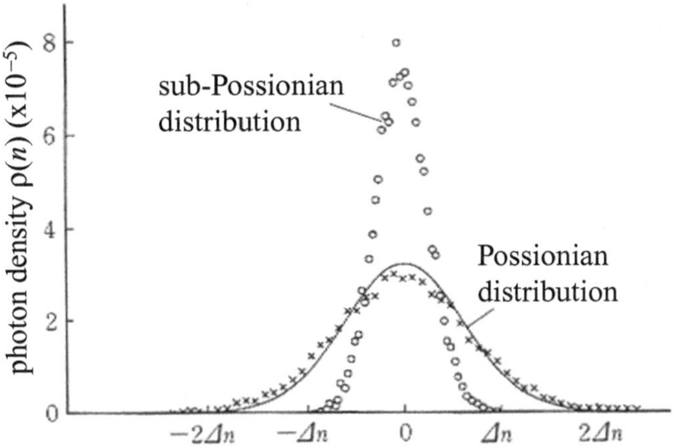

Fig. 1.9. Circles show a sub-Poissonian distribution of the photon-number squeezed state. Crosses show a Poissonian distribution observed in the coherent state for comparison [10]

schematically in Fig. 1.8. The sub-Poissonian distributions obtained both theoretically and experimentally in the photon-number squeezed state are given in Fig. 1.9.

Next, a practical advantage of the photon-number squeezed state over the quadrature phase squeezed state will be discussed. When a quadrature phase state is generated from a coherent state $|\alpha\rangle$, the parameters $|\nu|$ and $|\mu|$ must be comparatively large, in order to get enough squeezing of the fluctuation of the q' component. This restriction can be understood from the following relationships:

$$(\varDelta\hat{q})^2 = \frac{1}{4}(|\mu| - |\nu|)^2 = \frac{|\mu| - |\nu|}{4(|\mu| + |\nu|)}, \tag{1.147}$$

$$\langle\langle\hat{a}^\dagger\hat{a}\rangle\rangle \equiv \langle\langle\hat{n}\rangle\rangle = |\alpha|^2 + |\nu|^2, \tag{1.148}$$

$$|\mu| = \cosh(2R), \qquad |\nu| = \sinh(2R). \tag{1.149}$$

In this case, the total photon number $\langle\langle\hat{n}\rangle\rangle$ must be large enough, as known from (1.148). Therefore, quite large electromagnetic energy proportional to $|\nu|^2$ must be consumed for the generation of the quadrature phase squeezed state. In the case of the generation of the photon-number squeezed state, on the other hand, all photons in the squeezed state can be utilized as the signal.

One of the remarkable features of optical measurements with laser light is that the measurement can be carried out under the quantum limit. This is because the quantum zero-point energy $\hbar\omega/2$ is much larger than the thermal noise in the region of optical frequencies. Therefore, the performance of the detection is determined by the quantum noise in the coherent state or the vacuum field, for example, for gravitational wave detection using a laser interferometer, optical fiber communication systems, and so on. If the squeezing is employed in such applications, it is possible to overcome the limit of quantum noise of the coherent state and the vacuum field by reducing the quantum noise of one observable which carries information with enhancement of the quantum noise of the conjugated observables.

Recently, squeezed light has been applied to clarify the mechanism of neural cells in the retinas of mammals. Neural cells generate an electric signal which gives a temporal pulsation under the stimulation of light. The statistical properties of the neural signal is composed of two probabilistic processes. One is the quantum fluctuation of the incident light, and the other the intrinsic fluctuation of the neural cells. If experiments on the optic nerve can be done with squeezed light, it is possible to determine the intrinsic randomness embedded in human vision.

2

Interaction Between the Electron and the Radiation Field

In the previous chapter we have quantized the radiation field in free space. Then we obtained a clear concept about the photon and found that there is an uncertainty relation between two conjugate physical variables. The uncertainty relation mediates the duality between the wave-like and particle-like nature of the radiation field. This appears clearly in the uncertainty between the photon number n and the phase ϕ of the radiation field. In this chapter we study spontaneous emission, resulting from quantization of the radiation field, and related phenomena including laser cooling of atoms. In Sect. 2.1 we first formulate the interaction Hamiltonian between the radiation field and the electron system. In Sect. 2.2, using these formulas we describe the absorption and emission processes of the light due to the electron (atom) systems, and understand the origin of the spontaneous emissions from excited atoms. Concerning stimulated emission we will discuss it in connection with laser oscillation in Chap. 4. In Sect. 2.3, we investigate how the natural width of the absorption or emission spectrum is characteristic of the spontaneous emission process. In Sect. 2.4 we find that the spontaneous emission is not characteristic of an atom and can be controlled artificially by controlling the geometrical structure of the radiation field. Superradiance, i.e., spontaneous emission from a coherent atomic system, will be discussed separately in Chap. 5. In Sect. 2.5, we study the laser cooling of neutral atoms, and Bose–Einstein and Fermi condensation of the atoms will be discussed in Sects. 2.6 and 2.7, respectively.

2.1 Electron–Radiation Coupled System

The electromagnetic field accelerates a charged particle, while the motion of a charged particle generates an electromagnetic field. Namely, owing to the interaction with the electromagnetic field, a charged particle absorbs or emits an electromagnetic field, changing its energy state. In this section we discuss quantum mechanically the electron–radiation interaction [11].

The Lorentz force $\boldsymbol{F}$ that acts on an electron moving in the electric field $\boldsymbol{E}$ and magnetic field $\boldsymbol{B}$ is

$$\boldsymbol{F} = -e(\boldsymbol{E} + \boldsymbol{v} \times \boldsymbol{B}), \tag{2.1}$$

where $\boldsymbol{v}$ is the velocity vector of an electron. Using the scalar potential ϕ and the vector potential $\boldsymbol{A}$, the electric and magnetic fields $\boldsymbol{E}$ and $\boldsymbol{B}$ are, respectively, rewritten by

$$\boldsymbol{E} = -\nabla\phi - \frac{\partial}{\partial t}\boldsymbol{A}, \qquad \boldsymbol{B} = \mathrm{rot}\boldsymbol{A}. \tag{2.2}$$

Then, the equation of motion for the electron is

$$m\dot{\boldsymbol{v}} = e\nabla\phi + e\dot{\boldsymbol{A}} - e(\dot{\boldsymbol{r}} \times \mathrm{rot}\boldsymbol{A}), \tag{2.3}$$

where m is the electron mass. In general, the equation of motion is derived from the Lagrangian $\mathcal{L}$ by means of the relations

$$\frac{d}{dt}\left(\frac{\partial \mathcal{L}}{\partial \dot{x}}\right) - \frac{\partial \mathcal{L}}{\partial x} = 0, \ldots, \tag{2.4}$$

where $\boldsymbol{v} = \dot{\boldsymbol{r}} \equiv (\dot{x}, \dot{y}, \dot{z})$ is the velocity vector. In order to derive (2.3) from (2.4), the Lagrangian $\mathcal{L}$ should be

$$\mathcal{L} = \frac{m}{2}(\dot{\boldsymbol{r}})^2 + e\phi - e\dot{\boldsymbol{r}} \cdot \boldsymbol{A}. \tag{2.5}$$

The momentum vector $\boldsymbol{p}$, canonically conjugate to $\boldsymbol{r}$, is given by

$$\boldsymbol{p} = \left(\frac{\partial \mathcal{L}}{\partial \dot{x}}, \frac{\partial \mathcal{L}}{\partial \dot{y}}, \frac{\partial \mathcal{L}}{\partial \dot{z}}\right) = m\dot{\boldsymbol{r}} - e\boldsymbol{A}. \tag{2.6}$$

Following the standard treatment of classical mechanics, the Hamiltonian $\mathcal{H}$ for an electron in the electromagnetic field $(\phi, \boldsymbol{A})$ is derived as

$$\mathcal{H} = \boldsymbol{p} \cdot \boldsymbol{v} - \mathcal{L} = \frac{m}{2}(\boldsymbol{v})^2 - e\phi = \frac{1}{2m}(\boldsymbol{p} + e\boldsymbol{A})^2 - e\phi. \tag{2.7}$$

The classical Hamiltonian is formally converted to quantum-mechanical form by replacing the momentum vector $\boldsymbol{p}$ with the operator $\boldsymbol{p} \longrightarrow -i\hbar\nabla$, and by rewriting the vector potential $\boldsymbol{A}$ in terms of the creation and annihilation operators $\hat{a}_\lambda^\dagger$ and $\hat{a}_\lambda$ for the photon mode λ (see (1.16)). The total quantum-mechanical Hamiltonian $\mathcal{H}$ is divided into two parts. One is the unperturbed Hamiltonian $\mathcal{H}_0$ and the other is the Hamiltonian $\mathcal{H}'$ that describes the interaction between an electron and the radiation field:

$$\mathcal{H} = \mathcal{H}_0 + \mathcal{H}', \tag{2.8}$$

$$\mathcal{H}_0 = \sum_j \frac{1}{2m}\boldsymbol{p}_j^2 + \mathcal{V}(\boldsymbol{r}_1, \boldsymbol{r}_2, \ldots, \boldsymbol{r}_N) + \sum_j \hbar\omega_\lambda\left(\hat{a}_\lambda^\dagger \hat{a}_\lambda + \frac{1}{2}\right), \tag{2.9}$$

$$\mathcal{H}' = \sum_j \frac{e}{2m}\{\boldsymbol{p}_j \cdot \boldsymbol{A}(\boldsymbol{r}_j) + \boldsymbol{A}(\boldsymbol{r}_j) \cdot \boldsymbol{p}_j\} + \frac{e^2}{2m}\sum_j \boldsymbol{A}(\boldsymbol{r}_j)^2. \tag{2.10}$$

By choosing the Coulomb gauge div$\boldsymbol{A}(\boldsymbol{r}_j) = 0$ in the first term of (2.10), we obtain

$$\boldsymbol{p}_j \cdot \boldsymbol{A}(\boldsymbol{r}_j) = -i\hbar\{\mathrm{div}\boldsymbol{A}(\boldsymbol{r}_j) + \boldsymbol{A}(\boldsymbol{r}_j) \cdot \nabla_j\} = \boldsymbol{A}(\boldsymbol{r}_j) \cdot \boldsymbol{p}_j. \tag{2.11}$$

Substituting (1.16) and (1.37) into (2.10), the first term of the interaction Hamiltonian, which governs the first-order optical process such as absorption and emission of light, is formulated as

$$\mathcal{H}'_{(1)} = \frac{e}{m}\sqrt{\frac{\hbar}{2\epsilon_0 V}} \sum_j \sum_\lambda \frac{1}{\sqrt{\omega_\lambda}} \{e^{i\boldsymbol{k}\cdot\boldsymbol{r}_j}(\boldsymbol{e}_\lambda \cdot \boldsymbol{p}_j)\hat{a}_\lambda + e^{-i\boldsymbol{k}\cdot\boldsymbol{r}_j}(\boldsymbol{e}_\lambda \cdot \boldsymbol{p}_j)\hat{a}^\dagger_\lambda\}, \tag{2.12}$$

where $\lambda \equiv (\boldsymbol{k}, \gamma)$ corresponds to the photon mode and γ denotes two independent polarizations of light. Note that the second term in (2.10), proportional to $\boldsymbol{A}^2$, does not contribute to the first-order optical process, but it will be important for the second-order or higher-order optical transition such as Rayleigh and Compton scattering.

2.2 Spontaneous and Stimulated Emissions

In order to describe quantum optics, as presented in Chap. 1, we have described both the atomic systems and the electromagnetic field by quantum mechanics. On the other hand, the semiclassical theory, where only the electronic systems are quantized, is sometimes useful to discuss quantitatively absorption of light and stimulated emission of light. It is also possible to describe the spontaneous emission in the framework of the semiclassical theory. However, using the idea of the photon that is deduced from quantization of the electromagnetic field, we deeply understand not only the spontaneous emission but also the relation between the absorption and stimulated emission. Moreover, we can straightforwardly calculate the Einstein A and B coefficients on the basis of quantum theory.

Consider the optical transition due to the perturbation $\mathcal{H}'_{(1)}$ from the initial state $\Psi_i = |a\rangle\,|n_1, n_2, \ldots, n_\lambda, \ldots\rangle$ to a final state $\Psi_f = |b\rangle|n_1, n_2, \ldots, n_\lambda, \cdots\rangle$, which are both an eigenstate of the unperturbed Hamiltonian $\mathcal{H}_0$. Here the initial $|a\rangle$ and final $|b\rangle$ states are eigenstates of the electronic system. Instead of the total wavefunction $\Psi(t)$, we introduce the following form:

$$\Psi(t) = \exp\left(-\frac{i\mathcal{H}_0}{\hbar}t\right)\Psi'(t). \tag{2.13}$$

Then we obtain the interaction representation:

$$i\hbar\frac{\partial\Psi'}{\partial t} = \tilde{\mathcal{H}}'(t)\Psi', \tag{2.14}$$

where

$$\tilde{\mathcal{H}}'(t) = \exp\left(\frac{i\mathcal{H}_0}{\hbar}t\right)\mathcal{H}'_{(1)}\left(-\frac{i\mathcal{H}_0}{\hbar}t\right). \tag{2.15}$$

Using the eigenfunction Ψ_n of the Hamiltonian $\mathcal{H}_0$, we expand a solution $\Psi'(t)$ of (2.14) as follows:

$$\Psi'(t) = \sum_n c_n(t)\Psi_n. \tag{2.16}$$

Let us choose as the initial state of the total system an eigenstate Ψ_i of the Hamiltonian $\mathcal{H}_0$. Thus we take a set $\{c_i(0) = 1,\ c_n(0) = 0 \text{ for } n \neq i\}$ as an initial condition. Substituting (2.16) into (2.14), multiplying the final state wavefunction Ψ_f^* on the right and integrating it with respect to the whole coordinates, we obtain a differential equation for the coefficient $c_n(t)$ as

$$i\hbar\dot{c}_f(t) = \sum_n \langle f|\,\tilde{\mathcal{H}}'(t)\,|n\rangle\, c_n(t). \tag{2.17}$$

As we may use the initial state for the coefficients on the right side of (2.17), the coefficient $c_f(t)$ is integrated as

$$c_f(t) = \langle f|\,\mathcal{H}'_{(1)}\,|i\rangle\,\frac{1 - e^{i(E_f - E_i)t/\hbar}}{E_f - E_i}. \tag{2.18}$$

Therefore, the transition probability W_{fi} to the final state $|f\rangle$ per unit time is

$$\begin{aligned} W_{fi} &= \frac{|c_f(t)|^2}{t} \\ &= \frac{2\pi}{\hbar}\sum_\lambda\left(\frac{e}{m}\right)^2\frac{\hbar}{2\epsilon_0 V\omega_\lambda}\left\{\begin{matrix} n_\lambda \\ n_\lambda + 1\end{matrix}\right\} \\ &\quad\times|\langle b|\sum_j e^{\pm i\boldsymbol{k}\cdot\boldsymbol{r}_j}(\boldsymbol{e}_\lambda\cdot\boldsymbol{p}_i)|a\rangle|^2\delta(E_f - E_i). \end{aligned} \tag{2.19}$$

In order to describe the absorption process, the set $\{n_\lambda,\ e^{+i\boldsymbol{k}\cdot\boldsymbol{r}_j}\}$ in the above equation is chosen, which comes from the first term of the interaction Hamiltonian $\mathcal{H}'_{(1)}$ in (2.12). The other set $\{n_\lambda + 1,\ e^{-i\boldsymbol{k}\cdot\boldsymbol{r}_j}\}$, originating from the second term, gives rise to the emission process. In the absorption process where the photon system loses one photon energy $\hbar\omega_\lambda$, energy conservation yields

$$E_f - E_i = E_b - \hbar\omega_\lambda - E_a, \tag{2.20}$$

where E_a and E_b are eigenenergies of the initial and final electronic states $|a\rangle$ and $|b\rangle$, respectively. The energy density $\rho(\omega)d\omega$ of the photon system is given by

$$\rho(\omega_\lambda)d\omega_\lambda = \hbar\omega_\lambda \frac{n_\lambda}{V} d\omega_\lambda. \tag{2.21}$$

When we consider the electronic transition between energy levels of an isolated atom, the spatial distribution of the electron wavefunction is estimated to be of the order of the Bohr radius (∼0.5Å). Further, if the wavenumber $k = 2\pi/\lambda$ of the radiation field is very small (the wavelength λ is of the order of thousands of angstroms for visible light), the first term of the Taylor expansion of $\exp(i\boldsymbol{k}\cdot\boldsymbol{r}_j) = 1 + i\boldsymbol{k}\cdot\boldsymbol{r}_j + \cdots$ predominates over the matrix element of the electronic part in (2.19). Then we have

$$\langle b|\, \boldsymbol{e}_\lambda \cdot \sum_j \boldsymbol{p}_j \,|a\rangle = im\omega_{ba} \langle b|\, \boldsymbol{e}_\lambda \cdot \sum_j \boldsymbol{r}_j \,|a\rangle = -\frac{im}{e}\omega_{ba}\boldsymbol{e}_\lambda \cdot \langle b|\, \boldsymbol{P} \,|a\rangle\,, \tag{2.22}$$

where $E_b - E_a \equiv \hbar\omega_{ba}$, and $\boldsymbol{P} \equiv -e\Sigma_j \boldsymbol{r}_j$ is the operator of the electric dipole moment of the electronic system. In this dipole approximation, the transition probability for the absorption process is written

$$\begin{aligned} W_{fi}^{(\mathrm{a})} &\equiv B_{ba}\rho(\omega_{ba}) \\ &= \frac{2\pi}{\hbar^2}\left(\frac{e}{m}\right)^2 \frac{\hbar}{2\epsilon_0 V} \sum_\lambda \frac{n_\lambda}{\omega_\lambda} m^2\omega_{ba}^2 |\langle b|\boldsymbol{e}_\lambda \cdot \sum_j \boldsymbol{r}_j|a\rangle|^2 \delta(\omega_{ba} - \omega_\lambda) \qquad (2.23) \\ &= \frac{\pi}{3\epsilon_0\hbar^2}|\boldsymbol{P}_{ba}|^2\rho(\omega_{ba}). \qquad (2.24) \end{aligned}$$

Here we used the relation

$$\langle b|\, \boldsymbol{e}_\lambda \cdot \boldsymbol{P} \,|a\rangle = \frac{1}{3}\,|\boldsymbol{P}_{ba}|^2\,,$$

because the dipole moment of the electronic system is randomly oriented with respect to the polarization vector $\boldsymbol{e}_\lambda$ of the electromagnetic field. Accordingly, we obtain Einstein's B coefficient as $B_{ba} = \pi|\boldsymbol{P}_{ba}|^2/3\epsilon_0\hbar^2$. For the emission process, we interchange a and b: The higher energy $|b\rangle$ is initial and the lower energy $|a\rangle$ is final. Following similar calculations, the transition probability in the dipole approximation is

$$\begin{aligned} W_{ji}^{(\mathrm{e})} &\equiv B_{ab}\rho(\omega_{ba}) + A_{ab} \\ &= \frac{2\pi}{\hbar^2}\left(\frac{e}{m}\right)^2 \frac{\hbar}{2\epsilon_0 V} \sum_\lambda \frac{n_\lambda + 1}{\omega_\lambda} m^2\omega_{ba}^2 |\langle a|\boldsymbol{e}_\lambda \cdot \sum_j \boldsymbol{r}_j|b\rangle|^2 \delta(\omega_\lambda - \omega_{ba}), \qquad (2.25) \\ &= \frac{\pi}{3\epsilon_0\hbar^2}|\boldsymbol{P}_{ba}|^2\rho(\omega_{ba}) + \frac{\omega_{ba}^3}{3\pi\varepsilon_0\hbar c^3}|\boldsymbol{P}_{ab}|^2. \qquad (2.26) \end{aligned}$$

Here the λ summation was converted to an ω_λ integral. We therefore obtain Einstein's A and B coefficients: $B_{ba} = B_{ab}$ and $A_{ab}/B_{ba} = \hbar\omega_{ba}^3/\pi^2c^3$. Note here that $A_{ab} = (4\omega_{ba}^3/3\hbar c^3)|\boldsymbol{P}_{ab}|^2$ and $B_{ba} = B_{ab} = (4\pi^2/3\hbar^2)|\boldsymbol{P}_{ab}|^2$ are in cgs units.

We consider the rate constant A_{ab} of the spontaneous emission in more detail. The spontaneous emission comes from the part 1 of the factor $n+1$ in the second equation of (2.25). This physically means that the atomic system can emit light of any mode in any direction even under the condition $\{n_\lambda = 0\}$, i.e., in an environment with no photon around the atomic systems. When we consider such a large volume as $V \gg \lambda^3$ for the quantization of the radiation field, the λ summation can be converted to an integral with respect to ω_λ as

$$\sum_\lambda \longrightarrow 2\frac{V}{(2\pi)^3}\int d^3\boldsymbol{k} = \frac{V}{\pi^2 c^3}\int \omega_\lambda^2 d\omega_\lambda, \tag{2.27}$$

taking $\{n_\lambda = 0\}$ in (2.25). Then, we obtain the rate constant A_{ab} of the spontaneous emission as

$$\begin{aligned} A_{ab} &= \frac{2\pi}{\hbar^2}\frac{\hbar}{2\varepsilon_0 V}\frac{V}{\pi^2 c^3}\int d\omega_\lambda \frac{\omega_\lambda^2}{\omega_\lambda}\omega_{ba}^2 \frac{1}{3}\left|\boldsymbol{P}_{ab}\right|^2 \delta(\omega_\lambda - \omega_{ba}) \\ &= \frac{\omega_{ba}^3}{3\pi\varepsilon_0 \hbar c^3}\left|\boldsymbol{P}_{ab}\right|^2 . \end{aligned} \tag{2.28}$$

Note that the factor 2 in (2.27) comes from two independent polarizations of light per wavevector, which is an abbreviation of wavenumber vector.

Although the phenomenon of *spontaneous emission* is closely related to quantization of the radiation field as mentioned above, Einstein predicted this spontaneous emission without the quantization of the radiation field [12, 13]. At the same time, he successfully derived Planck's formula for thermal radiation as well as the so-called Einstein A and B coefficients. Here we briefly sketch Einstein's theory. Consider the radiation field interacting with an atomic systems in thermal equilibrium. We are interested in a pair of energy levels a and b of each atom. Figure 2.1 is a schematic energy diagram, where N_a and N_b are numbers of atoms in the a and b states, respectively.

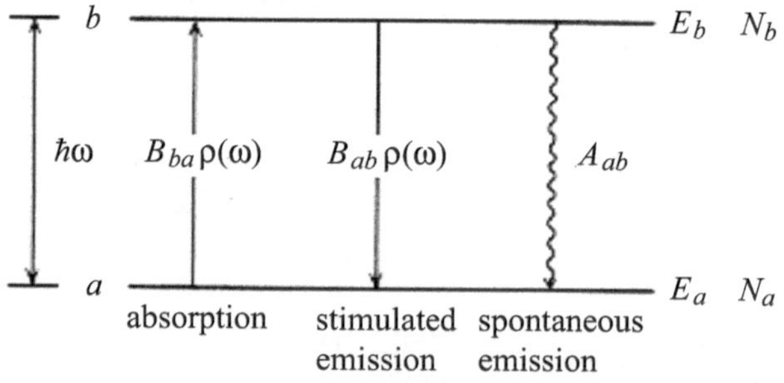

Fig. 2.1. Einstein obtained his A and B coefficient assuming thermal equilibrium between the electronic system (a, b) and the radiation field $\rho(\omega)$

There are three possible processes, as drawn in Fig. 2.1. Thus the rate equation for N_a is given by

$$\frac{dN_a}{dt} = -\frac{dN_b}{dt} = A_{ab}N_b - B_{ba}N_a\rho(\omega) + B_{ab}N_b\rho(\omega), \tag{2.29}$$

where $\rho(\omega)$ is the energy density of the radiation field. The second term in (2.29) illustrates the upward transition from a to b, absorbing the light with frequency ω, which results in a decrease of N_a. On the other hand, the third term describes the increase of N_a due to the downward transition from b to a, by stimulated emission in the presence of the relevant radiation field. This results in amplification of this field. It is evident that these two processes alone cannot maintain thermal equilibrium. In order to guarantee the thermal equilibrium of the atomic system, the first term corresponding to the spontaneous emission should be introduced. Thermal equilibrium means the stationary condition $dN_a/dt = 0$. Thus we obtain

$$\rho(\omega) = \frac{A_{ab}}{(N_a/N_b)B_{ba} - B_{ab}}. \tag{2.30}$$

Here (a) energy conservation should be kept for the exchange process between the atomic system and the radiation field, so that $E_b - E_a = \hbar\omega$. (b) The atomic system itself is in thermal equilibrium, so that $N_b/N_a = \exp\{-(E_b - E_a)\beta\}$, where $\beta \equiv 1/k_BT$. (c) In the limit of $T \to \infty$ ($\beta \to 0$), the two conditions $\rho(\omega) \to \infty$ and $N_b/N_a \to 1$ are required, leading to $B_{ba} = B_{ab}$. Furthermore, (d) under the condition $k_BT \gg \hbar\omega$, the energy density of the radiation field obeys the Rayleigh–Jeans formula:

$$\rho(\omega) \to \frac{\omega^2}{\pi^2 c^3} k_B T. \tag{2.31}$$

On the other hand, under the condition $k_BT \gg \hbar\omega$, relation (2.30) becomes $\rho(\omega) \to A_{ab}k_BT/B_{ba}\hbar\omega$. Comparing this result with (2.31), we obtain

$$\frac{A_{ab}}{B_{ba}} = \frac{\hbar\omega^3}{\pi^2 c^3}. \tag{2.32}$$

Substitution of (2.32) into (2.30) gives the well-known Planck formula for the thermal radiation:

$$\rho(\omega) = \frac{\hbar\omega^3}{\pi^2 c^3}\frac{1}{e^{\hbar\omega\beta} - 1}. \tag{2.33}$$

Next let us derive Planck's formula in terms of the quantum theory of radiation. In thermal equilibrium the energy distribution $\rho(\omega)$ for the thermal radiation is described by

$$\rho(\omega) = N(\omega)\hbar\omega \langle n_\omega \rangle, \tag{2.34}$$

where $N(\omega)$ is the density of states per unit volume for a photon $\hbar\omega$, and $\langle n_\omega \rangle$ is the thermal distribution of this photon at a temperature T. The density of states per unit volume for the radiation field confined within the volume $V = L^3$ is obtained as

$$N(\omega)d\omega = \frac{2}{V}\frac{V}{(2\pi)^3}d^3\boldsymbol{k} = \frac{k^2}{\pi^2}dk = \frac{\omega^2}{\pi^2 c^3}d\omega. \tag{2.35}$$

Here we have used the dispersion relation of the photon $\omega = ck$. The periodic boundary condition gives the relation

$$\boldsymbol{k} = \frac{2\pi}{L}(n_x, n_y, n_z) \qquad (n_x, n_y, n_z = 0, \pm 1, \pm 2, \ldots) \tag{2.36}$$

for the wavevector $\boldsymbol{k}$. If L is much larger than the wavelength λ of the light, $\boldsymbol{k}$ can be regarded as continuous numbers. When this approximation is not satisfied, suppression or enhancement of the spontaneous emission is expected, as will be discussed in Sect. 2.4. Since the photon obeys Bose–Einstein statistics, the thermal population $\langle n_\omega \rangle$ must be

$$\langle n_\omega \rangle = \frac{\sum n e^{-E_n\beta}}{\sum e^{-E_n\beta}} = \frac{1}{e^{\hbar\omega\beta} - 1}. \tag{2.37}$$

Here $E_n = \hbar\omega(n+1/2)$ is the eigenenergy of the photon mode ω (see Sect. 1.1). Substituting (2.35) and (2.37) into (2.34) we obtain Planck's formula:

$$\rho(\omega) = \frac{\omega^2}{\pi^2 c^3}\hbar\omega\frac{1}{e^{\hbar\omega\beta} - 1}.$$

It is emphasized that the energy distribution $\rho(\omega)$ of the radiation field, in general, is not restricted to the thermal distribution of (2.33). For example, laser systems have characteristic distributions, quite different from the thermal distribution, as will be discussed in Chap. 3.

2.3 Natural Width of a Spectral Line

In the previous section, the rate of spontaneous emission was calculated as radiative decay into a large number of photon modes in a large volume. Significantly, the finite lifetime τ restricts the resolution of the energy measurement in the emission spectrum [14]. Since an excited atom decays almost within its characteristic lifetime, the energy $\hbar\omega_{ba}$ between the electronic levels is observed only with an uncertainty

$$\hbar\Delta\omega_{ba} = \frac{\hbar}{\tau}. \tag{2.38}$$

It follows that the spontaneously emitted light is not monochromatic but has a spectral width, proportional to $1/\tau$. In this section we will consider this problem.

Supposing a two-level system with $(a,\ b)$, the downward transition from the excited state b to the lower state a emits a photon $\hbar\omega$, with an initial state $|b, n_\lambda = 0\rangle \equiv |b0\rangle$ and a final state $|a, n_\lambda = 1\rangle \equiv |a\lambda\rangle$ in (2.17). The coefficient $c_n(t)$ obeys the differential equation

$$i\hbar\dot{c}_{a\lambda} = \langle a\lambda|\, \mathcal{H}'_{(1)}\, |b0\rangle\, e^{i(\omega_a-\omega_b+\omega_\lambda)t} c_{b0}, \tag{2.39}$$

where the subscripts $a\lambda$ and $b0$ denote $|a, n_\lambda = 1\rangle$ and $|b, n_\lambda = 0\rangle$, respectively. On the other hand, by interchanging the initial and final states, we obtain

$$i\hbar\dot{c}_{b0} = \langle b0|\, \mathcal{H}'_{(1)}\, |a\lambda\rangle\, e^{i(\omega_b-\omega_a-\omega_\lambda)t} c_{a\lambda}. \tag{2.40}$$

Under the initial condition $\{c_{b0}(0) = 1,\ c_{a\lambda}(0) = 0\}$, we may assume $c_{b0}(t)$ to obey an exponential decay as $c_{b0}(t) = \exp(-\gamma t)$. Then we obtain

$$c_{a\lambda}(t) = \langle a\lambda|\, \mathcal{H}'_{(1)}\, |b0\rangle\, \frac{1 - e^{i(\omega_\lambda-\omega_0+i\gamma)t}}{\hbar\,(\omega_\lambda - \omega_0 + i\gamma)}, \tag{2.41}$$

where $\hbar\omega_0 \equiv E_b - E_a$, i.e., $\omega_0 = \omega_b - \omega_a$. Substituting (2.41) into the right side of (2.40), we obtain

$$\begin{aligned} -i\hbar\gamma &= \sum_\lambda \left|\langle a\lambda|\, \mathcal{H}'_{(1)}\, |b0\rangle\right|^2 \frac{1 - e^{i(\omega_0-\omega_\lambda-i\gamma)t}}{\hbar\,(\omega_0 - \omega_\lambda - i\gamma)} \\ &\to -\frac{i\pi}{\hbar} \sum_\lambda \left|\langle a\lambda|\, \mathcal{H}'_{(1)}\, |b0\rangle\right|^2 \delta(\omega_0 - \omega_\lambda). \end{aligned} \tag{2.42}$$

Because $\omega_0 t \gg 1$ and $\omega_0 \gg \gamma$, we have put $\gamma = 0$ on the right side of the equation, and neglected the real part for simplicity. The lifetime $\tau = 1/2\gamma$ due to the spontaneous emission is then obtained in agreement with the result in the previous section as

$$\begin{aligned} 2\gamma &= \frac{2\pi}{\hbar^2} \sum_\lambda \left|\langle a\lambda|\, \mathcal{H}'_{(1)}\, |b0\rangle\right|^2 \delta(\omega_0 - \omega_\lambda) \\ &= \frac{2\pi}{\hbar^2} \int \frac{d\Omega}{4\pi} \left|\langle a\lambda_0|\, \mathcal{H}'_{(1)}\, |b0\rangle\right|^2 N(\omega_0) = A_{ab}. \end{aligned} \tag{2.43}$$

It is evident that the A coefficient governs the lifetime of the spontaneous emission.

Using (2.41), the intensity $I(\omega)$ of the emitted light is written by

$$I(\omega)d\omega = \hbar\omega N(\omega)d\omega \int \frac{d\Omega}{4\pi} \left|c_{a\lambda}(\infty)\right|^2 = \frac{\gamma}{\pi} \frac{\hbar\omega d\omega}{(\omega - \omega_0)^2 + \gamma^2}. \tag{2.44}$$

It follows that, when we measure the energy difference $E_b - E_a = \hbar\omega_0$ between the electronic states, the spectral resolution is limited to $\Delta E = \hbar\gamma$. It is emphasized that the last equation of (2.44) has a Lorentzian form with central frequency ω_0. The full width 2γ at half maximum is called the *natural width of a spectral line.* In addition to the natural width, the linewidths of the absorption and emission spectra are more broadened by collision and molecular vibrations in gas systems or by lattice vibrations in solids. Furthermore, the spectra are also affected by the Doppler shift in a gas system or by inhomogeneous broadening in solids. These additional effects could be controlled. For example, the collision is greatly reduced at low temperatures and low pressure. It was believed that, in contrast to the additional effects, the natural width 2γ was hard to control. However, suppression of spontaneous emission, which will be discussed in the next section, demonstrates that the natural width 2γ is also controllable.

Let us discuss the contribution of the real part in (2.42). The real part originates from the interaction of the electronic system with zero-point vibration of the electromagnetic field and gives rise to the energy shift. This is called the *Lamb shift.* According to Dirac's theory, the $2S_{1/2}$ and $2P_{1/2}$ states of a hydrogen atom are energetically degenerate with respect to each other. However, Lamb and Rutherford found that the S state energy was higher than the P state energy by 1050 MHz. This difference comes from the effect of the higher order electromagnetic interaction on an electron in the $2S_{1/2}$ state being different from that in the $2P_{1/2}$ state. Tomonaga succeeded in explaining quantitatively the experimental results in terms of renormalized theory.

2.4 Suppression and Enhancement of the Spontaneous Emission

Einstein introduced the spontaneous emission to maintain the thermal equilibrium between the radiation field and the atomic system. It had been believed that spontaneous emission was a fundamental phenomenon and the lifetime of the excited atom due to spontaneous emission was a property inherent to each atom. However, the spontaneous emission rate is not a fundamental property of the isolated atom but a property of the coupled system consisting of the atom and the radiation field. The most significant feature is irreversibility of the process. This is due to the fact that an excited atom has an almost unlimited number of radiative channels having the same photon energy, as discussed in Sect. 2.3. On the other hand, for the excited atom in a resonator, the length of which is comparable to or smaller than the wavelength of light, its coupling to the radiation field could be markedly modified, leading to suppression or enhancement of the spontaneous emission rate. In this section we consider these phenomena.

2.4.1 Suppression of the Spontaneous Emission from Cs Atoms

Consider a resonator consisting of coupled mirrors with a separation d. The radiation mode with wavelength $\lambda \leq 2d$ is allowed to persist in this cavity (resonator), because the resonator mode must have a node on both mirror surfaces. For $\lambda > 2d$, on the other hand, spontaneous emission should be greatly suppressed and its lifetime should be infinite. Hulet, Hilfer and Kleppner [15] demonstrated the suppression of spontaneous emission in the coupled mirrors. They measured the spontaneous emission intensity arising from downward electronic transition from the high Rydberg state of Cs atoms with $n = 22$ and $|m| = 21$ to that with $n = 21$ and $|m| = 20$. From the experimental point of view, this transition has great advantages: (a) The wavelength of the emission light is about 0.45 mm. The resonator effect becomes clear for such a long-wavelength radiation field. (b) The lifetime is 450 µs, and is long enough to measure the change of the lifetime as a function of the wavelength to the cavity size.

Their experimental setup is composed of three parts. In the first stage a hot Cs atom is excited to the high Rydberg state through multiphoton processes, using a combination of a microwave and dye laser. In the second stage, the excited Cs atom enters the resonator and travels along two mirror planes. The resonator is made of a pair of gold-coated metal mirrors and the separation d of these two mirrors is fixed at

$$d = 230.1\,\mu\text{m} = 1.02\frac{\lambda_0}{2}.$$

Instead of changing d, the wavelength λ of the transition $\{n = 22, m = 21\} \to \{n = 21, m = 20\}$ is varied by applying a high voltage between the two electrodes of the metal mirrors. The maximum electric field applicable to the resonator is $E = 2600\,\text{V/cm}$. This can induce the second-order Stark shift of $\Delta\lambda/\lambda = 0.04$, leading to the wavelength modulation of $\lambda/2d = 0.98 \sim 1.02$. The third stage is to measure the number of Cs atoms in the excited $\{n = 22, |m| = 21\}$ state. Figure 2.2 shows the number of excited-state atoms measured in the third stage as a function of $\lambda/2d$. The number of excited atoms strikingly increases above $\lambda/2d = 1$, demonstrating the suppression of the spontaneous emission. Below $\lambda/2d = 1$, on the other hand, the spontaneous emission takes place even in the resonator, because the excited Cs atom possibly decays into the resonator modes. By analyzing the data, the lifetime is 20 times longer than 450 µs measured in free space. It is worth noting that the remarkable drop of the signal above $\lambda/2d = 1.01$ is not related to the enhancement of the spontaneous emission. The transition from the high Rydberg state $\{n = 22, |m| = 21\}$ to the ionic level is induced by the high electric field and, as a result, the number of excited atoms is greatly reduced.

2.4.2 Suppression and Enhancement of Spontaneous Emission

In this subsection, we show the alteration of visible spontaneous emission of atoms coupled to the degenerate modes of a confocal resonator [16]. This is due to a change in the density of coupling radiation modes. The partial emission rate into the resonator modes is enhanced by a factor of 19 when

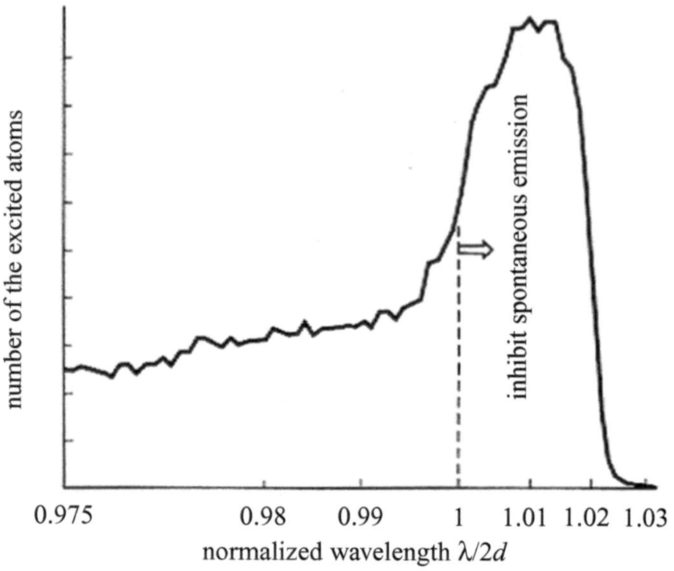

Fig. 2.2. The number of atoms staying in the excited state after passing through the cavity with separation d [15]

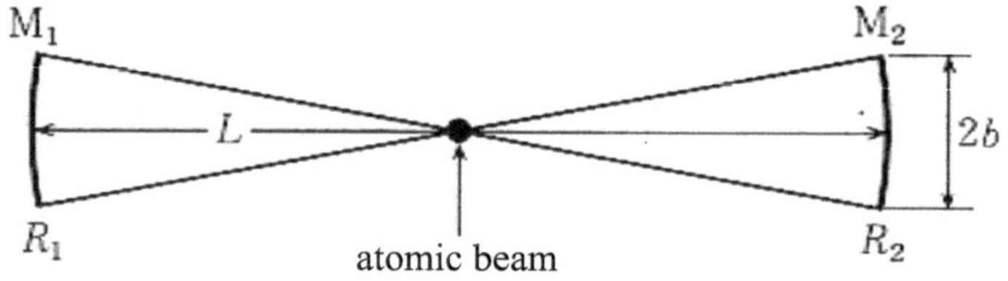

Fig. 2.3. The confocal resonator. The atomic and laser beams cross around its focal point

the resonator is tuned to the atomic transition frequency and is inhibited by a factor of 42 when it is detuned. In this experiment, the resonator linewidth is greater than the linewidth of the atomic transition, and the atomic sample is of negligible optical thickness as shown in Fig. 2.3. Consider an atom near the center of a confocal resonator of length L composed of mirrors M_1 and M_2 of reflectivities R_1 and R_2, both with aperture diameter $2b$ as shown in Fig. 2.3.

The atom illuminates the cavity with dipole radiation, producing a series of reflected and transmitted waves. The radiated power is obtained by adding together the multiple contributions of these transmitted waves. The ratio of γ, the spontaneous emission rate into the cavity, to γ_{sp}, the free-space rate into the same solid angle, is then given by

$$\frac{\gamma}{\gamma_{\mathrm{sp}}} = \frac{1}{1-R}\frac{1}{1+[1/(1-R)]^2\sin^2 kL}, \tag{2.45}$$

where L is the distance between the mirrors, $R=\sqrt{R_1R_2}$, $k = 2\pi/\lambda$ and $1 - R \ll 1$. It follows from (2.45) that the spontaneous emission should be greatly suppressed as $\gamma_{\mathrm{inh}} = (1-R)\gamma_{\mathrm{sp}} \ll \gamma_{\mathrm{sp}}$ for $kL = 2n\pi \pm \pi/2$, even if $\lambda \ll L$, that is, $kL \gg 2\pi$. Here n is integer. On the other hand, the spontaneous emission should be enhanced as $\gamma_{\mathrm{enh}} = \gamma_{\mathrm{sp}}/(1-R) \gg \gamma_{\mathrm{sp}}$ for $kL = n\pi$. Thus, by varying the distance L so as to make $k\Delta L$ of the order of π, suppression and enhancement of the spontaneous emission is expected to appear periodically.

For the atomic dipole transition with polarization perpendicular to the cavity axis, $\gamma_{\mathrm{sp}} = (3/8\pi)\Gamma_{\mathrm{sp}}\Delta\Omega$, where Γ_{sp} is the total free-space spontaneous emission rate and $\Delta\Omega = 8\pi b^2/L^2$ the solid angle subtended by both cavity mirrors. Therefore the total emission rate Γ is given by

$$\Gamma = \Gamma_{\mathrm{sp}}\left[1 + \left(\frac{\gamma}{\gamma_{\mathrm{sp}}} - 1\right)\frac{3}{8\pi}\Delta\Omega\right], \tag{2.46}$$

where $\gamma_{\mathrm{sp}} = (3/8\pi)\Gamma_{\mathrm{sp}}$. Γ has a maximum $\Gamma_{\mathrm{enh}} = \Gamma_{\mathrm{sp}}[1+(1-R)^{-1}3\Delta\Omega/8\pi]$ for $kL = n\pi$, whereas it has a minimum $\Gamma_{\mathrm{inh}} = \Gamma_{\mathrm{sp}}[1 - 3\Delta\Omega/8\pi]$ for $kL = 2n\pi \pm \pi/2$.

The experimental arrangement is shown in Fig. 2.4. An atomic beam of Yb is intercepted by a beam from a cw dye laser tuned to the $^1\mathrm{S}_0{-}^3\mathrm{P}_1$ transition of Yb with $\Gamma_{\mathrm{sp}} = 1.1 \times 10^6\mathrm{s}^{-1}$ at 556 nm as shown in Fig. 2.4(a). The excited atoms are positioned at the center of a confocal mirror resonator, and are confined to a region of size approximately 1.5 mm along the resonator. The laser is linearly polarized perpendicular to the resonator axis and is tuned to the ^{174}Yb isotopic component of the line.

Since this isotope has zero nuclear spin and is well resolved from the other components, only the single transition between $^1\mathrm{S}_0$ and $^3\mathrm{P}_1$ states of Yb is excited. A piezodevice (PZT), on which the mirror M_2 is mounted, varies the distance L of the resonator. The results are shown in Fig. 2.5. It is evident that the spontaneous emission rate varies periodically with cavity tuning frequency. When the beam stopper is inserted into the resonator, the emission rate is almost constant. The value is nearly equal to Γ_{sp}. By analyzing the data, they obtained $\gamma_{\mathrm{sp}}/\gamma_{\mathrm{inh}} = 42$ and $\gamma_{\mathrm{enh}}/\gamma_{\mathrm{sp}} = 19$. The former value agrees well with the theoretical value $\gamma_{\mathrm{sp}}/\gamma_{\mathrm{inh}} = (1-R)^{-1} \approx 2/(T_1 + T_2) = 43.3 \pm 2.0$, estimated for $T_1 = 2.8 \pm 0.1\%$, $T_1 = 1.8 \pm 0.1\%$, $L = 5.00$ cm and $2b = 4$ mm, where T_1 and T_2 are the transmittance of M_1 and M_2, respectively. On the other hand, the latter value is smaller than the theoretical value, probably owing to Doppler broadening of the emission spectrum and imperfection of the mirrors. Finally, it is worth noting that the suppression of the spontaneous emission greatly reduces the threshold of the laser oscillation, arising from the stimulated emission. This technique will enable one to make a single-atom laser (or maser) and a zero-threshold laser.

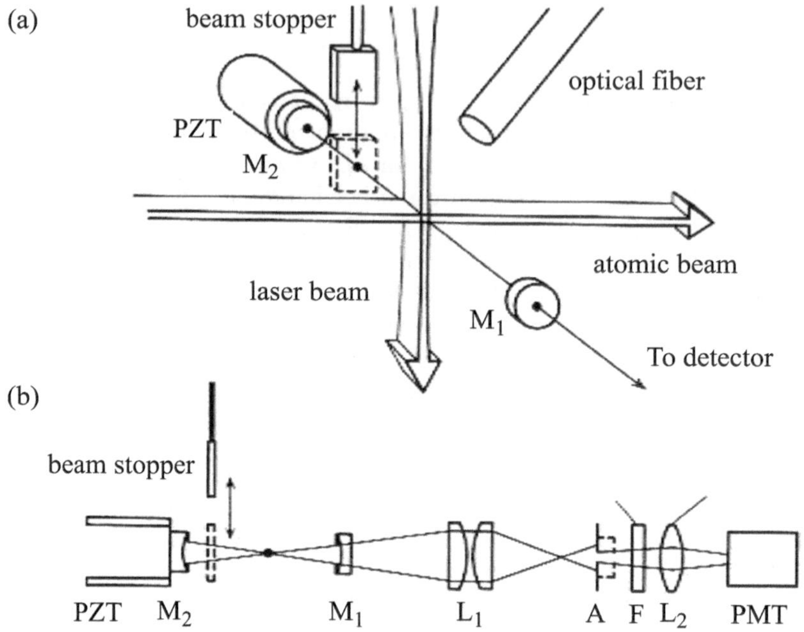

Fig. 2.4. Experimental apparatus. (**a**) Atomic-beam excitation geometry, showing the relative orientation of the atomic and laser beams, confocal resonator mirrors M_1 and M_2, the moveable beam stopper, and the optical fiber bundle. (**b**) On-axis optical configuration, showing the positions of the imaging lens (L_1), adjustable aperture (A), the image of M_1 by L_1, the interference filter (F), lens (L_2) and photomultiplier tube (PMT) [16]

2.5 Laser Cooling of an Atomic System

While semiconductor laser arrays or optical fiber lasers are being used to process metallic systems on one hand, laser systems are possible, on the other hand, to cool down an atomic gas to several tens of nanodegrees, i.e., to the order of 10^{-8} K. Such a low temperature has never been realized by conventional cooling methods. This has been made possible by making the best use of the electron–radiation interaction. As a result, both Bose–Einstein and Fermi condensation of neutral atoms have been realized as will be shown in Sect. 2.6. The system of neutral atoms has two kinds of freedom, i.e., internal and external degrees. Atom–photon interactions induce two types of effects: dissipative (or absorptive), on the one hand, and reactive (or dispersive), on the other hand. These two effects come mainly from the internal degree of atomic freedoms but the atom–photon interactions are shown to be possible to take off the kinetic energy of these atoms into the reservoir of the radiation field, i.e. the external degree of atoms. These processes of laser cooling are introduced in this section.

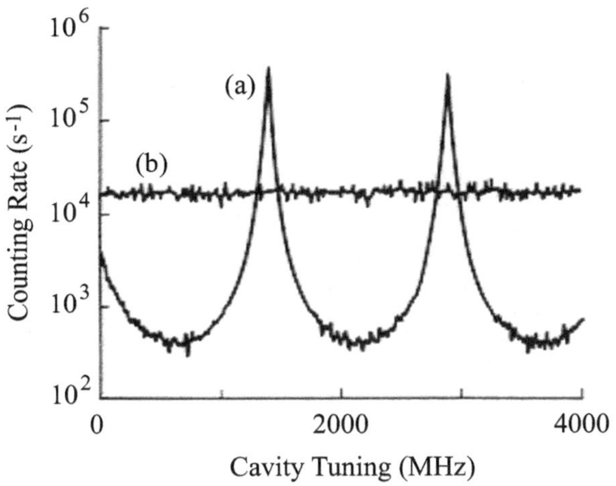

Fig. 2.5. Photon counting rate for light transmitted through the cavity mirror, as a function of cavity tuning. Curves (**a**): light emitted through the center of the mirror. Curves (**b**): normalized counting rate with the cavity blocked, showing the free-space rate into the same aperture [16]

2.5.1 Doppler Cooling

The absorption and emission spectra of an atomic gas with finite temperature are inhomogeneously broadened by the Doppler effect. The spectral width is proportional to the square-root of T, the gas temperature. We denote the frequency of the absorption peak by ω_A. When we pump the atomic system at the frequency ω_L below ω_A but within the inhomogeneously broadened width, the average frequency of spontaneous emission is ω_A so that the kinetic energy $\hbar(\omega_A - \omega_L)$ is taken off into the reservoir of the radiation field as an average per single cycle of the absorption and spontaneous emission. The atomic gas, e.g., of Sr with $T = 800\,\mathrm{K}$, is almost stopped within 0.3 ms and 4 cm by Doppler cooling because the acceleration caused by taking off the kinetic energy is as large as $10^7\,\mathrm{m/s}^2$, i.e., 10^5 times lager than the acceleration of gravity.

Interesting effects can also be obtained by combining the effects of two counterpropagating laser waves [17]. This cooling process results from a Doppler-induced imbalance between two radiation pressure forces with opposite directions. The two counterpropagating laser waves have the same intensity and the same frequency and they are slightly detuned to the red side of the atomic frequency ($\omega_L < \omega_A$). For an atom at rest, the two radiation pressure forces exactly balance each other and the net force is equal to zero. For a moving atom, the apparent frequencies of the two laser waves are Doppler shifted. The counterpropagating wave gets closer to resonance and exerts a stronger radiation pressure force than the copropagating wave which gets farther from resonance. The net force is thus opposite to the atomic velocity v and can be written for small v as $F = -\alpha v$ where α is a friction coefficient.

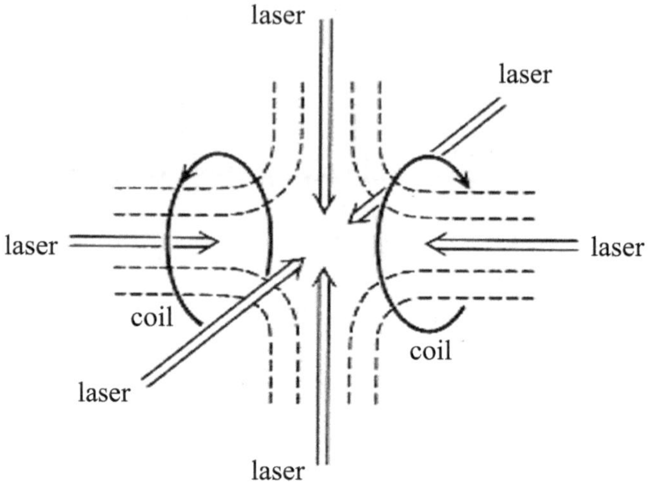

Fig. 2.6. Magnetooptical trap

By using three pairs of counterpropagating laser waves along three orthogonal directions, as shown in Fig. 2.6, one can damp the atomic velocity in a very short time, on the order of a few microseconds, achieving what is called an "optical molasses" [18].

A pair of circular coils around the x-axis can induce a quadrupolar magnetic field around the origin when the electric currents are opposite in the direction, so that this system can trap neutral atoms. Six laser beams are irradiating the atomic system with detuning in the lower frequency side and with opposite polarity in each of the x-, y- and z-axes. The cooling limit due to the present process comes from recoiling kinetic energy so that the available temperature kT is estimated to be $\hbar\Gamma/2$ with the spectral width Γ due to spontaneous emission. This is 240 μK for the Na atomic system and 125 μK for Cs. Chu et al., however, have observed that the Na atomic system reached 40 μK, far below the estimated limit temperature 240 μK [18, 19]. This mystery was resolved as another cooling mechanism contributing simultaneously. This is explained in the following.

2.5.2 Polarization-Gradient Cooling

In the previous subsection on Doppler cooling, we discussed separately the manipulation of internal and external degrees of freedom. In fact, however, there exist cooling mechanism resulting from an interplay between the internal degree spin and external degrees of freedom, and between dispersive and dissipative effects. In this subsection, one of these mechanisms, "polarization-gradient cooling", or the so-called "Sisyphus cooling" mechanism, is shown to lead to temperatures much lower than Doppler cooling [20].

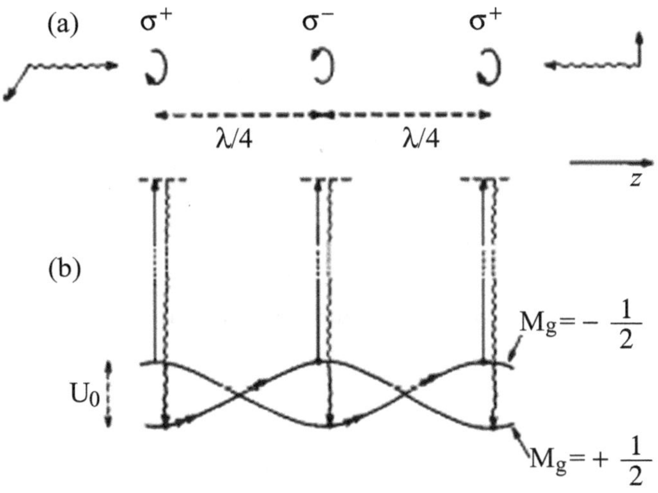

Fig. 2.7. Sisyphus cooling. Laser configuration formed by two counterpropagating plane waves along the z-axis with orthogonal linear polarizations (**a**). The polarization of the resulting electric field is spatially modulated with a period $\lambda/2$. Every $\lambda/4$, it changes from σ^+ to σ^- and vice versa. For an atom with two ground-state Zeeman sublevels $M_g = \pm 1/2$, the spatial modulation of the laser polarization results in correlated spatial modulations of the light shifts of these two sublevels and of the optical pumping rates between them (**b**). Because of these correlations, a moving atom runs up potential hills more frequently than down (double arrows of **b**) [20]

Most atoms, in particular alkali atoms, have a Zeeman structure in the ground state. Since the detuning used in laser cooling experiments is not too large compared to Γ, both different light shifts and optical pumping transitions exist for the various Zeeman sublevels in the ground state. Furthermore, the laser polarization varies in space so that the frequency shifts of the Zeeman sublevels and optical pumping rates are position dependent. It is shown here how the combination of these various effects can lead to a very efficient cooling mechanism.

Consider the laser configuration of Fig. 2.7(a), consisting of two couter-propagating plane waves along the z-axis, with orthogonal linear polarizations and with the same frequency and the same intensity. Then the polarization of the total field of the two waves changes from σ^+ to σ^- and vice versa every $\lambda/4$.

In between, it is elliptical or linear. As shown in Fig. 2.7(b), we consider the simple case where the atomic ground state has an angular momentum $J_g = 1/2$. Then two Zeeman sublevels $M_g = \pm 1/2$ undergo different frequency shifts of the Zeeman sublevels, depending on the laser polarization, so that the Zeeman degeneracy in zero magnetic field is removed. This gives the energy diagram of Fig. 2.7(b) showing spatial modulations of the Zeeman

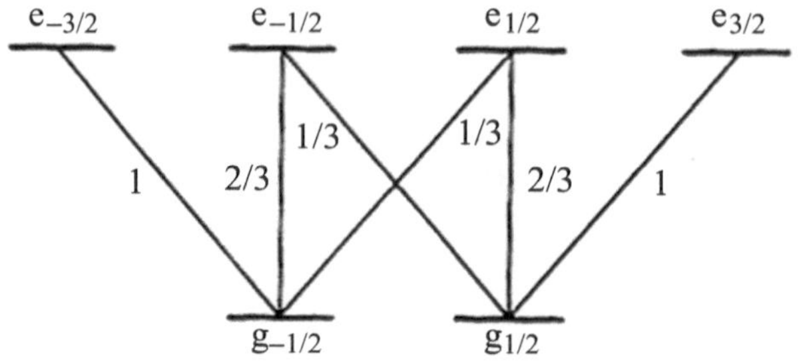

Fig. 2.8. Clebsch–Gordan coefficients for the transition between the ground state $J_g = 1/2$ and the excited state $J_e = 3/2$ [20]

splitting between the two sublevels with a period $\lambda/2$. This spatial dependence is evaluated in terms of the energy diagram of the Na atom with the ground state angular momentum $J_g = 1/2$ and the excited state $J_e = 3/2$ and with Clebsch–Gordan coefficients drawn in Fig. 2.8.

This is because the dispersive energy due to the two waves stabilizes each level $M_g = \pm 1/2$ by

$$\Delta(M_g) = \Omega^2 \frac{\delta}{\Gamma^2 + 4\delta^2} \tag{2.47}$$

with $\delta \equiv \omega_L - \omega_A$ and the relevant Rabi frequency Ω.

If the detuning δ is not too large compared to Γ, there is real absorption of photons by the atom and subsequent spontaneous emissions. Note here that on the centreal part in Fig. 2.7 with σ_- polarity only the photoabsorption from $M_g = 1/2$ to $M_e = -1/2$ is possible and the spontaneous emission from $M_e = -1/2$ to $M_g = -1/2$ is dominant. As a result, here the kinetic energy U_0 can be removed from an atomic system for an absorption-spontaneous emission. In the next step, part of the remaining kinetic energy is being changed into the potential energy U_0 by moving from the σ_- region to the σ_+ region by $\lambda/4$. Here the absorption from $M_g = -1/2$ to $M_e = 1/2$ is induced and subsequently the spontaneous emission into $M_g = 1/2$ is dominant so that more kinetic energy U_0 is removed to the radiation field. Finally when the remaining kinetic energy becomes less than U_0, these atoms are trapped within the potential wells.

As a result, this Sisyphus cooling leads to temperature T_{sis} such that $k_B T_{\text{sis}} \simeq U_0 \simeq \hbar\Omega^2/4\delta$ when $4|\delta| > \Gamma$. At low intensity, the light shift U_0 is much smaller than $\hbar\Gamma$. This explains why Sisyphus cooling leads to temperatures much lower than those achievable with Doppler cooling. One cannot, however, decrease the laser intensity indefinitely. The recoil due to the spontaneous emitted phonons, which has been neglected in the previous discussion, increased the kinetic energy of the atom by an amount on the order of E_R with $k = 2\pi/\lambda$ and λ the laser wavelength:

$$E_R = \frac{\hbar^2 k^2}{2M}. \tag{2.48}$$

The effective temperature which Sisyphus cooling can reach is limited by $T_R = E_R/k_B$. The value of T_R ranges from a few hundred nanokelvin for alkalis to a few microkelvin for helium.

2.6 Bose–Einstein Condensation in a Gas of Neutral Atoms

Bose–Einstein condensation (BEC) and fermionic condensation are ubiquitous phenomena which play significant roles in condensed matter, atomic, nuclear, and elementary particle physics, as well as astrophysics. The most striking feature of BEC is a macroscopic population in the ground state of the system at finite temperature. The study of these quantum condensations may also advance our understanding of superconductivity and superfluidity in more complex systems as well as the cross-over between BEC and the Bardeen–Cooper–Schrieffer (BCS) state. Laser cooling methods introduced in the previous section provided an effective approach towards very low temperatures of a gas of atoms, but have so far been limited to phase-space densities typically 10^5 times lower than required for BEC. The combination of laser cooling with evaporative cooling was a prerequisite for obtaining BEC in alkali atoms. The BEC of rubidium [21], lithium [22], and sodium [23] was reported independently by the group of Colorado, Texas, and MIT, respectively, within a few months in 1995.

Evaporative cooling requires an atom trap which is tightly confining and stable. The tightest confinement in a magnetic trap is achieved with a spherical quadrupole potential. The atoms, however, are lost from this trap due to nonadiabatic spin flips as the atoms pass near the center, where the field rapidly changes direction. This region constitutes a "hole" in the trap of micrometer dimensions. The Colorado group developed the time orbiting potential (TOP) trap which suppressed this trap loss, but at the cost of lower confinement. On the other hand, the MIT group suppressed the trap loss by adding a repulsion potential around the zero of the magnetic field, literally "plugging" the hole. This was accomplished by tightly focusing an intense blue-detuned laser that generated a repulsive optical dipole force. The total (dressed-atom) potential is a combination of the magnetic quadrupole trapping potential, the repulsive potential of the plug, and the effective energy shifts due to the rf, as shown in Fig. 2.9. At the point where atoms are in resonance with the rf, the trapped state undergoes a spin flip, crossing with the untrapped states. Over 7 s, the rf frequency was swept from 30 MHz to the final value below 1 MHz, while the field gradient was first increased to 550 G/cm and then 180 G/cm. Thus the Na atoms with higher temperature are evaporated, reducing the atomic temperature.

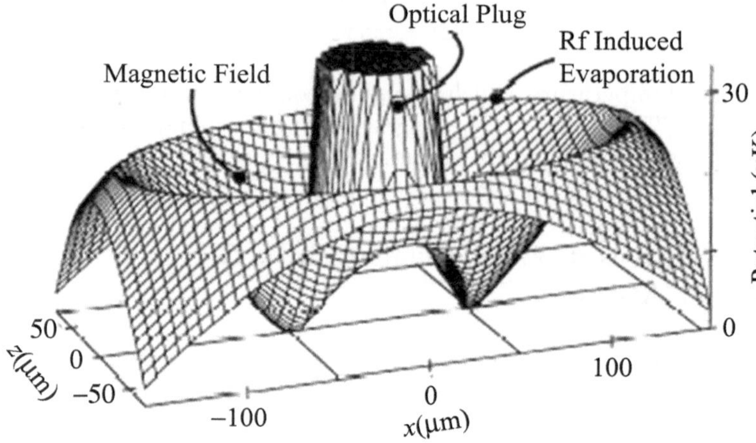

Fig. 2.9. Adiabatic potential due to the magnetic quadrupole field, the optical plug, and the rf. This cut of the three-dimensional potential is orthogonal to the propagation direction (y) of the blue-detuned laser. The symmetry axis of the quadrupole field is the z-axis [23]

The temperature and total number of atoms were determined using absorption imaging of the probe light pumping to the $F = 2$ state from the $F = 1$ state of Na atom. Above the critical frequency 0.7 MHz for the final frequency the distribution was perfectly spherical as expected for a thermal uncondensed cloud. When, however, the final frequency was lowered below this value, an elliptical core signal increased in intensity, whereas the broad spherical cloud became less intense. The elliptical core cloud, reflecting the potential shape, is assigned to be due to the Bose condensate, while the broad spherical cloud is due to the normal fraction. This result is drawn in Fig. 2.10 [22].

Here the kinetic energy of the condensed atoms is estimated around 1 nK, much less than the zero-point energy of the trap, 35 nK, and the internal energy of 120 nK. The number density in these condensates is estimated to be $4 \times 10^{14}\,\mathrm{cm}^{-3}$.

A few months before publication of the MIT paper on BEC of Na atoms, the Colorado group [21] reported the BEC of Rb atoms. This group also used the combination of laser cooling and evaporative cooling but additionally the time orbiting potential (TOP) magnetic trap to plug the "hole." This is a superposition of a large spherical quadrupole field and a small uniform transverse field that rotates at 7.5 kHz. This arrangement results in an effective average potential that is an axially symmetric, three-dimensional (3D) harmonic potential providing a tight and stable configuration during evaporation.

After laser cooling and TOP trapping, they obtained the Rb atomic system with the number density $2 \times 10^{10}\,\mathrm{cm}^{-3}$ and 90 μK. Evaporative cooling was performed by decreasing values of the rf evaporating frequency into a value ν_{ev} in Fig. 2.11, resulting in a corresponding decrease in the sample

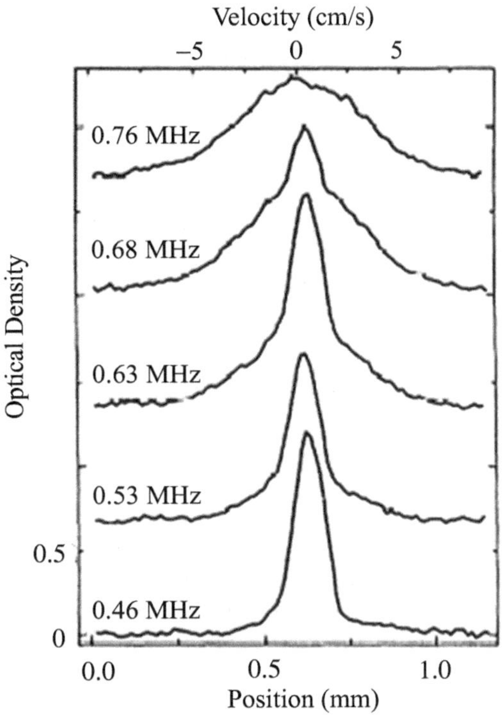

Fig. 2.10. Optical density as a function of position along the z-axis for progressively lower values of the final rf frequency. There are vertical cuts through time-of-flight images like those in Fig. 2.9. For $\nu_{\rm rf} < 0.7\,$MHz, they show the bimodal velocity distributions characteristic of the coexistence of a condensed and uncondensed fraction of Na atoms. The top four plots have been offset vertically for clarity [23]

temperature and an increase in phase-space density. As shown in Fig. 2.11, a sharp increase in the peak density was observed at a value of $\nu_{\rm ev} = 4.23\,$MHz. This increase is expected at the BEC transition. Below the transition, i.e., at the smaller value of $\nu_{\rm ev} \le 4.23\,$MHz, there is a two-component cloud, with a dense central condensate surrounded by a diffuse, noncondensed fraction as shown in Fig. 2.11. At $\nu_{\rm ev} = 4.25\,$MHz just above the BEC transition, the number density is estimated to be $2.6 \times 10^{12}\,{\rm cm}^{-3}$ and the temperature is 170 nK. These values were calculated for the sample in the unexpanded trap. However, after the adiabatic expansion stage, the atoms are still in good thermal equilibrium, but the temperatures and densities are greatly reduced. The 170 nK temperature is reduced to 20 nK, and the number density is reduced from $2.6 \times 10^{12}\,{\rm cm}^{-3}$ to $1 \times 10^{11}\,{\rm cm}^{-3}$.

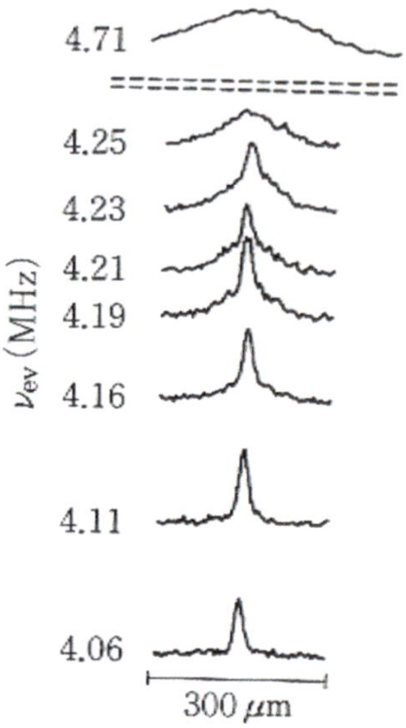

Fig. 2.11. Horizontal sections taken through the velocity distribution at progressively lower values of ν_{ev} show the appearance of the condensate fraction [21]

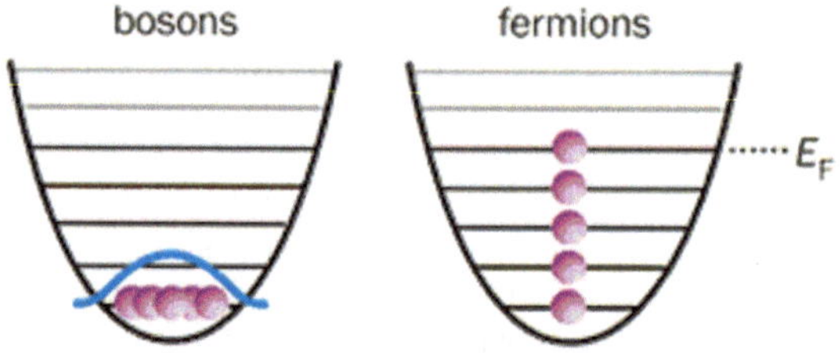

Fig. 2.12. At absolute zero, gaseous boson atoms all end up in the lowest energy state. Fermions, in contrast, fill the available states with one atom per state shown here for a one-dimensional harmonic confining potential. The energy of the highest filled state at $T = 0$ is the Fermi energy E_F

2.7 Condensates of a Fermionic Gas

Fermions and bosons are very different at the quantum level. Obeying Pauli's exclusion principle, identical fermions cannot occupy the same quantum state at the same time. Bosons, however, can share quantum states. At ultralow temperatures, bosons will eagerly fall into a single quantum state to form a Bose–Einstein condensate, whereas fermions tend to fill energy states from the lowest up, with one particle per quantum state as shown in Fig. 2.12. At high temperatures, in contrast, bosons and fermions spread out over many states with, on average, much less than one atom per state.

The quantum behavior emerges gradually as the fermion gas is cooled below the Fermi temperature $T_F = E_F/k_B$, where E_F is the Fermi energy. T_F, which is typically less than 1 μK for atomic gases, marks the crossover from the classical to the quantum regime. When the system temperature is reduced furthermore, the weakly interacting pairs of fermionic atoms begin to act collectively and finally collapse into a coherent many-body state which is called a Bardeen–Cooper–Schrieffer state (BCS) . This concept originates in the conventional superconductor due to the condensation of electron pairs. In the case of noncharged pairs like pairs of liquid helium-3 atoms, this state is accompanied with superfluidity.

On the other hand, two fermionic atoms are sometimes tightly bound into a molecule. These molecules are bosonic so that these can collapse into a common ground state, known as a Bose–Einstein condensate (BEC), below the critical temperature of an order of 10^{-5} K [24–26]. In 1998 Ketterle and his coworkers first demonstrated the technique that allows the interactions between ultracold atoms to be controlled, thus making a true fermionic condensate possible. This technique takes advantage of resonant scattering between atoms, and allows the strength as well as the sign of the interactions between the atoms to be tuned with an external magnetic field. This phenomenon, which is known as a Feshbach resonance, arises when the kinetic energy of a pair of colliding atoms that have one particular spin orientation is close to the kinetic energy corresponding to a quasibound pair of atoms with a different spin configuration.

This magnetic-field Feshbach resonance provides the means for controlling both the strength of cold atom interactions, characterized by the s-wave scattering length a, as well as whether they are, in the mean-field approximation, effectively repulsive ($a > 0$) or attractive ($a < 0$). For magnetic-field detuning on the $a > 0$ (BEC) side of the resonance there exists an extremely weakly bound molecular state whose binding energy depends on the detuning from the Feshbach resonance. On the side of $a < 0$, condensation of Cooper pairs is expected for an atomic gas. Therefore it is possible to observe the BCS–BEC crossover regime by changing the magnetic field so as to cover both sides of the Feshbach resonance. Here we distinguish fermionic condensates, i.e., condensation of Cooper pairs, from the BEC extreme where there remains no fermionic degree of freedom because all fermions are bound into bosonic molecules. The fermionic condensation was confirmed for the $a < 0$, or BCS, side of the Feshbach resonance by Regal, Creiner and Jin [27]. They trapped and cooled a dilute gas of the fermionic isotope ^{40}K, which has a total atomic spin $f = 9/2$ in its lowest hyperfine ground state and thus ten available Zeeman spin states $|f, m_f\rangle$. A trapped gas of ^{40}K atoms is evaporatively cooled to quantum degeneracy and then a magnetic-field Feshbach resonance is used to control the atom–atom interactions. The location of this resonance was precisely determined from low-density measurements of molecular dissociation as $B_0 = 202.10 \pm 0.07$ G. Here the magnetic-field dependence of the s-wave scattering length a is drawn in Fig. 2.13.

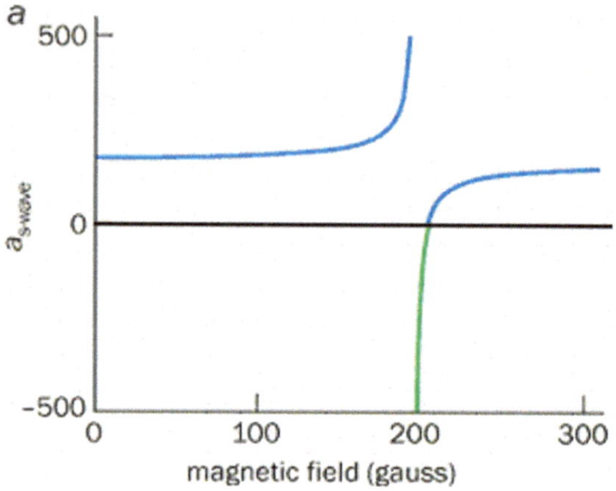

Fig. 2.13. The "scattering length" of s-wave collisions of ultracold ^{40}K atoms, $a_{s\text{-wave}}$, as a function of applied magnetic field. The effective interactions between the colliding atoms are attractive if $a_{s\text{-wave}} < 0$, and repulsive if $a_{s\text{-wave}} > 0$. The collision cross-section is proportional to the square of $a_{s\text{-wave}}$. A magnetic field of about 200 gauss would create the large, attractive interaction required for the atoms to form Cooper pairs [24]

Experiments are initiated by preparing atoms in a nearly equal, incoherent mixture of the $|9/2, -7/2\rangle$ and $|9/2, -9/2\rangle$ spin states at a low temperature $T/T_F \sim 0.1$. The two kinds of fermions are scattered so efficiently that thermal equilibrium can be reached more rapidly than in the case of a single kind. In order to investigate the BCS–BEC crossover regime, this ultracold two-component atom gas is prepared at a magnetic field far above the Feshbach resonance. The magnetic field is then slowly lowered, at typically 10 ms/G, to a value B_{hold} near the Feshbach resonance $B_0 = 202.1$ G. This sweep is slow enough to allow the atoms to collide effectively in the trap. As Fig. 2.13 shows, the BEC of the molecules is realized for $\Delta B \equiv B_h - B_0 < 0$ with the scattering length $a > 0$ in [24–26]. A Fermi gas of ^{40}K is cooled initially below $T < T_F = 0.17\,\mu$K, and the magnetic field B is swept down to $\Delta B = -0.56$ G. Here the BEC of the molecules is confirmed. The fermionic ^{40}K atoms are condensed in the BCS state $\Delta B \equiv B_h - B_0 > 0$, where the scattering length is $a < 0$. In order to prove the BCS condensation, the fermionic atoms are pairwise projected adiabatically onto the molecules in the molecular regime $\Delta B \equiv B_h - B_0 < 0$, and the momentum distribution of the resulting molecular gas is measured. After a total of typically 17 ms of expansion, the molecules are selectively detected by optical absorption images and analyzed as a two-component function for a condensate and noncondensed molecules. The measured condensate fraction N_0/N is plotted in Fig. 2.14 as a function of the magnetic-field detuning from the resonance $\Delta B = B_{\text{hold}} - B_0$. The data in Fig. 2.14 were taken for a Fermi gas initially at $T/T_F = 0.08$ and for two different holding times at B_{hold}. Condensation is observed on both the BCS

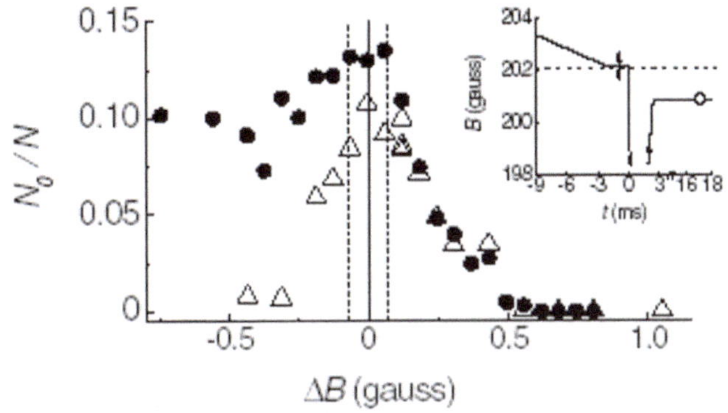

Fig. 2.14. Measured condensate fraction as a function of detuning from the Feshbach resonance $\Delta B = B_{\text{hold}} - B_0$. Data here were taken for $t_{\text{hold}} = 2\,\text{ms}$ (•) and $t_{\text{hold}} = 30\,\text{ms}$ (△) with an initial cloud at $T/T_F = 0.08$ and $T_F = 0.35\,\mu\text{K}$ [24]

Fig. 2.15. Time of flight images showing the fermionic condensate for $\Delta B = 0.12$, 0.25, and 0.55 G (left to right) on the BCS side of the resonance. The original atom cloud starts at $T/T_F = 0.07$, and the resulting fitted condensate fractions are $N_0/N = 0.10$, 0.05, and 0.01 (left to right) [24]

($\Delta B > 0$) and BEC ($\Delta B < 0$) sides of the resonance. The condensate on the BCS side of the Feshbach resonance has a relatively long lifetime ($> 30\,\text{ms}$) as denoted by triangles in Fig. 2.14. For the BEC side of the resonance, no condensate is observed for $t_{\text{hold}} = 30\,\text{ms}$ except very near the resonance. Figure 2.15 displays some examples of time-of-flight absorption images for the fermionic condensate on the BCS side.

3

Statistical Properties of Light

A laser is a light source with temporal and spatial coherence. In this chapter, we discuss the statistical properties of light including coherence and related topics. The degree of optical coherence can be defined quantitatively as the magnitude of correlation of the electromagnetic field at two points separated spatially and temporally. In Sect. 3.1, we give the definition of the first- and the second-order correlation functions. In order to understand the physical meanings of the correlation functions, the first- and second-order correlation functions are calculated for (a) coherent light, (b) thermal radiation (chaotic light), and (c) the photon-number squeezed state (nonclassical light). The first-order correlation function is observed as the contrast of interference fringes formed by the light that has passed through two slits or two pinholes as demonstrated by Young's experiment. The second-order correlation function can be measured from the correlation of light intensity at two separate times as a function of temporal distance, and the experiment was first conducted by Hanbury-Brown and Twiss. The second-order correlation function includes the information of the quantum properties of radiation that is undetectable from first-order effects such as fringes. Photons in chaotic light such as thermal radiation tend to bunch, and on the other hand, photons in photon-number squeezed states tend to be antibunching. Coherent light shows the property intermediate between these two. Photons in the antibunched state of light tend to arrive at the detector at regular interval. Such a property is not available from classical light.

In Sect. 3.2, we discuss the theory of photon counting. The probability distribution of the photon number counted in a fixed period will be calculated for several types of photon states. Both from measurements of the photon-count distribution and the second-order correlation function, the coherence time (correlation time) τ_c can be obtained. The statistical characteristics of light generated by laser oscillation show how chaotic light evolves into coherent light as the population inversion grows, although the details of the mathematical description of laser oscillation will be given in Chap. 4.

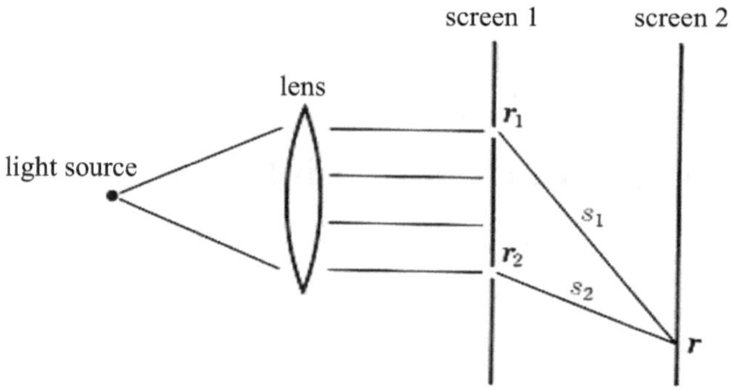

Fig. 3.1. Schematic of experiment to observe Young's interference pattern

3.1 The Degree of Coherence and Correlation Functions of Light

The characteristics of optical beams observed in interferometric experiments are usually expressed by the concept of coherence. Two beams are called coherent if interference is observed when two beams overlap at the same temporal and spatial position. Young's experiment shows the interference between beams from two slits or pinholes. The visibility of the fringe is determined by the first-order coherence of light, and it is described by the first-order correlation function of the electric field at two spatial and temporal positions. On the other hand, the second-order correlation function gives information on the fluctuation of light itself. The first experiment was demonstrated by Hanbury-Brown and Twiss. The concept of second-order coherence is introduced, because it reflects the quantum properties of light that cannot be obtained from first-order coherence. In this section, we discuss how the statistical characteristics of coherent laser light and thermal radiation are described by the first- and second-order coherence and correlation functions.

Figure 3.1 depicts a schematic diagram of Young's experiment. The light from a point source is assumed to be collimated by a lens system and it passes through the two slits located at $\boldsymbol{r}_1$ and $\boldsymbol{r}_2$ on screen 1, and then an interference pattern will be formed on screen 2. The electric field at position $\boldsymbol{r}$ and time t on screen 2 can be expressed as a superposition of the electric fields $\boldsymbol{E}(\boldsymbol{r}_1 t_1)$ and $\boldsymbol{E}(\boldsymbol{r}_2 t_2)$ observed at $\boldsymbol{r}_1$ and $\boldsymbol{r}_2$ at time $t_1 = t - s_1/c$ and $t_2 = t - s_2/c$, respectively:

$$\boldsymbol{E}(\boldsymbol{r}t) = u_1 \boldsymbol{E}(\boldsymbol{r}_1 t_1) + u_2 \boldsymbol{E}(\boldsymbol{r}_2 t_2). \tag{3.1}$$

Here, s_1 and s_2 are the distance from the slits to the detecting position $\boldsymbol{r}$, c is the velocity of light, and the amplitudes u_1 and u_2 are inversely proportional to s_1 and s_2 and are dependent on the size and shape of the slits. The light intensity at $\boldsymbol{r}$ can be expressed as

$$
\begin{aligned}
I(\boldsymbol{r}t) &= \left(\frac{\epsilon_0 c}{2}\right)|\boldsymbol{E}(\boldsymbol{r}t)|^2 \\
&= \left(\frac{\epsilon_0 c}{2}\right)\{|u_1|^2|\boldsymbol{E}(\boldsymbol{r}_1t_1)|^2 + |u_2|^2|\boldsymbol{E}(\boldsymbol{r}_2t_2)|^2 \\
&\quad +2\mathrm{Re}[u_1^* u_2 \boldsymbol{E}^*(\boldsymbol{r}_1t_1)\boldsymbol{E}(\boldsymbol{r}_2t_2)]\}. \qquad (3.2)
\end{aligned}
$$

Because the integration time for detection is usually much longer than the optical coherence time τ_c, we detect the intensity averaged over a long period when a cw light source is used. Due to ergodic theory, the temporal average can be interpreted as an ensemble average over the statistical distribution. As the interference in Young's experiment originates from the last term of (3.2), the interference effect is expressed by the following first-order correlation function:

$$
\langle \boldsymbol{E}^*(\boldsymbol{r}_1t_1)\boldsymbol{E}(\boldsymbol{r}_2t_2)\rangle = \lim_{T\to\infty}\frac{1}{T}\int_0^T \boldsymbol{E}^*(\boldsymbol{r}_1t_1)\boldsymbol{E}(\boldsymbol{r}_2t_1+t_{21})dt_1, \qquad (3.3)
$$

where $t_{21} \equiv t_2 - t_1$. By using this correlation function, the degree of first-order coherence $g^{(1)}(\boldsymbol{r}_1t_1,\, \boldsymbol{r}_2t_2)$ is defined as follows:

$$
g^{(1)}(\boldsymbol{r}_1t_1,\, \boldsymbol{r}_2t_2) \equiv g_{12}^{(1)} = \frac{|\langle \boldsymbol{E}^*(\boldsymbol{r}_1t_1)\boldsymbol{E}(\boldsymbol{r}_2t_2)\rangle|}{\{\langle|\boldsymbol{E}(\boldsymbol{r}_1t_1)|^2\rangle\langle|\boldsymbol{E}(\boldsymbol{r}_2t_2)|^2\rangle\}^{1/2}}. \qquad (3.4)
$$

When the light beam shows $g_{12}^{(1)} = 1$ for any different spatial and temporal positions, the light beam has first-order coherence. When $g_{12}^{(1)} = 0$, it is completely incoherent. If $g_{12}^{(1)}$ shows a value between 0 and 1, the beam is said to have first-order partial coherence. The classical electromagnetic wave and laser light have first-order coherence, because we can show that $g_{12}^{(1)} = 1$ in Sect. 3.1.1. In the case of thermal radiation, we assume a single cavity mode of thermal radiation selected by a very narrow band-pass filter. In this case, even though the selected mode is subjected to random fluctuation, the light observed at two spatial and temporal positions always shows the first-order coherence ($g_{12}^{(1)} = 1$) as shown in Sect. 3.1.2. The optical coherence time τ_c is given by the relation $\tau_c = 1/\gamma$ for light with a Lorentzian spectral shape and damping constant γ. From this point of view, Young's experiment can be a method to measure the interference effect utilizing the first-order coherence when the condition $|t_1 - t_2 - (s_1 - s_2)/c| \ll \tau_c$ is satisfied. Actual lasers have various coherence characteristics intermediate between those of thermal radiation and classical electromagnetic waves as shown in Chap. 4. The difference among them cannot be distinguished by first-order coherence, but can be clarified by using higher order coherence.

The experiment of Hanbury-Brown and Twiss is a method to measure the second-order coherence of light, and thus it is possible to detect the quantum optical properties of light. The importance of their experiment should be recognized not only because of their results, but also because it opened a new page in the field of quantum optics [28]. A diagram of the experiment is shown in Fig. 3.2.

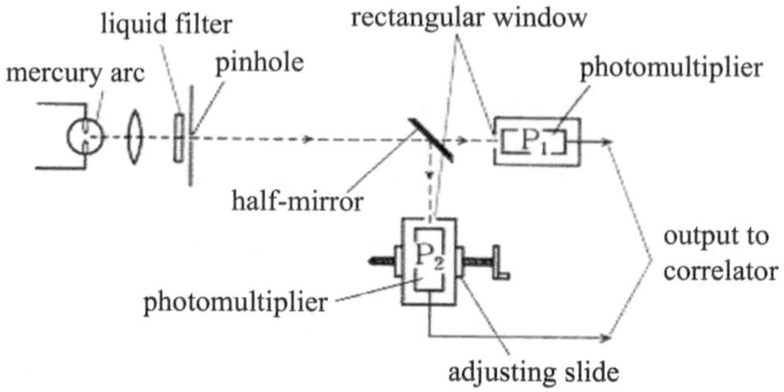

Fig. 3.2. Measurement of intensity correlation by Hanburg-Brown and Twiss [28]

The emission at 435.8 nm from a mercury arc lamp was selected by a filter. The beam was divided into two parts with equal intensity by a half-mirror. The intensities of the two beams were detected by photomultiplier tubes, and the product of the amplified fluctuation of the two outputs was observed by correlator. The long time integration of the signal gives the second-order coherence, the definition of which will be given later. The temporal positions of the observation for two beams can be varied by changing the spatial position of one photomultiplier tube P_2. Then, the signal intensity can be given by a function of two temporal positions as

$$\langle\{I(\boldsymbol{r}t_1) - I\}\{I(\boldsymbol{r}t_2) - I\}\rangle = \langle I(\boldsymbol{r}t_1)I(\boldsymbol{r}t_2)\rangle - I^2. \tag{3.5}$$

Here, $I = \langle I(\boldsymbol{r}t)\rangle$ and the bracket originally means an ensemble average due to the ergodicity of physical quantities, but it can be replaced with the temporal average as shown by (3.3).

The second-order intensity correlation function can be defined generally as

$$\langle I(\boldsymbol{r}_1t_1)I(\boldsymbol{r}_2t_2)\rangle = \left(\frac{\epsilon_0 c}{2}\right)^2 \langle \boldsymbol{E}^*(\boldsymbol{r}_1t_1)\boldsymbol{E}^*(\boldsymbol{r}_2t_2)\boldsymbol{E}(\boldsymbol{r}_2t_2)\boldsymbol{E}(\boldsymbol{r}_1t_1)\rangle.$$

Therefore, the formula (3.5) is known to be the special case of this form, that is, $\boldsymbol{r}_1 = \boldsymbol{r}_2$. The degree of coherence of the radiation field at two spatial and temporal positions $(\boldsymbol{r}_1t_1)$ and $(\boldsymbol{r}_2t_2)$ can be defined by the following expression:

$$g^{(2)}(\boldsymbol{r}_1t_1, \boldsymbol{r}_2t_2; \boldsymbol{r}_2t_2, \boldsymbol{r}_1t_1) = \frac{\langle \boldsymbol{E}^*(\boldsymbol{r}_1t_1)\boldsymbol{E}^*(\boldsymbol{r}_2t_2)\boldsymbol{E}(\boldsymbol{r}_2t_2)\boldsymbol{E}(\boldsymbol{r}_1t_1)\rangle}{\langle|\boldsymbol{E}(\boldsymbol{r}_1t_1)|^2\rangle\langle|\boldsymbol{E}(\boldsymbol{r}_2t_2)|^2\rangle}. \tag{3.6}$$

In the following part, this quantity will be described briefly as $g_{12}^{(2)}$, and the angle bracket $\langle\cdots\rangle$ means the ensemble average as before. If $g_{12}^{(1)} = 1$ and $g_{12}^{(2)} = 1$ are simultaneously satisfied, the light at two spatial and temporal positions has second-order coherence.

In order to obtain the quantum mechanical counterparts of the first and second-order correlation functions, the electric fields $\boldsymbol{E}^*$ and $\boldsymbol{E}$ must be replaced by the corresponding operators $\hat{\boldsymbol{E}}^+$ and $\hat{\boldsymbol{E}}^-$. These are the first and second terms of (1.38) which can be represented as

$$\hat{\boldsymbol{E}}^+ = i\sqrt{\frac{1}{2V}}\sum_\lambda e_\lambda\sqrt{\frac{\hbar\omega_\lambda}{\epsilon_0}}e^{i\boldsymbol{k}\cdot\boldsymbol{r}}\hat{a}_\lambda,$$

$$\hat{\boldsymbol{E}}^- = -i\sqrt{\frac{1}{2V}}\sum_\lambda e_\lambda\sqrt{\frac{\hbar\omega_\lambda}{\epsilon_0}}e^{-i\boldsymbol{k}\cdot\boldsymbol{r}}\hat{a}_\lambda^\dagger. \tag{3.7}$$

In quantum theory, the expression $\langle\cdots\rangle$ means the expectation value for the considered quantum state if the quantum system of light is in a pure state, and it means the statistical average if it is in a mixed state. In the latter case, the definition is given as $\langle\cdots\rangle = \mathrm{Tr}[\rho(\cdots)]$. Calculations based on this average will be performed in the following subsections.

When all of photon annihilation operators $\hat{\boldsymbol{E}}^+$ are located at the right side of all the photon generating operators $\hat{\boldsymbol{E}}^-$, the ordering is called the normal order. In order to define the first- and second-order correlation functions as normally ordered operators in the framework of quantum mechanics, this ordering indicates the fact that detection of a photon accompanies the annihilation of a photon, and a detector responds only when a photon arrives at the detector position. Hence, the first- and second-order correlation functions or the degree of first- and second-order coherence correspond to the absolute square of the probability amplitude for annihilating one or two photons, respectively. Now, let us calculate the optical coherence for the coherent state, thermal radiation, and the photon-number state of light.

3.1.1 Coherent State

The first-order correlation function of a single-mode coherent state $|\alpha\rangle$ can be expressed as

$$\begin{aligned}\langle\hat{\boldsymbol{E}}^-(\boldsymbol{r}_1t_1)\hat{\boldsymbol{E}}^+(\boldsymbol{r}_2t_2)\rangle &\equiv \left(\frac{\hbar\omega}{2\epsilon_0V}\right)e^{i\omega(t_1-t_2)-i\boldsymbol{k}\cdot(\boldsymbol{r}_1-\boldsymbol{r}_2)}\langle\alpha|\hat{a}^\dagger\hat{a}|\alpha\rangle \\ &= \frac{\hbar\omega}{2\epsilon_0V}|\alpha|^2e^{i\omega(t_1-t_2)-i\boldsymbol{k}\cdot(\boldsymbol{r}_1-\boldsymbol{r}_2)}.\end{aligned} \tag{3.8}$$

The degree of first-order coherence is found to be $g_{12}^{(1)} = 1$ by substituting (3.8) into (3.4). Substitution of the result

$$\langle\hat{\boldsymbol{E}}^-(\boldsymbol{r}_1t_1)\hat{\boldsymbol{E}}^-(\boldsymbol{r}_2t_2)\hat{\boldsymbol{E}}^+(\boldsymbol{r}_2t_2)\hat{\boldsymbol{E}}^+(\boldsymbol{r}_1t_1)\rangle = \left(\frac{\hbar\omega}{2\epsilon_0V}\right)^2|\alpha|^4, \tag{3.9}$$

and (3.8) into (3.6) gives $g_{12}^{(2)} = 1$.

The quantum mechanical degree of nth order coherence is defined by the following formula:

$$g^{(n)}(\boldsymbol{r}_1t_1,\ \ldots,\boldsymbol{r}_nt_n;\boldsymbol{r}_{n+1}t_{n+1},\ \ldots,\ \boldsymbol{r}_{2n}t_{2n})$$
$$= \frac{|\langle\hat{\boldsymbol{E}}^-(\boldsymbol{r}_1t_1)\cdots\hat{\boldsymbol{E}}^-(\boldsymbol{r}_nt_n)\hat{\boldsymbol{E}}^+(\boldsymbol{r}_{n+1}t_{n+1})\cdots\hat{\boldsymbol{E}}^+(\boldsymbol{r}_{2n}t_{2n})\rangle|}{\{\langle\hat{\boldsymbol{E}}^-(\boldsymbol{r}_1t_1)\hat{\boldsymbol{E}}^+(\boldsymbol{r}_1t_1)\rangle\cdots\langle\hat{\boldsymbol{E}}^-(\boldsymbol{r}_{2n}t_{2n})\hat{\boldsymbol{E}}^+(\boldsymbol{r}_{2n}t_{2n})\rangle\}^{1/2}}. \quad (3.10)$$

Then, the degrees of first- and second-order coherence defined by (3.4) and (3.6) are the special cases of the general form (3.10). For the coherent state $|\alpha\rangle$, it can be shown that $g^{(n)} = 1$ $(n \geq 1)$, in other words, the coherent state has coherences of all orders.

Next, let us consider the degree of coherence for the case of a multimode coherent state expressed by

$$|\{\alpha_{\boldsymbol{k}}\}\rangle \equiv |\alpha_1\rangle|\alpha_2\rangle\cdots|\alpha_{\boldsymbol{k}}\rangle\cdots.$$

From the definition of $\hat{\boldsymbol{E}}^+$ and the relation given by

$$a_{\boldsymbol{k}'}|\{\alpha_{\boldsymbol{k}}\}\rangle = \alpha_{\boldsymbol{k}'}|\{\alpha_{\boldsymbol{k}}\}\rangle,$$

the state $|\{\alpha_{\boldsymbol{k}}\}\rangle$ is proven to be an eigenstate of $\hat{\boldsymbol{E}}^+$ as shown by

$$\hat{\boldsymbol{E}}^+(\boldsymbol{r}t)|\{\alpha_{\boldsymbol{k}}\}\rangle = \boldsymbol{\varepsilon}(\boldsymbol{r}t)|\{\alpha_{\boldsymbol{k}}\}\rangle, \quad (3.11)$$

where

$$\boldsymbol{\varepsilon}(\boldsymbol{r}t) = i\sum_{\boldsymbol{k}}\sqrt{\frac{\hbar\omega_{\boldsymbol{k}}}{2\epsilon_0 V}}\boldsymbol{e}_{\varepsilon\boldsymbol{k}}\alpha_{\boldsymbol{k}}\exp[-i(\omega_{\boldsymbol{k}}t - \boldsymbol{k}\cdot\boldsymbol{r})].$$

The numerator and denominator in (3.10) are evaluated to be equal so that the degrees of nth order coherence become unity as

$$g^{(n)}(\boldsymbol{r}_1t_1,\cdots,\boldsymbol{r}_nt_n;\boldsymbol{r}_{n+1}t_{n+1},\cdots,\boldsymbol{r}_{2n}t_{2n}) = 1 \quad (3.12)$$

for all n. Therefore, the multimode coherent state $|\{\alpha_{\boldsymbol{k}}\}\rangle$ is also quantum mechanically coherent in all orders. Experimentally, any complex interference effects produced by the light in these states or the correlation functions of these radiation fields are the same as those predicted for classical electromagnetic waves. Then, the states $|\{\alpha_{\boldsymbol{k}}\}\rangle$ are called coherent states.

3.1.2 Thermal Radiation

The density operator $\hat{\rho}$ for the thermal radiation field at temperature T with a single mode $\hbar\omega$ can be expanded by photon-number states $|n\rangle$ as

$$\begin{aligned}\hat{\rho} &= \sum_{n=0}^{\infty} P_n |n\rangle\langle n| \\ &= \{1 - \exp(-\beta\hbar\omega)\} \sum_{n=0}^{\infty} \exp(-\beta n\hbar\omega) |n\rangle\langle n|. \end{aligned} \tag{3.13}$$

Here, $\beta \equiv 1/k_B T$. At a finite temperature, P_n, the probability that n photons are excited in the mode can be expressed as

$$\begin{aligned} P_n &= \frac{\exp(-\beta E_n)}{\sum_{n=0}^{\infty} \exp(-\beta E_n)} = \frac{\exp(-\beta n\hbar\omega)}{\sum_{n=0}^{\infty} \exp(-\beta n\hbar\omega)} \\ &= \exp(-\beta n\hbar\omega)\{1 - \exp(-\beta\hbar\omega)\}, \end{aligned} \tag{3.14}$$

where the eigenenergy is given by the relation $E_n = \hbar\omega(n + 1/2)$. By using the mean number of thermally excited photons

$$\overline{n} = \langle n \rangle \equiv \mathrm{Tr}(\hat{\rho}\hat{a}^\dagger\hat{a}) = \frac{1}{\exp(\hbar\omega\beta) - 1}, \tag{3.15}$$

the density operator $\hat{\rho}$ for a single mode can be expressed as follows:

$$\hat{\rho} = \sum_{n=0}^{\infty} \frac{\overline{n}^n}{(1 + \overline{n})^{1+n}} |n\rangle\langle n|. \tag{3.16}$$

Because all modes consisting of the thermal radiation are independent of each other, the density operator $\hat{\rho}$ for total thermal radiation system can be given in the same way as

$$\hat{\rho} = \sum_{\{n_{\boldsymbol{k}}\}} P_{\{n_{\boldsymbol{k}}\}} |\{n_{\boldsymbol{k}}\}\rangle\langle\{n_{\boldsymbol{k}}\}|. \tag{3.17}$$

Here,

$$P_{\{n_{\boldsymbol{k}}\}} = \prod_{\boldsymbol{k}} \frac{(\overline{n}_{\boldsymbol{k}})^{n_{\boldsymbol{k}}}}{(1 + \overline{n}_{\boldsymbol{k}})^{1+n_{\boldsymbol{k}}}}, \tag{3.18}$$

and $\{n_{\boldsymbol{k}}\}$ represents a photon-number state for multimode.

In order to evaluate the degree of first-order coherence $g_{12}^{(1)}$ for this state, the fisrt-order correlation function must be calculated first. That is,

$$\begin{aligned} \langle \boldsymbol{E}^*(\boldsymbol{r}_1 t_1)\boldsymbol{E}(\boldsymbol{r}_2 t_2)\rangle &= \mathrm{Tr}\{\hat{\rho}\boldsymbol{E}^*(\boldsymbol{r}_1 t_1)\boldsymbol{E}(\boldsymbol{r}_2 t_2)\} \\ &= \sum_{\{n_{\boldsymbol{k}}\}} P_{\{n_{\boldsymbol{k}}\}} \frac{\hbar\omega_{\boldsymbol{k}}}{2\epsilon_0 V} n_{\boldsymbol{k}} e^{i\omega_{\boldsymbol{k}}(t_1 - t_2) - i\boldsymbol{k}\cdot(\boldsymbol{r}_1 - \boldsymbol{r}_2)} \\ &= \sum_{\boldsymbol{k}} \frac{\hbar\omega_{\boldsymbol{k}}}{2\epsilon_0 V} \overline{n}_{\boldsymbol{k}} e^{i\omega_{\boldsymbol{k}}\tau}, \end{aligned} \tag{3.19}$$

where $\tau \equiv t_1 - t_2 - \boldsymbol{k} \cdot (\boldsymbol{r}_1 - \boldsymbol{r}_2)/\omega_{\boldsymbol{k}}$. By substituting (3.19) into (3.4), the expression for $g_{12}^{(1)}$ is given by

$$g_{12}^{(1)} = \frac{\left| \sum_{\boldsymbol{k}} \overline{n}_{\boldsymbol{k}} \omega_{\boldsymbol{k}} \exp(i\omega_{\boldsymbol{k}} \tau) \right|}{\sum_{\boldsymbol{k}} \overline{n}_{\boldsymbol{k}} \omega_{\boldsymbol{k}}}. \tag{3.20}$$

Let us consider a typical example where an emission spectrum has a Lorentzian shape with a width of 2γ:

$$\overline{n}_{\boldsymbol{k}} \omega_{\boldsymbol{k}} \propto \frac{\gamma}{(\omega_0 - \omega_{\boldsymbol{k}})^2 + \gamma^2}. \tag{3.21}$$

Such a distribution can be obtained by filtering the photon flux with the bosonic distribution given by (3.18). In this case, the correlation function

$$g_{12}^{(1)} = \exp(-\gamma|\tau|) \tag{3.22}$$

can be obtained by substituting (3.21) into (3.20), and transforming summation into integration, under the approximation that $\omega_0 \gg \gamma$. If a Gaussian distribution

$$\overline{n}_{\boldsymbol{k}} \omega_{\boldsymbol{k}} \propto \exp\left[-\frac{(\omega_{\boldsymbol{k}} - \omega_0)^2}{2\delta^2} \right] \tag{3.23}$$

is given by filtering of thermal radiation, a similar calculation gives the correlation function as

$$g_{12}^{(1)} = \exp\left(-\frac{1}{2}\delta^2 \tau^2 \right). \tag{3.24}$$

The degrees of first-order coherence $g_{12}^{(1)}$ are depicted as functions of $\tau \equiv t_1 - t_2 - \boldsymbol{k} \cdot (\boldsymbol{r}_1 - \boldsymbol{r}_2)/\omega_{\boldsymbol{k}}$ in Fig. 3.3.

Next, let us show the calculation of the degree of second-order coherence $g_{12}^{(2)}$. The electric field is expressed by the Fourier expansion of all modes as

$$\hat{\boldsymbol{E}}^+(\boldsymbol{r}t) = \sum_{\boldsymbol{k}} \hat{\boldsymbol{E}}_{\boldsymbol{k}}^+ e^{i\boldsymbol{k}\cdot\boldsymbol{r} - i\omega_{\boldsymbol{k}} t}. \tag{3.25}$$

Substituting this expression into the numerator of (3.6), the correlation function is reduced to the following expression

$$\begin{aligned} &\langle \hat{\boldsymbol{E}}^-(\boldsymbol{r}_1 t_1) \hat{\boldsymbol{E}}^-(\boldsymbol{r}_2 t_2) \hat{\boldsymbol{E}}^+(\boldsymbol{r}_2 t_2) \hat{\boldsymbol{E}}^+(\boldsymbol{r}_1 t_1) \rangle \\ &= \sum_{\boldsymbol{k}_1 \boldsymbol{k}_2 \boldsymbol{k}_3 \boldsymbol{k}_4} \langle \hat{\boldsymbol{E}}_{\boldsymbol{k}_1}^- \hat{\boldsymbol{E}}_{\boldsymbol{k}_2}^- \hat{\boldsymbol{E}}_{\boldsymbol{k}_3}^+ \hat{\boldsymbol{E}}_{\boldsymbol{k}_4}^+ \rangle \exp[i(-\boldsymbol{k}_1 \cdot \boldsymbol{r}_1 + i\omega_1 t_1 \\ &\quad -i\boldsymbol{k}_2 \cdot \boldsymbol{r}_2 + i\omega_2 t_2 + i\boldsymbol{k}_3 \cdot \boldsymbol{r}_2 - i\omega_3 t_2 + i\boldsymbol{k}_4 \cdot \boldsymbol{r}_1 - i\omega_4 t_1)]. \end{aligned} \tag{3.26}$$

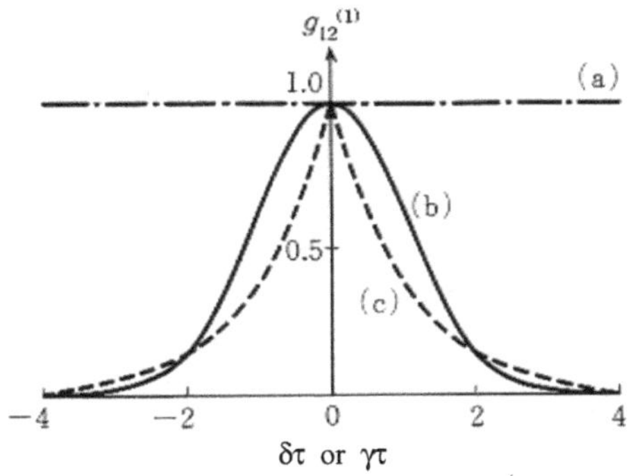

Fig. 3.3. First-order correlation $g_{12}^{(1)}(\tau)$ for (**a**) a stable classical wave, and for chaotic light with (**b**) a Gaussian frequency distribution, and (c) a Lorentzian frequency distribution. $\tau \equiv t_1 - t_2 - \boldsymbol{k}\cdot(\boldsymbol{r}_1 - \boldsymbol{r}_2)/\omega_{\boldsymbol{k}}$

Because the relation

$$\langle X\rangle = \sum_{\boldsymbol{k}} \sum_{\{n_{\boldsymbol{k}}\}} P_{\{n_{\boldsymbol{k}}\}} |\{n_{\boldsymbol{k}}\}\rangle\langle\{n_{\boldsymbol{k}}\}|X \tag{3.27}$$

is satisfied for any electric field operator X, (3.26) gives finite results only for two cases, that is, ($\boldsymbol{k}_1 = \boldsymbol{k}_3$ and $\boldsymbol{k}_2 = \boldsymbol{k}_4$) and ($\boldsymbol{k}_1 = \boldsymbol{k}_4$ and $\boldsymbol{k}_2 = \boldsymbol{k}_3$). For the latter case, the phase factor in (3.26) is 1, and for the former case it is given by $\exp(i\omega_1\tau)\cdot\exp(-i\omega_2\tau)$. Then, we can obtain the following relationship for the first- and second-order coherence:

$$g_{12}^{(2)} = \frac{\left|\sum_{\boldsymbol{k}} \overline{n}_{\boldsymbol{k}}\omega_{\boldsymbol{k}} \exp(i\omega_{\boldsymbol{k}}\tau)\right|^2}{\left(\sum_{\boldsymbol{k}} \overline{n}_{\boldsymbol{k}}\omega_{\boldsymbol{k}}\right)^2} + 1 = \left(g_{12}^{(1)}\right)^2 + 1. \tag{3.28}$$

From this expression, the second-order coherence $g_{12}^{(2)}$ of thermal radiation at $\tau = 0$ is known to be 2. The physical meaning of this fact ($g_{12}^{(2)} > 1$) is that when one photon from thermal radiation is observed at time t_1, the probability to observe another photon at time $t_2 = t_1 + \tau$ is larger than 1 if $\tau < \tau_c$. This shows the existence of temporal bunching of photons in the thermal radiation field. This is the most pronounced difference from the coherent state where $g_{12}^{(1)} = 1$, $g_{12}^{(2)} = 1$, $\cdots$ and $g^{(n)} = 1$.

The difference between second-order coherence in coherent and chaotic light has also been confirmed experimentally [29]. As an example, we introduce an experiment done by using light with a Gaussian frequency distribution. In this experiment, photon numbers m_1 and m_2 are measured during short periods δt_1, and δt_2, respectively where both periods are separated by τ. The

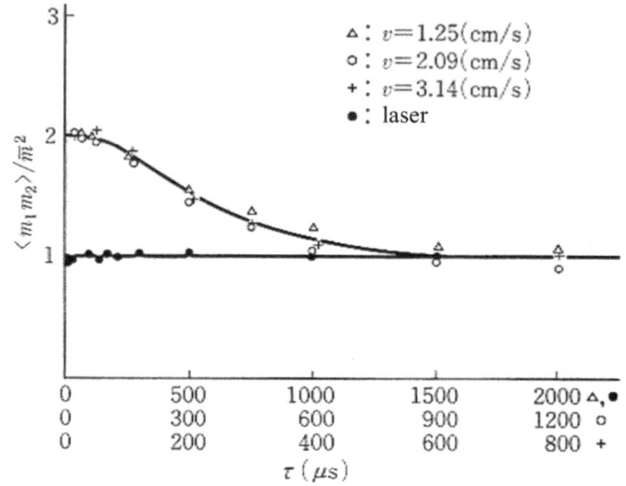

Fig. 3.4. Second-order correlation $g_{12}^{(2)}(\tau) = \langle m_1 m_2 \rangle / \overline{m}^2$ for laser light (•) and chaotic light induced by frosted glass rotating with different velocity v. Solid lines denote the theoretical result, and the points are the observed results [29]

correlation between photon numbers is obtained as a function of temporal distance τ as done in the experiment of Hanbury-Brown and Twiss. If $\delta t_1 = \delta t_2$ and τ is much shorter than the coherence time τ_c, the correlation $\langle m_1 m_2 \rangle$ gives the degree of second-order coherence $g_{12}^{(2)}$ of (3.6) for the same spatial position. Hence, when the average of counted photons during the short period $\delta t_1 = \delta t_2$ is $\overline{m}_1$, $g_{12}^{(2)}$ is given as

$$g_{12}^{(2)}(\tau) = \frac{\langle m_1 m_2 \rangle}{\overline{m}^2} . \tag{3.29}$$

Arecchi et al. [29] obtained $g_{12}^{(2)} = 1$ for a monochromatic laser beam as shown in Fig. 3.4, and this result corresponded to the coherence properties predicted from (3.12) in Sect. 3.1.1. The light source with a Gaussian frequency distribution can be obtained by passing the beam through a rotating frosted glass. The degree of second-order coherence $g_{12}^{(2)}(\tau)$ measured for this light is also shown in Fig. 3.4. We can easily find the exact confirmation of the theoretical prediction that $g_{12}^{(2)}(\tau) \to 2$ for $\tau \ll \tau_c$, and $g_{12}^{(2)}(\tau) \to 1$ for $\tau \gg \tau_c$.

The degrees of coherence higher than third order become important in higher order nonlinear optical processes. For example, in the process of third-harmonic generation described in Chap. 6, three photons are absorbed simultaneously from a light beam and one photon of the third-harmonic wave is emitted. The rate of the events is proportional to the third-order coherence $g^{(3)}$ of the incident beam. Therefore, the intensity of third harmonics gives information about the third-order coherence of the light.

3.1.3 Photon-Number State and Antibunching Characteristics

Correlation functions of the single-mode photon-number state $|n\rangle$ can be calculated from the definition (3.4) and (3.6), and the electric field correlations are given by

$$\begin{aligned}\langle \boldsymbol{E}^*(\boldsymbol{r}_1 t_1)\boldsymbol{E}(\boldsymbol{r}_2 t_2)\rangle &= \langle n|\boldsymbol{E}^*(\boldsymbol{r}_1 t_1)\boldsymbol{E}(\boldsymbol{r}_2 t_2)|n\rangle \\ &= \frac{\hbar\omega}{2\epsilon_0 V} n \exp(i\omega\tau), \end{aligned} \qquad (3.30)$$

$$\langle n|\boldsymbol{E}^*(\boldsymbol{r}_1 t_1)\boldsymbol{E}^*(\boldsymbol{r}_2 t_2))\boldsymbol{E}(\boldsymbol{r}_2 t_2)\boldsymbol{E}(\boldsymbol{r}_1 t_1)|n\rangle = \left(\frac{\hbar\omega}{2\epsilon_0 V}\right)^2 n(n-1). \qquad (3.31)$$

Substituting these results into (3.4) and (3.6), $g_{12}^{(1)}$ and $g_{12}^{(2)}$ can be expressed as

$$\begin{aligned} g_{12}^{(1)} = 1, \quad g_{12}^{(2)} &= \frac{n-1}{n} = 1 - \frac{1}{n} \quad (n \geq 1) \\ &= 0 \quad (n = 0). \end{aligned} \qquad (3.32)$$

If we can prepare the light beam with a well-defined photon number, the experiment of Hanbury-Brown and Twiss will show the correlation of $g_{12}^{(2)} - 1 = -1/n < 0$. This negative value means the antibunching of photons, and it is a unique quantum effect of photons because such an effect is unpredictable from classical coherence theory. This is a property of the photon-number squeezed state described in Sect. 1.4.2. In other words, if one photon in a photon number state is observed at time t_1, the probability of arrival of another photon during the next τ_c is smaller than average.

For general cases, n and $n(n-1)$ appearing in the correlation functions of the electric field (3.30) and (3.31) for single-mode light beams can be replaced by the corresponding expectation values, $\overline{n}$ and $\overline{n^2} - \overline{n}$, respectively. Hence, the degrees of coherence can be given by

$$g_{12}^{(1)} = 1, \quad g_{12}^{(2)} = \frac{\overline{n^2} - \overline{n}}{\overline{n}^2}. \qquad (3.33)$$

Finally, we briefly describe the relation between the second-order coherence and photon-counting characteristics. When the photon-count is given by a Poisson distribution as for the coherent state, the variance of the photon number is $(\Delta n)^2 \equiv \overline{n^2} - \overline{n}^2 = \overline{n}$ as shown in the next section. In this case, we obtain $g_{12}^{(2)} - 1 = 0$ from (3.33). On the other hand, we obtain $g_{12}^{(2)} - 1 > 0$ in a super-Poisson distribution in which the photon number spreads broader than the Poisson distribution, such as $(\Delta n)^2 \equiv \overline{n^2} - \overline{n}^2 > \overline{n}$, and this situation is found in chaotic bunching light. When $g_{12}^{(2)} - 1 < 0$ is observed, the photon count has a sub-Poisson distribution where the photon number has a narrower distribution than the Poisson distribution, as $(\Delta n)^2 \equiv \overline{n^2} - \overline{n}^2 < \overline{n}$. This state corresponds to an antibunching photon-number squeezed state. It is understood that the coherent state forms a border between bunching and antibunching states.

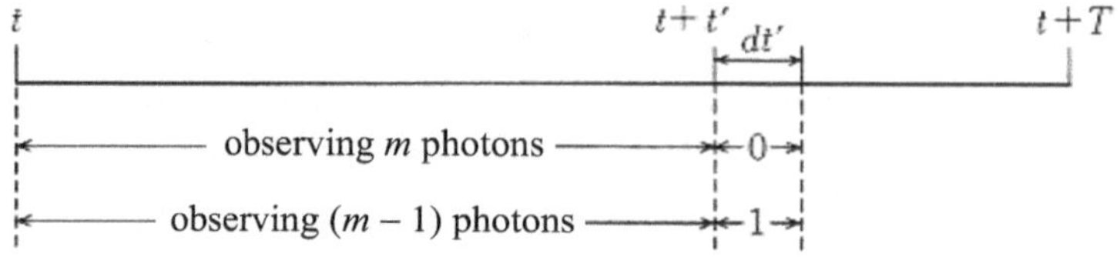

Fig. 3.5. Schematic to obtain the photon-number counting function $P_m(t, T)$, the probability of observing m photons between t and $t + t' + dt'$

3.2 Photon-Count Distribution

The statistical distribution $P_m(T)$ of photon number m observed during time T is called the photon-count distribution. In this section, we discuss the relation between the properties of a light source and the results obtained by the photon-counting measurement.

It is possible to measure the number of incident photons by means of photoemission in a phototube. In experiments, the number of incident photons in a time duration T is measured by controlling a shutter located in front of the tube. The statistical distribution of the photon number can be obtained by repeating the experiment many times (typically more than 10^4 times). The light beam is assumed to be in a steady state. The rate of photoemission is proportional to the beam intensity, and it is determined by the average of the product of two electric field operators $\hat{\boldsymbol{E}}^-\hat{\boldsymbol{E}}^+$ defined in (3.7). Now, we use the envelope of the classical magnitude $I(t)$ as the intensity of the incident electromagnetic field. If the probability that the light beam causes photoemission from an atom in a phototube is $p(t)$, the count between t and $t+dt$ is $p(t)dt$. The interval dt must be long enough to apply transition probability theory for photoemission, but must be short enough in order to eliminate the probability that more than two photoelectrons are counted during dt. In this case, $p(t)$ is proportional to the beam intensity $I(t)$:

$$p(t)dt = \zeta I(t)dt. \tag{3.34}$$

Here, ζ means the factor determined by the experimental conditions such as quantum efficiency of the phototube, and it depends on the matrix element for photoemission, the atomic density of the cathode metal, and so on.

Let $P_m(t, T)$ denote the probability that m photons are counted between time t and $t + T$. Then, $P_m(T)$ can be defined as the average of $P_m(t, T)$ for starting time t. Choose time $t + t'$ between t and $t + T$ as shown in Fig. 3.5.

The probability of observing m photons between t and $t + t' + dt'$ can be expressed as

$$P_m(t, t' + dt') \tag{3.35}$$

by definition. This probability can be considered as the summation of two kinds of contribution, that is, (1) the probability for observing m photons between t and $t + t'$ and no photon during the following dt' as

$$P_m(t,\, t')\{1 - p(t')dt'\}, \tag{3.36}$$

and (2) the probability to observe $m-1$ photons between t and $t+t'$ and one photon during dt' as

$$P_{m-1}(t,\, t')p(t')dt'. \tag{3.37}$$

We assume that the probability to observe two photons during dt' is negligible. Then (3.35) is equivalent to the summation of (3.36) and (3.37), and the following equation is given:

$$\frac{dP_m(t,\, t')}{dt'} = \zeta I(t')\{P_{m-1}(t,\, t') - P_m(t,\, t')\}. \tag{3.38}$$

The probability for observing m photons $P_m(t,\, t')$ is connected to the probability functions for other smaller m values by simultaneous differential equations. We start to discuss the case of $m = 0$. Because the first equation with $m = 0$ does not have a term with $m-1$, it is expressed by the following simple form:

$$\frac{dP_0(t,\, t')}{dt'} = -\zeta I(t')P_0(t,\, t'). \tag{3.39}$$

The probability to observe no photon $P(t,\, t')$ must be unity if the time interval is zero. This restriction gives the initial condition as

$$P_0(t,\, 0) = 1. \tag{3.40}$$

Equation (3.39) can be solved formally under the initial condition (3.40) as

$$P_0(t,\, T) = \exp\left[-\zeta \int_t^{t+T} I(t')dt'\right]. \tag{3.41}$$

By introducing the average intensity $I(t,\, T) \equiv \int_t^{t+T} I(t')dt'/T$, the solution of (3.38) can be generally given under the initial condition $P_m(t,\, 0) = 0$ $(m \neq 0)$ as

$$P_m(t,\, T) = \frac{\{\zeta I(t,\, T)T\}^m}{m!} \exp\{-\zeta I(t,\, T)T\}. \tag{3.42}$$

This result can be proven by mathematical induction. The average on the initial time t can be replaced by an average on ensembles as

$$P_m(T) \equiv \langle P_m(t,\, T)\rangle = \left\langle \frac{\{\zeta IT\}^m}{m!} \exp\{-\zeta IT\}\right\rangle. \tag{3.43}$$

Because we assume that the intensity I of the incident beam does not depend on time t, $I(t,\, T)$ is a constant. By giving the constant number $\overline{m} = \zeta I(t,\, T)T$ as the averaged photon-count, it is found that the distribution of the photon count $P_m(T)$ has a Poisson distribution:

$$P_m(T) = \frac{\overline{m}^m}{m!} \exp(-\overline{m}). \tag{3.44}$$

The fluctuation of the photon count from the averaged value $\overline{m}$ can be given by the standard deviation Δm or $(\Delta m)^2$. In the case of a Poisson distribution, $(\Delta m)^2 = \overline{m}$. This type of fluctuation is called particle fluctuation because it is a result of the fact that energy is subtracted from light beam detection only by a multiple of the single photon energy. A classical electromagnetic wave and quantum mechanical coherent light are subjected to a Poisson distribution.

Next, let us consider the opposite extreme from coherent light, that is, chaotic light. Chaotic light can be obtained from discharge lamps, filaments, or thermal cavities. Chaotic light has a random electric field amplitude and phase modulation, and the time scale of the random fluctuation is determined by Doppler broadening of the spectrum or sometimes by collision broadening. This characteristic time is called the coherence time τ_c, and it is of the order of the reciprocal of the frequency broadening of light. The order of frequency broadening of the thermal radiation is almost the same as that of optical frequencies, but it is much narrower for the case of discharge lamps. In spite of such differences, the statistical properties of both light beams have similar chaotic features. If the observing time T is much longer than the coherence time τ_c, $I(t, T)$ does not depend on t, and is determined by a Poisson distribution given by (3.44).

In order to find properties that are different for coherent and chaotic light, let us consider the case where the counting time T is much shorter than the coherence time ($T \ll \tau_c$). In this limit, the instantaneous intensity $I(t)$ is assumed to be a constant in the counting duration. Then $I(t, T)$ is approximated as $I(t)$. The distribution of the intensity can be obtained by applying the theory of stochastic processes to the random electric field fluctuation and phase modulation. The result is represented as

$$p[I(t)] = \frac{1}{I} \exp\left[-\frac{I(t)}{I}\right],$$

where I means the averaged light intensity. The details of the calculation are described in the following paragraphs.

From the ergodic hypothesis for a steady state light source, the photon counting distribution $P_m(T)$ given by (3.43) can be expressed as the average over the probability distribution of intensity:

$$P_m(T) = \frac{1}{I} \int dI(t) \exp\left[-\frac{I(t)}{I}\right] \frac{\{\zeta I(t) T\}^m}{m!} \exp[-\zeta I(t) T] \tag{3.45a}$$

$$= \frac{\overline{m}^m}{(1+\overline{m})^{m+1}}. \tag{3.45b}$$

Here, $\overline{m} = \zeta I T$ means the average value of counted photons. This expression is the same as the photon counting distribution of the single-mode light obtained by the filtering of thermal radiation as given by (3.16). Therefore, it is satisfied not only for a single-mode beam but also for any chaotic light under the condition that $T \ll \tau_c$.

In general, the variance of the photon-count distribution can be written as

$$(\Delta m)^2 = \overline{m^2} - (\overline{m})^2 = \sum_{m=0}^{\infty} m^2 P_m(T) - \left\{ \sum_{m=0}^{\infty} m P_m(T) \right\}^2 . \tag{3.46a}$$

The average of m and m^2 can be calculated from (3.45) as

$$\overline{m} = \sum_{m=0}^{\infty} m P_m(T) = \langle \zeta I(t) T \rangle = \zeta I T$$

$$\overline{m^2} = \sum_{m=0}^{\infty} m^2 P_m(T) = \langle \{\zeta I(t) T\}^2 \rangle + \langle \zeta I(t) T \rangle .$$

Then, the final result is represented as

$$(\Delta m)^2 = \overline{m} + \zeta^2 T^2 \{ \langle I(t)^2 \rangle - I^2 \} . \tag{3.46b}$$

The first term is the particle fluctuation specific for a Poisson distribution, and the second term is known as the wave fluctuation originating from random variation of the instantaneous intensity.

Until now, we have considered the photon-count distribution from the classical intensity $I(t)$ and stochastic processes. In the last half of this section, we will show that a calculation based on quantum theory also gives the same results for coherent and chaotic light. A quantum mechanical expression for the photon count was derived by Kelley and Kleiner [30]. The quantum counterpart of the classical expression shown by (3.43) is given by

$$P_m(T) = \mathrm{Tr} \left[\rho \hat{\mathrm{N}} \frac{\{\zeta \hat{I}(T) T\}^m}{m!} \exp\{-\zeta \hat{I}(T) T\} \right] , \tag{3.47}$$

where

$$\hat{I}(T) = \frac{2\epsilon_0 c}{T} \int_0^T \hat{\boldsymbol{E}}^-(\boldsymbol{r}t) \hat{\boldsymbol{E}}^+(\boldsymbol{r}t) dt . \tag{3.48}$$

The letter $\boldsymbol{r}$ is the position of the detector, $\hat{\mathrm{N}}$ is the normal-ordering operator which makes the electric field operators $\hat{\boldsymbol{E}}^+$ and $\hat{\boldsymbol{E}}^-$ be arranged in normal order. This is a mathematical representation of the fact that the detection process accompanies the annihilation of photons. In the following paragraph, we will show that the photon-count distribution of several photon states can be derived by using the quantum-mechanical expression (3.47).

Let us consider the case for a single-mode light beam, where the photon-counting distribution (3.47) can be reduced to the following simple form

$$P_m(T) = \mathrm{Tr}\left\{\rho \hat{\mathrm{N}} \frac{(\xi \hat{a}^\dagger \hat{a})^m}{m!} \exp(-\xi \hat{a}^\dagger \hat{a})\right\}, \tag{3.49}$$

because of the relation that $\hat{\boldsymbol{E}}^-(\boldsymbol{r}_1 t_1)\hat{\boldsymbol{E}}^+(\boldsymbol{r}_1 t_1) = (\hbar\omega/2\epsilon_0 V)\hat{n}$. In (3.49), $\xi = \zeta c\hbar\omega T/V$ means the quantum yield of the detector, i.e., a measure of the detector efficiency. The density operator ρ can be expanded with photon number states $|n\rangle$ as

$$\rho = \sum_{n=0}^{\infty} P_n |n\rangle\langle n|. \tag{3.50}$$

By substituting (3.50) into (3.49), expanding $\exp(-\xi \hat{a}^\dagger \hat{a})$, and arranging the operators into normal ordering by $\hat{\mathrm{N}}$, the following probability distribution can be obtained:

$$\begin{aligned} P_m(T) &= \sum_{n=0}^{\infty} P_n \frac{\xi^m}{m!} \langle n| \sum_{l=0}^{\infty} (-1)^l \frac{\xi^l}{l!} (\hat{a}^\dagger)^{m+l} \hat{a}^{m+l} |n\rangle \\ &= \sum_{n=0}^{\infty} P_n \frac{\xi^m}{m!} \sum_{l=0}^{n-m} (-1)^l \frac{\xi^l}{l!} \frac{n!}{(n-m-l)!} \\ &= \sum_{n=m}^{\infty} P_n \frac{n!}{m!(n-m)!} \xi^m (1-\xi)^{n-m}. \end{aligned} \tag{3.51}$$

Each term follows a binomial distribution, and it reflects the quantum mechanical features of the photon. Here, the letter ξ means the probability of photon detection during the counting time T. When there are n photons in a mode, the probability to count m photons is $(\xi)^m$, and the probability for not counting $n-m$ photons is $(1-\xi)^{n-m}$. The factor ${}_nC_m = n!/m!(n-m)!$ represents the fact that we cannot distinguish one photon from another.

Next, consider single-mode coherent light $|\alpha\rangle$. According to the relation $|\alpha|^2 = \overline{n}$, we can obtain

$$P_n = \langle n|\rho|n\rangle = \frac{|\alpha|^{2n}}{n!} \exp(-|\alpha|^2) = \frac{\overline{n}^n}{n!} \exp(-\overline{n}). \tag{3.52}$$

Substituting this expression into (3.51), we find that a Poisson distribution with average value $\xi\overline{n}$ can be obtained for the coherent light as shown in the following:

$$\begin{aligned}
P_m(T) &= \sum_{n=m}^{\infty} \frac{\overline{n}^n}{n!} e^{-\overline{n}} \frac{n!}{m!(n-m)!} \xi^n (1-\xi)^{n-m} \\
&= \frac{(\xi\overline{n})^m}{m!} \exp(-\overline{n}) \left[\sum_{n=m}^{\infty} \frac{\{\overline{n}(1-\xi)\}^{n-m}}{(n-m)!} \right] \\
&= \frac{(\xi\overline{n})^m}{m!} \exp(-\xi\overline{n}). \qquad (3.53)
\end{aligned}$$

A chaotic single-mode beam is available by filtering chaotic light like thermal radiation. In this case, the diagonal matrix element P_n is obtained from the basis of photon number states as

$$P_n = \frac{\overline{n}^n}{(1+\overline{n})^{1+n}}, \qquad (3.54)$$

which is also known from (3.16), where $\overline{n} \equiv \sum_{n=0}^{\infty} nP_n$. If (3.54) is substituted into the formula (3.51), the result is given by

$$\begin{aligned}
P_m(T) &= \sum_{n=m}^{\infty} \frac{\overline{n}^n}{(1+\overline{n})^{1+n}} \frac{n!}{m!(n-m)!} \xi^m (1-\xi)^{n-m} \\
&= \frac{(\xi\overline{n})^m}{(1+\overline{n})^{m+1}} \sum_{n=m}^{\infty} \left\{ \frac{\overline{n}(1-\xi)}{1+\overline{n}} \right\}^{n-m} \frac{n!}{m!(n-m)!}. \qquad (3.55)
\end{aligned}$$

By using the relation $x = 1 - (1+\xi\overline{n})/(1+\overline{n})$ in the following

$$\frac{1}{(1-x)^{p+1}} = \sum_{n=0}^{\infty} \frac{(n+p)!}{n!p!} x^n, \qquad (3.56)$$

the distribution of (3.55) is reduced to a final simple form as

$$P_m(T) = \frac{(\xi\overline{n})^m}{(1+\xi\overline{n})^{1+m}}. \qquad (3.57)$$

In this chaotic case, the distribution has the same form as (3.45b) but using the average of $\xi\overline{n}$.

Photon-count distribution depends strongly on the quantitative relationship between the counting time T and the coherence time τ_c. Figure 3.6 clearly shows the transition from the expression of (3.45b) for $T \ll \tau_c$ to the Poisson distribution (3.44) [31]. The coherence time of chaotic light can be determined by comparing the experimental results for $P_m(T)$ with the theoretical prediction.

With the conventional spectroscopic methods using a monochromator, the resolution of the frequency is usually 10^{10} Hz or more. If a Fabry–Pérot interferometer is employed, we can resolve the line spectra with width from 10^7

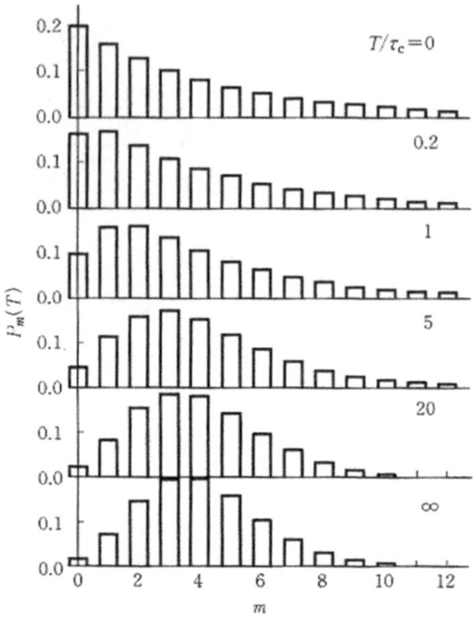

Fig. 3.6. Photon number distribution function $P_m(T)$ for chaotic light for the average photon number $\overline{m} = 4$. $P_m(T) = \{\overline{m}/(1+\overline{m})\}^m$ for $T/\tau_c = 0$ changes into $P_m(T)$ = Poisson distribution for $T/\tau_c \to \infty$ [31]

to 10^{12} Hz. The method using intensity fluctuation introduced in this chapter can be applied to spectroscopy. Because the resolution is limited by the temporal response of fast detectors (typically 10^{-9} s), it is really a complementary method adequate for measurements of spectra with linewidth narrower than 10^8 Hz.

4

Laser Oscillation

The advantages of lasers over conventional light sources are high luminance, superior monochromaticity, and excellent directionality. In other words, lasers have good optical coherence, that is, ideal interfering ability, both in temporal and spatial regions. These features are strongly related to the oscillation mechanism of lasers. Laser light can be generated by oscillation and amplification of a selected mode with the stimulated emission of electromagnetic radiation. In order to obtain the stimulated emission, population inversion must be formed between two states in a system. Population inversion means the situation that a higher energy state is more populated than a lower energy state. Therefore this would be impossible to achieve without external excitation. In this chapter, we discuss the principle of laser oscillation on the basis of quantum mechanics. In Sect. 4.1, we start from a model consisting of two-level systems, and describe the oscillation and amplification processes in the context of quantum mechanical interaction between the electromagnetic wave and the atomic system in a cavity. The photon number in an oscillating mode increases with the degree of population inversion and the behavior can be understood macroscopically as a second-order phase transition. The high luminance of laser light is shown as the drastic increase of photon number in the relevant mode above the threshold value of the external pumping. Theoretical studies of the statistical properties of the photon below and above the threshold reveal that the qualitative transition from chaotic to coherent light occurs at the same time. In Sect. 4.2, we will discuss the phase fluctuation of laser light, the characteristics of which are observed as the high monochromaticity and good directionality of lasers. In a coherent state where the photon number $\overline{n}$ in the oscillating mode is much larger than unity, phase diffusion caused by fluctuation is shown to be very slow. The small phase diffusion constant is understood as high monochromaticity in the spectral region. Sometimes, it is possible to make the phase diffusion time longer than 1 minute, and in this case, the light beam behaves as an ideal classical wave with a well-defined phase in that time scale. In Sect. 4.3, we explain several examples of lasers, one of which is the ruby laser which was the first to be

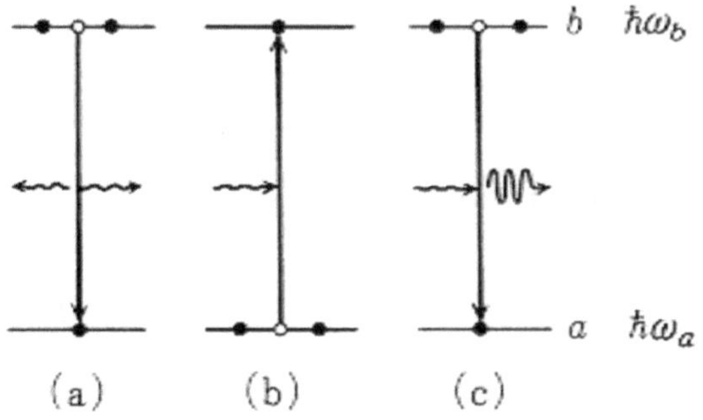

Fig. 4.1. Schematic diagram of (**a**) spontaneous emission, (**b**) optical absorption, and (**c**) stimulated emission due to the electronic transition in a two-level system consisting of levels a and b

demonstrated in the optical region. Next, we introduce the Ti-sapphire and alexandrite laser as typical tunable lasers. We also describe the mechanism of several gas lasers and recent progress in the field of short-wavelength generation.

4.1 Laser Light and its Statistical Properties

The great advantages of lasers are the temporal and spacial coherence as well as the high brightness. The two kinds of coherence are equivalent to good monochromaticity and directionality, as known from Fourier transformation. In this section, we first discuss the mechanism of laser oscillation in the context of the interaction between the radiation field and the atomic system. Next, the statistical properties of the laser beam will be calculated. The results show how the laser light approaches the ideal coherent state described in Chap. 1 with increasing pump intensity. We also give the theoretical results on the photon-counting statistics or photon correlation in laser light.

The interaction between two-level systems and the electromagnetic field includes three types of elementary processes as described in Chap. 2. The schematic diagrams for these processes are shown in Fig. 4.1, that is, spontaneous emission (Fig. 4.1a), absorption (Fig. 4.1b) and stimulated emission (Fig. 4.1c). Stimulated emission, shown in Fig. 4.1(c), is a process in which the electromagnetic field is amplified by the transition of an electron from state b to state a. When this process overcomes (a) spontaneous emission and (b) absorption, the electromagnetic mode starts to oscillate. Based on these principles, the phenomenon is named LASER which is an acronym for Light Amplification by Stimulated Emission of Radiation.

In thermal equilibrium, optical absorption is usually stronger than stimulated emission since the atoms are more populated in the lower level a (Fig. 4.1b). But, once atoms are more populated in the higher level b, stimu-

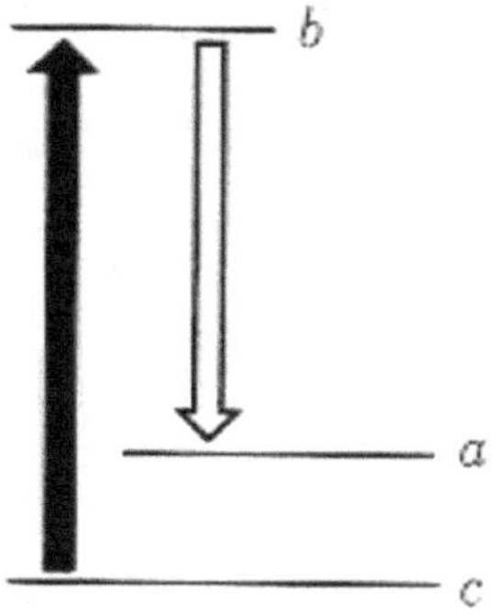

Fig. 4.2. Three-level model for laser oscillation. Population inversion between level a and b is formed by pumping from the lower level c to an excited state b

lated emission overcomes absorption. In other words, in order to obtain laser oscillation, population inversion must be achieved between energy levels a and b.

Among the several methods to achieve population inversion, the most popular way is to continue optical pumping from the lowest level c to an excited level b in a three-level system depicted in Fig. 4.2. In this case, the population inversion is formed between levels a and b as shown in Fig. 4.1(c).

Another condition to obtain laser oscillation is that the stimulated emission from levels b to a overcomes the spontaneous emission between the same levels. Using the Einstein coefficients A and B described in Chap. 2, this condition can be expressed as $B\rho(\omega) > A$, where $\rho(\omega)$ is the energy density of the radiation field with angular frequency ω. Because the coefficients A and B are constants specific to atoms contained in a volume larger than the wavelength, as described in Sect. 2.4, the energy density $\rho(\omega)$ of the radiation must be large in order to satisfy the condition $B\rho(\omega) > A$. Therefore, the laser medium with three-level system is usually contained in a Fabry–Pérot or ring cavity as shown in Figs. 4.3(a) and (b). When the medium is contained in such cavities, the electromagnetic wave generated by the stimulated emission is reflected by high-reflection mirrors and returned to the medium again. Repeated amplification works as positive feedback and the energy will be built up inside the cavity.

Before lasing starts, the radiation is distributed continuously over the modes of a broad frequency and wavenumber range. But, several discrete modes in the frequency and wavevector (direction) domains grow selectively as a result of the feedback, and most of the radiation energy is concentrated in a small number of modes. In the following part of this section, we describe the mechanism of laser oscillation and the statistical properties of laser light on the basis of quantum mechanics.

Since laser oscillation is a result of nonlinear interaction between the atomic system and the radiation field, the evolution of both systems must

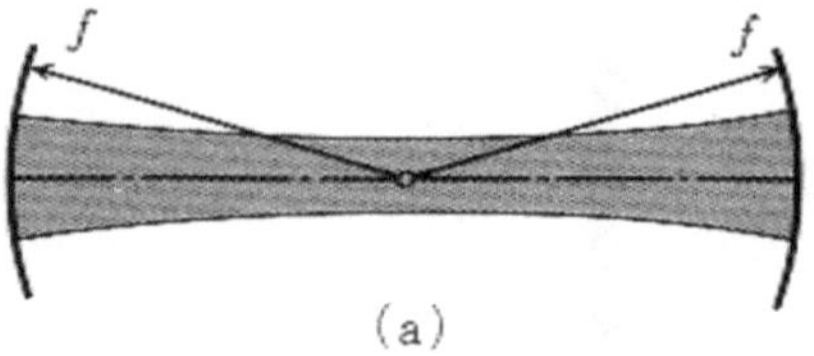

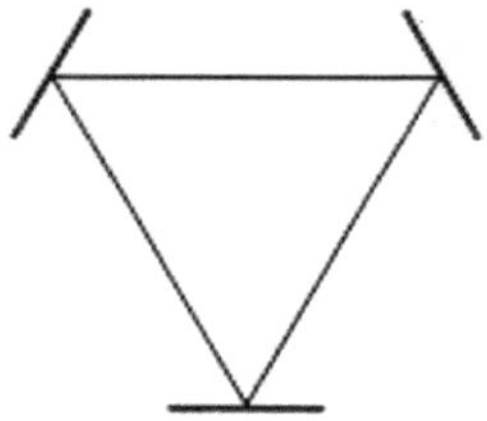

Fig. 4.3. Examples of laser cavities. (**a**) Fabry–Pérot cavity composed of two confocal mirrors, and (**b**) ring cavity with three mirrors. There are many other variations according to the physical properties of the laser media and the required laser performance

be investigated simultaneously. We employ the three-level system as a model of the laser medium. Currently, most commercial lasers consist of media with four-level systems. However, the theoretical description of the typical three-level case can be easily expanded to the four-level case, so we discuss mainly the three-level system in this chapter. Lasers are driven by an external excitation source pumping the atomic system into an excited state b. For a complete description, the numbers of populations N_a and N_b for the lower a and higher b levels, respectively, and the photon-number distribution P_n for the electromagnetic mode must be incorporated. Here and hereafter we confine ourselves to the case of a single cavity mode for simplicity. These physical variables depend on the pumping rate r, and the decaying factor C for the considered mode, as well as the light and matter interaction term. In practice, the pumping is mainly made by an external light beam or electric discharge. Using these parameters, rN atoms are pumped into the state b continuously per second where N denotes the number of atoms. If the laser beam is obtained continuously from one mirror with transmissivity T, the decaying factor can be given by $C = (c/2L)T$ where $2L$ is the round-trip length of the cavity. When the laser medium has a finite absorption coefficient $\alpha(\omega)$ at the oscillating frequency ω, the absorption loss $\alpha(\omega)c$ must be added to the decaying rate C. The damping constant for the atomic system is on the order of the spontaneous emission rate from the levels a or b to the ground state c, and in typical cases the values $2\gamma_a \simeq 2\gamma_b$ are about $3 \times 10^7\,\mathrm{s}^{-1}$. On the other hand, the photon loss constant C in a cavity is estimated to be about $10^6\,\mathrm{s}^{-1}$ for the case of gas lasers. Therefore, the damping of atomic systems is one or two orders faster than that of the radiation field. Since the temporal evolution of the atomic part follows the evolution of the photon part, the behavior of both parts can be separated adiabatically. In such cases, we can do the calculation in two steps.

For the first step, the motion of the atomic system will be calculated under the steady photon-number distribution P_n. The behavior of the atomic system can be described by the induced electric dipole represented by the off-diagonal components ρ_{ab} and ρ_{ba} of the density matrix, and also by the population given by the diagonal components ρ_{aa} and ρ_{bb}. According to perturbation theory, these components are formulated as follows:

$$\dot{\rho}_{ba} = -(i\omega_{ba} + \gamma_{ba})\rho_{ba} - \frac{i}{\hbar}\mathcal{H}'_{ba}(t)(\rho_{bb} - \rho_{aa}), \tag{4.1a}$$

$$\dot{\rho}_{bb} = rP_n - \frac{i}{\hbar}(\mathcal{H}'_{ba}\rho_{ab} - \rho_{ba}\mathcal{H}'_{ab}) - 2\gamma_b\rho_{bb}, \tag{4.1b}$$

$$\dot{\rho}_{aa} = \frac{i}{\hbar}(\rho_{ab}\mathcal{H}'_{ba} - \mathcal{H}'_{ab}\rho_{ba}) - 2\gamma_a\rho_{aa}. \tag{4.1c}$$

Here, $\mathcal{H}'$ means the first-order interaction Hamiltonian between the atomic system and the radiation field with angular frequency ω given by (2.12) and we assume that the excited state is generated with the rate rP_n. Because the transverse off-diagonal damping factor $\gamma_{ab} = 1/T_2$ (transverse relaxation) is generally much larger than the values of the diagonal damping factors $2\gamma_a$ and $2\gamma_b$ (longitudinal relaxation), the off-diagonal component with the angular frequency ω can be obtained under the adiabatic approximation and putting $\dot{\rho}_{ba} = -i\omega\rho_{ba}$ as

$$\rho_{ab} = \frac{\mathcal{H}'_{ba}(t)(\rho_{bb} - \rho_{aa})}{\hbar(\omega - \omega_{ba} + i\gamma_{ba})}. \tag{4.2}$$

Also the equations of motion of the diagonal parts can be obtained by substituting this expression into (4.1b) and (4.1c). In this first step, the distributions of the atomic system $R_n^a \equiv \rho_{aa}(n)$ and $R_n^b \equiv \rho_{bb}(n)$ were given as functions of the photon number n. In the next step, we must substitute R_n^a and R_n^b into the equations of the photon systems in order to investigate the photon dynamics. The evolution of photons is determined by the light–matter interaction and the loss of photons inside cavities. Here, R_n^a (R_n^b) is the probability that n photons exist in the mode and an atom does at the state $a\,(b)$. Therefore, NR_n^a/P_n means the average number of atoms existing in the state a when there are n photons. The total probability of an atom existing in the state $a\,(b)$ can be written as the summation over all photon number n:

$$\sum_n R_n^{a(b)} = \frac{N_{a(b)}}{N}.$$

Then, let us study the temporal evolution of R_n^b and R_{n+1}^a. Both states have almost the same energy, and are coupled via exchange of one photon as expressed by the following rate equations:

$$\frac{dR_n^b}{dt} = rP_n - 2\gamma_b R_n^b + gR_{n+1}^a(n+1) - gR_n^b(n+1), \tag{4.3a}$$

$$\frac{dR_{n+1}^a}{dt} = -2\gamma_a R_{n+1}^a - gR_{n+1}^a(n+1) + gR_n^b(n+1). \tag{4.3b}$$

Here, the pumping increases the number of atoms existing in the state b with the rate rP_n. The damping rate can be expressed as $\gamma_{ba} = \gamma_a + \gamma_b$ by neglecting the purely transversal damping part. In this case, the coupling constant g can be obtained from (4.2) as

$$g = \frac{e^2 \omega_{ba} |\boldsymbol{r}_{ab}|^2}{3\epsilon_0 V \hbar(\gamma_a + \gamma_b)}, \tag{4.4}$$

under the resonant condition $\omega = \omega_{ab}$. The average over the polarization direction was taken into account as done in (2.24). From now on, we consider the behavior in the typical time scale of gas lasers that is longer than the longitudinal relaxation time $T_1 \equiv 1/\gamma_a,\ 1/\gamma_b \sim 10^{-8}$s, but shorter than the decay time of the radiation mode $\sim 10^{-6}$s, under which scale both the atomic system and photon distribution are in equilibrium. By replacing the left-hand sides of (4.3a) and (4.3b) with zero, the solution in the quasisteady state is given as the following:

$$R_n^b = \frac{rP_n\{g(n+1) + 2\gamma_a\}}{4\gamma_a\gamma_b + 2(\gamma_a + \gamma_b)g(n+1)}, \tag{4.5}$$

$$R_{n+1}^a = \frac{rP_n g(n+1)}{4\gamma_a\gamma_b + 2(\gamma_a + \gamma_b)g(n+1)}. \tag{4.6}$$

The addition of (4.3a) and (4.3b) gives the following relation:

$$rP_n = 2\gamma_b R_n^b + 2\gamma_a R_{n+1}^a, \tag{4.7}$$

and the summation over n gives:

$$rN = 2\gamma_b N_b + 2\gamma_a N_a. \tag{4.8}$$

This equation means that N_a and N_b are determined by the balance between the pumping rate rN from the third level c to the highest level b and the decaying rates from the levels b or a to the level c. Now, we introduce a new parameter $\beta \equiv 2\gamma_a\gamma_b/(\gamma_a + \gamma_b)g$. We estimate the values for typical lasers of volume $V = 2 \times 10^{-5}\text{m}^3$, where the number of atoms contained in the cavity is $N = 2 \times 10^{20}$. In this case, values of the parameters are $g \sim 0.5\text{s}^{-1}$ and $\beta \sim 3 \times 10^7$ for the transition frequency $\omega_{ba} \sim 3 \times 10^{15}$ Hz. If $\beta \gg n$, the second term $2(\gamma_a + \gamma_b)g(n+1)$ in the denominators of (4.5) and (4.6) are negligible, compared to the first of them, $4\gamma_a\gamma_b$. Therefore, R_n^b and R_{n+1}^a can be approximated as follows:

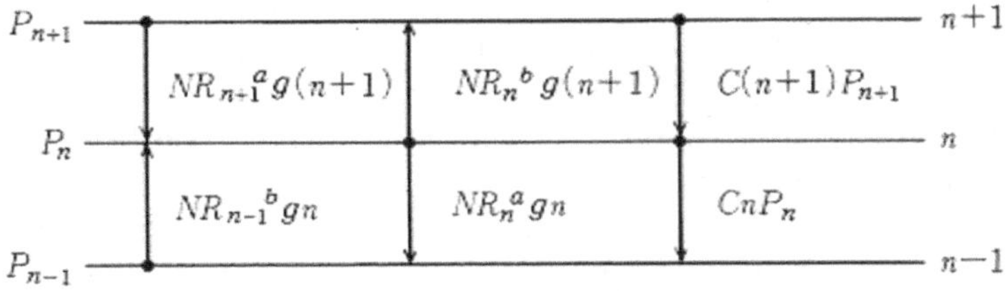

Fig. 4.4. Ladder diagram representing the change of photon-number distribution P_n caused by a transition in the atomic state and cavity loss C. The rates R_n^b and R_n^a are given by (4.5) and (4.6), respectively

$$R_n^b \simeq \frac{rP_n}{2\gamma_b} \qquad (n \ll \beta), \tag{4.9}$$

$$R_{n+1}^a \simeq \frac{rP_n g(n+1)}{4\gamma_a\gamma_b} \simeq R_n^b \frac{g(n+1)}{2\gamma_a} \qquad (n \ll \beta). \tag{4.10}$$

On the other hand, for the case of $n \gg \beta$, we obtain

$$R_n^b \simeq R_{n+1}^a \simeq \frac{rP_n}{2(\gamma_a + \gamma_b)} \qquad (n \gg \beta). \tag{4.11}$$

In this case, the distributions of the a and b levels are almost equal, and that corresponds to the saturation effect.

In the second step, the rate equation for P_n can be derived from the expressions for R_n^b and R_{n+1}^a, (4.5) and (4.6), and the ladder diagram describing the photon-number distribution P_n (shown in Fig. 4.4):

$$\frac{dP_n}{dt} = -\frac{A\beta P_n(n+1)}{\beta+n+1} + \frac{A\beta P_{n-1}n}{\beta+n} - CnP_n + C(n+1)P_{n+1}. \tag{4.12}$$

Here, $A \equiv Nrg/2\gamma_b$. The last two terms describe the rate for the $(n+1)$ photon state P_{n+1} to change into the n-photon state by losing one photon, as given by $C(n+1)P_{n+1}$, and the decreasing rate of P_n with the transition from the n-photon state to the $(n-1)$ photon state as given by $-CnP_n$. We can obtain P_n in the steady state by directly solving (4.12), but the same result is more easily given by considering the balancing of photon states shown in Fig. 4.4, the details of which are given in the following paragraph.

When the balancing is satisfied between the n and $(n-1)$ levels in the photon-number ladder shown in Fig. 4.4, the following expression can be obtained:

$$NR_{n-1}^b gn - NR_n^a gn - CnP_n = 0. \tag{4.13}$$

Substituting R_n^b and R_{n+1}^a given in (4.5) and (4.6), into (4.13), and using the parameters $A \equiv Nrg/2\gamma_b$ and $\beta \equiv 2\gamma_a\gamma_b/(\gamma_a+\gamma_b)g$, we obtain the following recurrence formula:

$$P_n = \frac{A\beta}{C(n+\beta)} P_{n-1} = \left(\frac{A\beta}{C}\right)^n \frac{\beta!}{(n+\beta)!} P_0 . \tag{4.14}$$

The probability P_0 of no photon existing in the cavity can be derived from the normalization condition

$$\sum_n P_n = P_0 \sum_n \left(\frac{A\beta}{C}\right)^n \frac{\beta!}{(n+\beta)!} = 1 . \tag{4.15}$$

The summation in (4.15) can be expressed by the confluent hypergeometric function $F(a, b; x)$ as

$$F\left(1, 1+\beta; \frac{A\beta}{C}\right) P_0 = 1 . \tag{4.16}$$

Here, the general form of the function is defined by the following expression

$$F(a, b; x) = \sum_n \frac{(a+n-1)!(b-1)!}{(b+n-1)!(a-1)!} \frac{x^n}{n!} . \tag{4.17}$$

Hence, P_n is given in a simpler form as

$$P_n = \left(\frac{A\beta}{C}\right)^n \frac{\beta!}{(n+\beta)! F(1, 1+\beta; A\beta/C)} . \tag{4.18}$$

The physical parameters characterizing laser oscillation can be calculated by using the photon number distribution P_n. For example, the mean photon number $\overline{n}$ in an electromagnetic mode can be given as a function of the intensity of pumping in the next equation:

$$\overline{n} = \sum_{n=0}^{\infty} n P_n = \sum_{n=1}^{\infty} (n+\beta-\beta) \left(\frac{A\beta}{C}\right)^n \frac{\beta!}{(n+\beta)!} P_0 = \frac{A\beta}{C} - \beta(1-P_0) . \tag{4.19}$$

As the pumping rate r is included in (4.19) as the form of $A/C \equiv rNg/(2\gamma_b C)$, we can discuss the laser characteristics by two parameters A/C and β. The mean photon number is plotted as a function of A/C in Fig. 4.5 when $\beta = 3 \times 10^7$ [32]. From this figure, we can understand that the photon number builds up drastically at the threshold, that is, $A/C = 1$. Laser oscillation corresponds to the situation that the photon number increases only in one selected mode. This is one reason why laser oscillation is interpreted as a second-order phase transition.

In the following couple of subsections, we investigate the photon statistics above and below the threshold.

4.1.1 Photon Statistics Below the Threshold ($A/C < 1$)

When $A/C < 1$, the second argument in the confluent hypergeometric function (4.17) is larger than the third. In this case, we can define y with $x = by$ and the function can be expanded in terms of b^{-1} because $b = 1 + \beta \gg 1$ [33]:

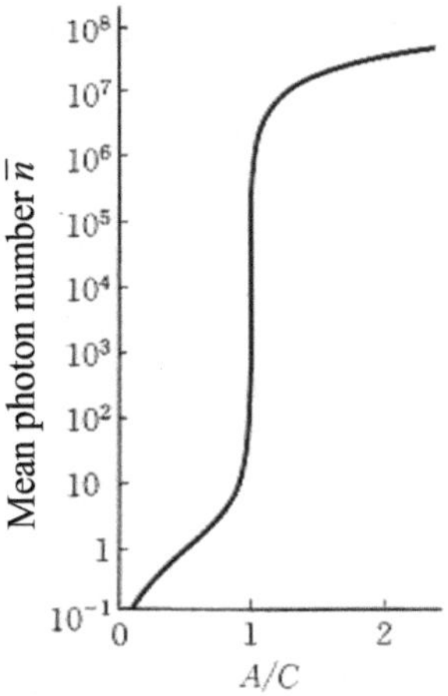

Fig. 4.5. Mean photon number $\overline{n}$ in a cavity mode as a function of pumping rate A/C. The parameter β is defined as $\beta = 2\gamma_a\gamma_b/(\gamma_a+\gamma_b)g$, and the value $\beta = 3\times 10^7$ is chosen in this case [32]

$$F(a,\, b;\, by) = \frac{1}{(1-y)^a}\left\{1 - \frac{a(a+1)}{2b}\left(\frac{y}{1-y}\right)^2 + O(|b|^{-2})\right\}. \quad (4.20)$$

In our case, because equation (4.20) can be written in the following form:

$$F\left(1,\, 1+\beta;\, \frac{A\beta}{C}\right) = \frac{1}{1-(A/C)}\left[1 - \frac{A/C}{\beta\{1-(A/C)\}^2} + O(\beta^{-2})\right], \quad (4.21)$$

the probability P_0 can be obtained by substituting this expression into (4.16). Then, we can derive the average photon number $\overline{n}$ from (4.19) by neglecting the terms of $O(\beta^{-2})$ as

$$\overline{n} = \frac{A/C}{1-(A/C)} \qquad (A/C < 1). \quad (4.22)$$

The expression of the photon-number distribution (4.18) can be approximated by the expansion of the confluent hypergeometric function and by applying Stirling's formula $(n+\beta)! \simeq \beta^n \beta!$ $(\beta \gg n)$ as

$$P_n = \left(\frac{A}{C}\right)^n \left(1 - \frac{A}{C}\right) = \frac{\overline{n}^n}{(1+\overline{n})^{n+1}} \qquad (A/C < 1). \quad (4.23)$$

This formula gives the chaotic photon distribution which was given for the thermal radiation in agreement with (3.16).

In other words, when lasers are driven below the threshold, a small number of photons ($\overline{n} \sim A/C$) with chaotic characteristics will exist in the cavity.

4.1.2 Photon Statistics Above the Threshold ($A/C > 1$)

In the opposite limit, we can also make an approximation for the confluent hypergeometric function in order to calculate P_0. By using the following expansion with $x \equiv A\beta/C(> b \equiv 1 + \beta)$ [33]:

$$F(a,\, b;\, x) = \frac{\Gamma(b)}{\Gamma(a)} e^{x} x^{a-b}\{1 + O(|x|^{-1})\}, \tag{4.24}$$

the final form can be obtained as:

$$F\left(1,\, 1+\beta;\, \frac{A\beta}{C}\right) = \beta! \exp\left(\frac{A\beta}{C}\right)\left(\frac{A\beta}{C}\right)^{-\beta}\left\{1 + O\left(\frac{C}{A\beta}\right)\right\}. \tag{4.25}$$

Because this value is much larger than 1, we know that $P_0 \ll 1$ from (4.16). Then, assuming that $P_0 \simeq 0$ in (4.19), the mean photon number in the lasing mode can be expressed as

$$\overline{n} = \beta\left(\frac{A}{C} - 1\right). \tag{4.26}$$

By substituting (4.26) into (4.18), the photon-number distribution P_n can be calculated as:

$$P_n = \left(\frac{A\beta}{C}\right)^{n+\beta} \frac{\exp(-A\beta/C)}{(n+\beta)!} = \frac{(\overline{n}+\beta)^{n+\beta}\exp(-\overline{n}-\beta)}{(n+\beta)!}. \tag{4.27}$$

In particular, when $A/C - 1 \gg 1$, that means $\overline{n} \gg \beta$, P_n can be approximated as

$$P_n = \frac{\overline{n}^n \exp(-\overline{n})}{n!} \quad (A/C - 1 \gg 1). \tag{4.28}$$

As known from this Poisson distribution, the photon statistics of laser light are found to be the same as that of the coherent state, in agreement with (1.52) with $\overline{n} \equiv |\alpha|^2$.

Next, we investigate the second-order correlation function $g_{12}^{(2)}(\tau)$ at $\tau = 0$. This can be expressed as the following, in the quantum-mechanical representation for a single-mode radiation field,

$$g_{12}^{(2)}(0) = \frac{\mathrm{Tr}\{\rho(\hat{a}^\dagger)^2\hat{a}^2\}}{\{\mathrm{Tr}(\rho\hat{a}^\dagger\hat{a})\}^2}. \tag{4.29}$$

The density matrix in (4.29) is defined as

$$\rho = \sum_{n=0}^{\infty} P_n |n\rangle\langle n|. \tag{4.30}$$

Therefore, the correlation function $g_{12}^{(2)}(0)$ can be obtained by employing (4.23) for P_n below the threshold value ($A/C < 1$), and (4.27) above the threshold of laser oscillation ($A/C > 1$). The results are expressed as

$$g_{12}^{(2)}(0) = \frac{\sum_n \left(\frac{A}{C}\right)^n \left(1 - \frac{A}{C}\right) \langle n|(\hat{a}^\dagger)^2 \hat{a}^2|n\rangle}{\left\{\sum_n \left(\frac{A}{C}\right)^n \left(1 - \frac{A}{C}\right) \langle n|\hat{a}^\dagger \hat{a}|n\rangle\right\}^2} \tag{4.31}$$

$$= \frac{2(A/C)^2/\{1-(A/C)\}^2}{(A/C)^2/\{1-(A/C)\}^2} = 2 \quad (A/C < 1), \tag{4.32}$$

$$g_{12}^{(2)}(0) = \frac{1}{\overline{n}^2} \sum_{n=0}^{\infty} \frac{(\overline{n}+\beta)^{n+\beta} \exp\{-(\overline{n}+\beta)\}}{(n+\beta)!} \langle n|(\hat{a}^\dagger)\hat{a}^2|n\rangle$$

$$= \frac{1}{\overline{n}^2} \exp\{-(\overline{n}+\beta)\} \sum_{n=2}^{\infty} \frac{n(n-1)(\overline{n}+\beta)^{n+\beta}}{(n+\beta)!}$$

$$= 1 + \frac{\beta}{\overline{n}^2} \quad (A/C > 1). \tag{4.33}$$

From these results, the photon sequence shows the bunching below the threshold ($A/C < 1$) because $g_{12}^{(2)}(0) - 1 = 1 > 0$. This is the characteristic of the chaotic light. On the other hand, increasing the average photon number $\overline{n} \equiv \beta(A/C - 1)$, $g_{12}^{(2)}(0) - 1 = \beta/\overline{n}^2$ approaches 0 above the threshold value. This shows that laser light approaches ideal coherent light above the threshold.

Now, we compare the theory with experimental results in Fig. 4.6 [34] which shows the results of the theory and experiment for the degree of coherence $g_{12}^{(2)}(0)$.

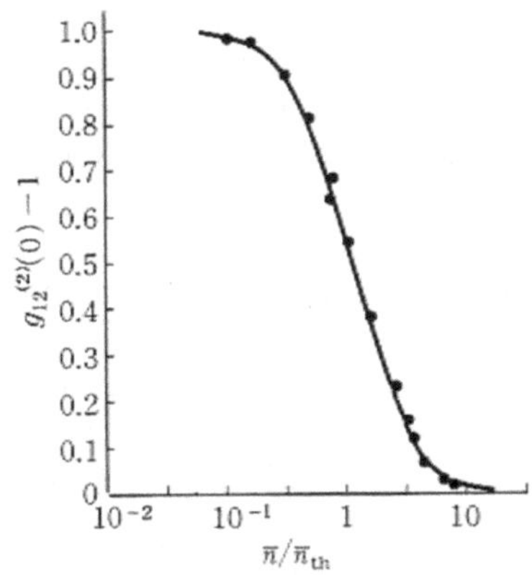

Fig. 4.6. The degree of second-order coherence as a function of mean photon number $\overline{n}$. The theoretical calculation is given by the solid curve, and experimental results by the closed circles [34]

In the experiment, the value was obtained as a function of the mean photon number $\overline{n}$ in a cavity; on the other hand, theoretical results were given by (4.18). The horizontal axis is normalized with the photon number at the threshold value $\overline{n}_{\mathrm{th}}$, and the vertical axis indicates the degree of second-order coherence $g_{12}^{(2)}(0) - 1$. The theoretical and experimental results for the correlation function $g_{12}^{(2)}$ correspond very well over a wide pumping range including the transition from incoherent to coherent light as well as two limiting cases described in Sects. 4.1.1 and 4.1.2, that is, $A/C \ll 1$ $(\overline{n}/\overline{n}_{\mathrm{th}} \ll 1)$ and $A/C \gg 1$ $(\overline{n}/\overline{n}_{\mathrm{th}} \gg 1)$, respectively.

In the beginning of this section, we noted that one of three important characteristics is high luminance. This property is reflected in the fact that a large number of photons exist in a single mode above the threshold of oscillation. The behavior can be understood intuitively from Fig. 4.5 and equation (4.26) showing the photon number $\overline{n} = \beta(A/C - 1)$ for $A/C > 2$. We also realize that the light has coherent characteristics with small photon-number fluctuation expressed by $g_{12}^{(2)} - 1$. In the next section, let us discuss theoretically the high monochromaticity and directionality, the second and the third properties of lasers.

4.2 Phase Fluctuations of Lasers

Laser oscillation resembles the second-order phase transition typically observed in ferromagnetic ordering as shown in Fig. 4.5. In ferromagnetic materials, the ordered phase appears below the Curie temperature T_{c}, and likewise laser oscillation, as an ordered phase, appears when the pumping rate is higher than its threshold $(A/C = 1)$. In the case of lasers, the order parameter is the photon number $\overline{n}$ in the mode contributing to laser oscillation. The pumping rate in the laser oscillation plays the role of temperature for the ferromagnetic phase transition. When the higher energy level b is more populated than the lower energy level a, it is called population inversion and this quasi-equilibrium situation can be described with the concept of negative temperature. The negative temperature is defined assuming that the two selected levels are populated according to a Boltzmann distribution, so that the temperature is necessarily negative when the population is inverted.

The direction of ordered magnetic moments in isotropic Heisenberg ferromagnetic materials cannot be determined by the Hamiltonian of the system, and they will have any directions with the same probability. But the direction of the magnetization diffuses into different directions with a time scale much longer than experimental scales. The electric field of a laser beam has a property similar to a Heisenberg ferromagnet, because its phase diffuses very slowly from the initial value into other equivalent phases.

In the discussion of laser oscillation in the former section, we show that the density matrix ρ of photons in lasers can be expanded in terms of the photon-number state $|n\rangle\langle n|$ as

$$\rho = \sum_n P_n |n\rangle\langle n|. \tag{4.34}$$

In this description, the photon system is characterized by $P_n = \langle n|\rho|n\rangle$ or diagonal components of the density matrix with the basis of photon-number state, but the information from the phase component does not appear in the expression. On the other hand, the electric field $\boldsymbol{E}$ in (1.38) includes a linear combination of generation and annihilation operators of the photon $\hat{a}^\dagger_\lambda$ and $\hat{a}_\lambda$. Then, the off-diagonal components of the density matrix $\langle n \pm 1|\rho|n\rangle$ are essential for the complete description of the electric field. In this section, the discussion will be expanded in order to include the phase motion of the radiation field. In the discussion, it will be found that the phase diffusion is suppressed above the laser threshold, and the spectral width of the laser becomes very narrow. Then finally the single-mode oscillation in a cavity will be established. This is the origin of the second and third important properties, i.e., the high monochromatiticity and the superior directionality of lasers.

Because the details of calculation are available in other books [35], we briefly outline the framework of the theory and results. Here, we assume that the initial density matrix of the coupled system consisting of photonic (a) and atomic (B) parts can be given by a direct product of both systems as:

$$\rho_{\mathrm{aB}}(0) = \rho_{\mathrm{a}}(0) \times \rho_{\mathrm{B}}(0). \tag{4.35}$$

The interaction between a photon and an atom is written as

$$\mathcal{V}_{\mathrm{aB}} = \hbar g \sum_i \sigma_i \hat{a}^\dagger + \mathrm{h.\,c.} \tag{4.36}$$

Here, σ_i means the operator to induce the transition from the excited state b to the lower energy state a in the ith atom, and the Hermite conjugated operator $\sigma_i^\dagger$ induces the reverse process. The equation of the density matrix for the photonic system $\rho(t)$ can be derived by taking the trace for the atomic system as

$$\rho(t) = \mathrm{Tr}_{\mathrm{B}} \rho_{\mathrm{aB}}(t). \tag{4.37}$$

Then, the following equation can be obtained by including perturbation terms up to forth order:

$$\begin{aligned}\dot{\rho} = &-\frac{1}{2}A(\rho\hat{a}\hat{a}^\dagger - \hat{a}^\dagger\rho\hat{a}) - \frac{1}{2}C(\rho\hat{a}^\dagger\hat{a} - \hat{a}\rho\hat{a}^\dagger) \\ &+\frac{1}{8}B[\rho(\hat{a}\hat{a}^\dagger)^2 + 3\hat{a}\hat{a}^\dagger\rho\hat{a}\hat{a}^\dagger - 4\hat{a}^\dagger\rho\hat{a}\hat{a}^\dagger\hat{a}] + \mathrm{h.\,c.}\end{aligned} \tag{4.38}$$

Here, the coefficients A and C have the same meaning as in the former section, and B is defined as C/β.

Next, we consider the expansion of the photonic density matrix $\rho(t)$ with coherent states $|\alpha\rangle$ of the single mode:

$$\rho(t) = \int d^2\alpha P(\alpha, t)|\alpha\rangle\langle\alpha|. \tag{4.39}$$

The representation of the coherent state $|\alpha\rangle$ contains information about the off-diagonal components of the photon-number state, that is, information on the phase, as shown by the equation $\langle\alpha|\hat{a}|\alpha\rangle = \alpha$. By substituting (4.39) into (4.38), the following Fokker–Planck equation can be obtained:

$$\frac{\partial}{\partial t}P(\alpha, t) = -\frac{1}{2}\left\{\frac{\partial}{\partial\alpha}[(A - C - B|\alpha|^2)\alpha P] + \text{c.c.}\right\} + A\frac{\partial^2 P}{\partial\alpha\partial\alpha^*}. \tag{4.40}$$

If the complex parameters of the coherent state α and α^* are expressed in polar form as $\alpha = r\exp(i\theta)$ and $\alpha^* = r\exp(-i\theta)$, the Fokker–Planck equation (4.40) can be transformed into the following equation:

$$\begin{aligned}\frac{\partial}{\partial t}P(r, \theta, t) &= -\frac{1}{2r}\frac{\partial}{\partial r}[r^2(A - C - Br^2)P(r, \theta, t)] \\ &\quad + \left(\frac{A}{4r^2}\frac{\partial^2}{\partial\theta^2} + \frac{A}{4r}\frac{\partial}{\partial r}r\frac{\partial}{\partial r}\right)P(r, \theta, t).\end{aligned} \tag{4.41}$$

This equation can be simplified by introducing the nondimensional parameters T, R, and a:

$$\begin{aligned}T &\equiv \sqrt{\frac{AB}{8}}t, \quad R \equiv \sqrt[4]{\frac{2B}{A}}r, \quad a \equiv \sqrt{\frac{2B}{A}}\left(\frac{A - C}{B}\right), \\ \frac{\partial P}{\partial T} &= -\frac{1}{R}\frac{\partial}{\partial R}\{R^2(a - R^2)P\} + \frac{1}{R}\frac{\partial}{\partial R}\left(R\frac{\partial P}{\partial R}\right) + \frac{1}{R^2}\frac{\partial^2}{\partial\theta^2}P.\end{aligned} \tag{4.42}$$

The time-independent solution of the equation can be obtained by the separation of variables. The part depending on θ can be solved as $e^{im\theta}$ where m is an integer. In the case that $m = 0$, the steady state solution can be given as the following:

$$P(R) = \frac{N}{2\pi}\exp\left(-\frac{1}{4}R^4 + \frac{1}{2}aR^2\right). \tag{4.43}$$

Here, the normalized constant N can be expressed as

$$\frac{1}{N} = \int_0^\infty R\exp\left(-\frac{1}{4}R^4 + \frac{1}{2}aR^2\right)dR. \tag{4.44}$$

Now, the photon-number distribution function $P(R)$ was derived instead of P_n in (4.18) where the discrete parameter n was used. Two types of expressions are connected by the relation $R^2 = \sqrt{2B/A}n$, and both expressions are equivalent in the limit $n \gg 1$. We can calculate the photon-number distribution $P_m(T)$ obtained in Sect. 3.2, and the second-order correlation function $g_{12}^{(2)}(0)$

by using this steady state solution. For the case that $a \gg 1$, the photon distribution is found to be Gaussian with the peak at $\overline{n} = (A-C)/B = \beta(A/C-1)$ and width $\Delta n = \sqrt{A/B} = \sqrt{A\beta/C}$, by substituting $R^2 = \sqrt{2B/A\overline{n}} = a$.

When we employ (4.43) for the R dependence and assume that $P(\theta, t) \sim \exp(\pm i\theta)$ by choosing $m = \pm 1$ as the θ dependence, the equation (4.42) is reduced to the following form:

$$\frac{\partial}{\partial t}P(\theta, t) = \frac{A}{4\overline{n}}\frac{\theta^2}{\partial\theta^2}P(\theta, t) = -\frac{A}{4\overline{n}}P(\theta, t). \tag{4.45}$$

Defining the diffusion constant for the phase θ as $D = A/2\overline{n}$, the solution can be given as

$$P(\theta, t) = \exp\left(-\frac{1}{2}Dt\right). \tag{4.46}$$

The expectation value of the electric field amplitude $E(t)$ can be calculated from the density matrix (4.39) and (3.11) as

$$\begin{aligned}\langle E(t)\rangle \equiv \mathrm{Tr}\{\rho(t)E(t)\} &= \frac{1}{2}\varepsilon\langle ae^{-i\omega t}\rho(t)\rangle + \mathrm{c.\,c.}\\ &= \langle E(0)\rangle \sin\omega t \exp\left(-\frac{1}{2}Dt\right). \end{aligned}\tag{4.47}$$

From the Fourier transformation of (4.47), we can obtain the power spectrum of laser light as

$$\begin{aligned}|E(\Omega)|^2 &= \left|\int_0^\infty dt\, e^{i\Omega t}\langle E(0)\rangle \sin\omega t \exp\left(-\frac{1}{2}Dt\right)\right|^2\\ &\simeq |\langle E(0)\rangle|^2 \frac{1}{(\Omega-\omega)^2 + (D/2)^2}. \end{aligned}\tag{4.48}$$

Here, we neglected the antiresonance term. Above the threshold value ($A/C > 2$), since the diffusion constant becomes very small because $D/2 = A/4\overline{n} < 10^{-2}s^{-1}$, the spectral width of the laser is found to be very small. If the spectral width $D/2$ becomes narrower than the mode separation in a cavity $2\pi c/L$ (L is the length of the cavity), single-mode oscillation can be achieved. These mathematically derived properties appear as the high monochromaticity and directionality of the laser beam.

Now, we consider a typical example in which the decay rate C of a cavity is comparable to $A \sim 10^6\,\mathrm{s}^{-1}$, and the mean photon number n is 3×10^7. Then the inverse of the phase diffusion constant $D/2$ ($= A/4\overline{n}$) is the order of 1 minute. Therefore, the phase diffusion constant is 10^7 times slower than the rate of stimulated emission (A) and the decay rate of the cavity (C). In this case, the phase of the light does not change during a period shorter than 1 min. This is the reason why the laser light behaves as an ideal classical electromagnetic field.

4.3 Several Examples of Lasers

4.3.1 Ruby Laser

Maiman succeeded in producing laser oscillation for the first time in 1960. Ruby for lasers is a crystalline sapphire (Al_2O_3) containing Cr^{3+} ions with a concentration of 0.01–0.03 weight %. Because the color of ruby for a laser is paler than jewel ruby, which typically contains 0.5% Cr^{3+}, it is called pink ruby. Since a Cr^{3+} ion substitutes a Al^{3+} ion, it is located at the center of a corundum structure composed of the six nearest neighbor O^{2-} ions. The Cr^{3+} ion feels approximately a cubic crystal field. In this field, the degenerate 3d level of the central ion splits into e_g and t_{2g} levels as shown in Fig. 4.7. Among these 3d orbitals, $t_{2g}(\xi, \eta, \zeta)$ orbitals are described by the wavefuctions distributing toward the direction of xy, yz, zx, and $e_g(u, v)$ orbitals have the dependence $2z^2 - x^2 - y^2$, and $\sqrt{3}(x^2 - y^2)$. Since the e_g orbitals expand in the direction of O^{2-} ions, e_g electrons are subjected to a stronger repulsion force. On the other hand, since 3d electrons in t_{2g} orbitals distribute in the space between oxygen ions, their energies are lower, as shown in Fig. 4.7. In the case of Cr^{3+}, three 3d electrons occupy t_{2g} orbitals, and the ground state forms a quartet $^4A_2$ according to Hund's rule [36].

There are several types of excited states. States with an electron configuration of $(t_{2g})^2e_g$ or $t_{2g}(e_g)^2$ are formed by the excitation of one or two t_{2g} electrons into e_g orbitals, respectively. Another type of excitation is formed by the transition of one electron in a t_g state into another t_g state changing the sign of the spin. By diagonalizing the interaction terms among 3d electron configurations, we can obtain $(3d)^3$ multiplet states as ^2E, $^2T_1$, $^2T_2$, and $^4A_2$ (these belong to $(t_{2g})^3$ configurations), and $^4T_2$, $^4T_1$, $^2A_1$, etc. (which belong to $(t_{2g})^2e_g$). The energies of all levels can be represented as a function of the intensity of the crystal field $10Dq$, and they are depicted in a Tanabe–Sugano chart as given in Fig. 4.8 [36]. Usually $10\,Dq$ is scaled by B, a matrix element of the configuration interaction. In the case of Cr^{3+}:Al_2O_3, the value of $10Dq/B$ is 2.5, so that excited states line up from the lower energy side in the order ^2E, $^2T_1$, $^4T_2$, $^2T_2$, $^4T_1$ $[(t_{2g})^2e_g]$, $^2A_1[(t_{2g})^2e_g]$, and others (see Fig. 4.8).

Experimental results of the absorption spectra of Cr^{3+}:Al_2O_3 given by Misu et al. are shown in Fig. 4.9. We find that the ^2E, $^2T_1$, and $^2T_2$ levels form

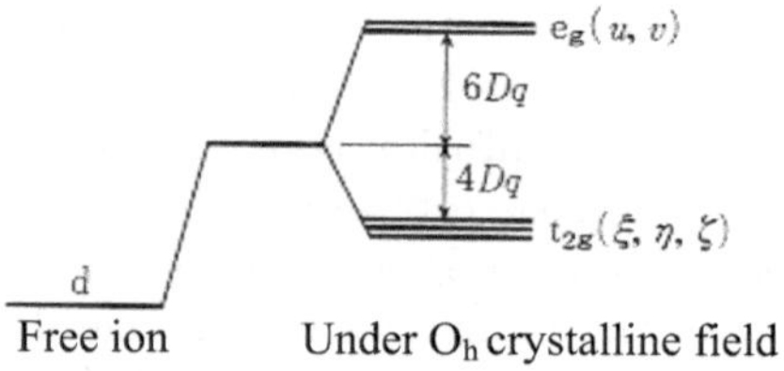

Fig. 4.7. Splitting of d-electron levels in a crystal field with O_h symmetry

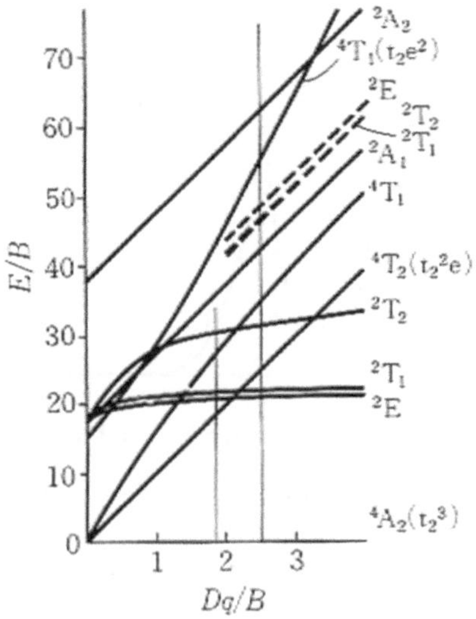

Fig. 4.8. Dependence of electronic energy of multiplet states $^{2S+1}$L of a $(3d)^3$ system (for a example, Cr^{3+}) in O_h symmetry on the crystal field parameter $10Dq$ [36]

weak and sharp line spectra, and on the contrary, that $^4T_2$ and $^4T_1[(t_{2g})^2e_g]$ appear as strong and broadband absorption peaks. The difference in these two cases can be explained also by the Tanabe–Sugano chart. The crystal field $10Dq$ varies with the change of distance between the Cr^{3+} ion and O^{2-} ions, and the distances are subjected to the influence of lattice vibrations. The energy dependence of the ^2E, $^2T_1$, and $^2T_2$ levels on the $10Dq$ values is very small around $10Dq/B = 2.5$ as shown in Fig. 4.8, so that they are not affected by lattice vibrations. On the other hand, the energy levels of $^4T_2$ and $^4T_1$ depend strongly on $10Dq$, so that the vibration makes their linewidth broader.

The notation of multiplets $^{2S+1}L_n$ defined for atoms or ions describes the total orbital angular momentum L and the total spin S of multi-electron systems. When the system is located in a cubic symmetry field, we use the irreducible representation of the cubic O_h group A_1, A_2, E, T_1, and T_2 instead of L. From Fig. 4.9, it is found that the transitions accompanied by the change of total spin S, such as $^4A_2 \rightarrow {}^2E$, $^2T_1$ and $^2T_2$, have sharp lines, and the transitions without spin change, such as $^4A_2 \rightarrow {}^4T_1$ and $^4T_2$, have broad line shapes. The narrow spectra of the former cases can be explained by the weaker effects from lattice vibrations as predicted from Fig. 4.8; and the origin of the weakness is that these are higher order transitions including the spin-flip process. Figure 4.10 depicts schematically the energy levels and their variation due to lattice vibrations in a configurational coordinate. Both types of transitions, that is, among the $(t_{2g})^3$ configuration and those of $(t_{2g})^3 \rightarrow (t_{2g})^2e_g$ are parity forbidden, so that these absorptions should

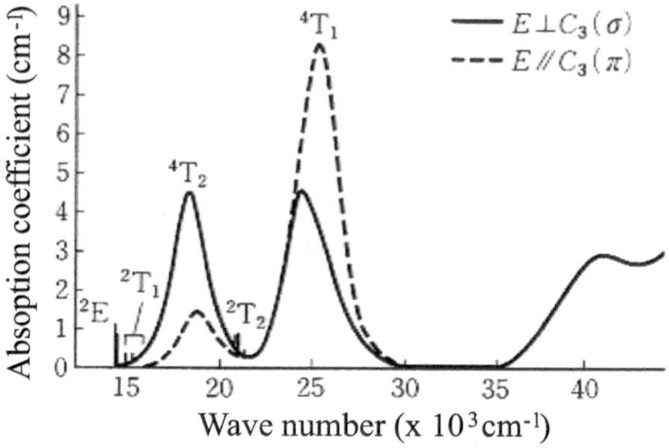

Fig. 4.9. Absorption spectrum of ruby. The data over 35 000 cm^{-1} were measured by light with random polarization. Only the results for σ polarization are shown for narrow spectral bands, and their width and intensity are influenced by the resolution of the apparatus. The spectrum was observed by A. Misu for 0.26 weight % Cr_2O_3 at room temperature [36]

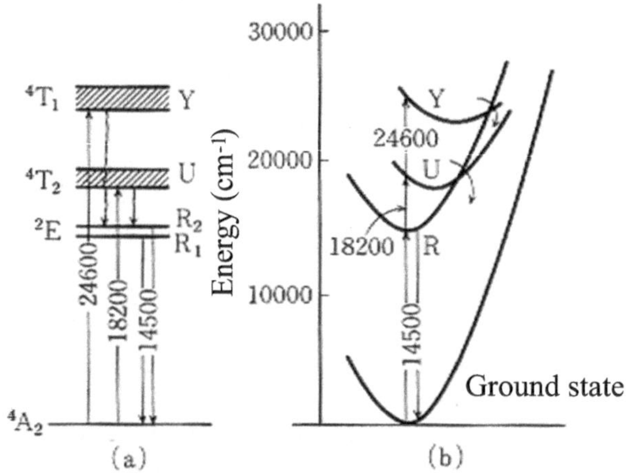

Fig. 4.10. (**a**) Energy diagram and (**b**) configuration coordinate curves for the Cr^{3+} ion in ruby. Population inversion is formed between the ^{2}E and $^4A_2$ levels by the excitation of highly absorptive $^4T_1$ and $^4T_2$ levels followed by relaxation to the ^{2}E (R_1, R_2) levels

not exist without perturbation. Next, we discuss the mechanism of electronic transitions observed in optical absorption.

Cr^{3+} ions are affected not only by the cubic O_h field due to the nearest neighbor O^{2-} ions, but also by the crystal field with three-fold symmetry from the second nearest Al^{3+} ions. Because the perturbation from the second crystal field breaks the inversion symmetry at the position of the Cr^{3+} ions, the t_{2g} or e_g states will be mixed with states of different parity. As a result, the electric dipole transitions from $^4A_2$ to $^4T_2$ or $^4T_1$ become partially allowed, and the oscillator strength of these transitions is estimated to be

$$f \sim f_{\text{allowed}} \left(\frac{\langle \mathcal{V}_{\text{odd}} \rangle}{\Delta E} \right)^2 \sim 10^{-4}. \tag{4.49}$$

Here, we used $\langle \mathcal{V}_{\text{odd}} \rangle \sim 10^3\,\text{cm}^{-1}$ for the magnitude of the crystal field and $\Delta E \sim 10^5\,\text{cm}^{-1}$ for the energy difference between mixing states. The oscillator strength of the dipole allowed transition f_{allowed} is the order of unity. In order to cause the spin forbidden transitions such as $^4A_2 \rightarrow {}^2E$, $^2T_1$, and $^2T_2$, the perturbation from the spin–orbit interaction $\mathcal{H}_{\text{so}}$ is required to flip spins, as well as the crystal field $\mathcal{V}_{\text{odd}}$. The oscillator strength of this transition can be estimated to be

$$f \sim f_{\text{allowed}} \left(\frac{\langle \mathcal{V}_{\text{odd}} \rangle}{\Delta E} \right)^2 \left(\frac{\langle \mathcal{H}_{\text{so}} \rangle}{\Delta E_0} \right)^2 \sim 10^{-7}. \tag{4.50}$$

Here, we assume that $\langle \mathcal{H}_{\text{so}} \rangle \sim 100\,\text{cm}^{-1}$, $\Delta E_0 \sim$ several thousand cm^{-1}. The lowest excited state ^2E is split into two lines R_1 and R_2. The oscillator strength of the lowest excited state R_1 is very weak (10^{-7}); then the emission lifetime is as long as 3 ms at room temperature, and 4.3 ms at 77 K.

It is possible to achieve laser oscillation utilizing the optical property of Cr^{3+} ions in Al_2O_3 described in the preceding paragraph. In order to make the population inversion between the states $^4A_2$ and R_1, electrons are at first excited from the ground to $^4T_2$ and $^4T_1$ states by a xenon flash lamp because the higher two levels have stronger optical absorption. As shown in Fig. 4.10, the absorption bands are located around 18 200 cm^{-1}, and 24 600 cm^{-1}. The excitation relaxes into the lowest excited state R_1 in the time scale of several μs which is three orders shorter than the lifetime of the R_1 level. Then the accumulation of excitation in the R_1 level results in a population inversion between the ground $^4A_2$ and R_1 states. The energy difference between the two levels is 14 500 cm^{-1} (λ_0 = 6943 Å), and the linewidth of the emission is 20 cm^{-1}. Laser oscillation is achieved by setting this laser medium into a Fabry–Pérot cavity and giving positive feedback as described in Sect. 4.1. A pulse energy of 10 J can be obtained from this system, and the maximum peak power of 100 MW is available by using Q-switching which will be described in the next chapter.

Unfortunately, ruby lasers are rarely used in current industries because of their practical inconvenience (one pulse available per minute). As an alternative to ruby lasers, Nd^{3+}:YAG lasers are widely used in many fields such as laser processing, harmonic generation, and other research and development purposes. In this laser, interconfigurational f-f transitions are used for laser action. Because the details of the mechanism can be found in other books [37], we do not discuss it further here.

4.3.2 Alexandrite and Ti-Sapphire Lasers

Chrysoberyl crystal $BeAl_2O_4$ can be artificially synthesized from a mixture of beryllium oxide (BeO) and aluminum oxide (Al_2O_3). When Cr^{3+} ions partially substitute Al^{3+}, the crystal is called alexandrite. The natural crystal was discovered in the Ural district of Russia in 1833, and it was named after Czar Alexander in those days. It is a famous jewel because it shows a green color in daylight, but the color changes into dark red under an electric light. Since this crystal also shows a broad light emission spectrum, it is used for the medium of tunable laser sources. Laser-active trivalent Cr^{3+} is located at O_h site, which is the same as the case of ruby, but the emission spectrum is much broader than that of ruby. This difference can be explained by the difference of the crystal field parameter Dq/B, which is 2.5 in ruby and 1.9 in alexandrite. As shown in Fig. 4.8, the lowest excited state of Cr^{3+} is $^4T_2[(t_{2g})^2e_g]$ instead of ^2E in ruby. The oscillator strength of the transition to this state is three orders stronger than that of the transition to ^2E, and because the transition energy is proportional to $10Dq$, the corresponding band is broader due to the variation of $10Dq$ caused by the lattice vibration of O^{2-} around Cr^{3+}. The experimental results in Fig. 4.11 show broad emission from $^4T_2$ and sharp emission from ^2E [38]. These characteristics can be understood in the same context as the case of ruby, as shown in Fig. 4.8. Lasing at 12 500–14 000 cm^{-1} has been obtained by using the emission from $^4T_2$. When it was excited by a flash lamp with energy of 150 J, a laser pulse of 1 J was obtained in the

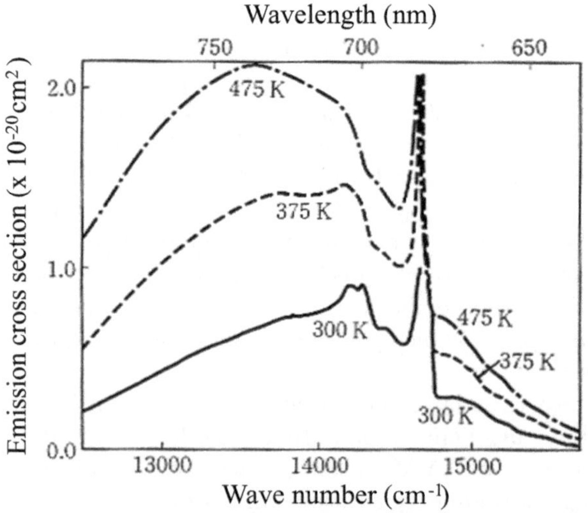

Fig. 4.11. Emission spectra from Cr^{3+} ions in alexandrite. The sharp line is the emission from the ^2E level, and the broadband emission in the longer wavelength region is due to the emission from the $^4T_2$ level [38]

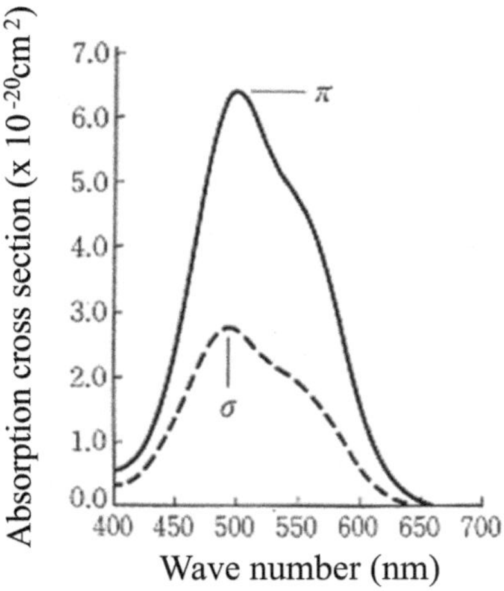

Fig. 4.12. Optical absorption spectra produced by the transition $^2T_2 \rightarrow {}^2E$ in Ti:Al_2O_3 [39]

range of 700–800 nm. One of the important advantages of this laser is that high power is easily obtained in spite of the higher oscillation threshold.

Ti^{3+}:Al_2O_3 is also practically used as a tunable solid state laser medium. Trivalent titanium ion has one 3d electron in the O_h crystal field formed by the surrounding octahedral O^{2-} ions. In this case, laser oscillation occurs by the optical transition $t_{2g} \rightarrow e_g$ ($^2T_2 \rightarrow {}^2E$, in configuration). The transition energy is proportional to the crystal field $10Dq$, and its oscillator strength f is much larger than that of ruby because of its spin allowed characteristics. But, the shorter lifetime (3 µs) requires higher pumping. Absorption spectra have broad peaks because of the strong dependence of the transition energy on the crystal field $10Dq$ (Figs. 4.12 and 4.13). The emission spectrum spreads over a broad range in the near-IR region (700–1000 nm) [39].

The broad emissions of these two laser media Ti^{3+}:Al_2O_3 and Cr^{3+}:$BeAl_2O_4$ are necessary for the generation of ultrashort laser pulses. The details of the principle and the method of ultrashort pulse generation will be described in the next chapter.

4.3.3 Other Important Lasers

The helium–neon (He-Ne) gas laser is one of the most widely used lasers. With this laser, 632.8 nm emission is obtained by the transition from the $(2p)^5 5s$ state to the $(2p)^5 3p$ state of the neon atom. The mechanism of this laser is depicted in Fig. 4.14. First, the He atom is excited by collision with high-velocity electrons and the energy is transferred from an excited He to a Ne atom. Population inversion is formed between the $3S[(2p^5)5s]$ state and the $2P[(2p)^5 3p]$ state of the neon atoms, and laser emission occurs due to the transition between them.

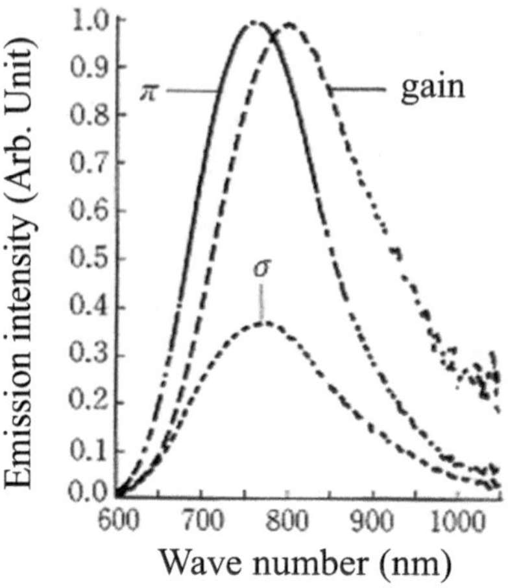

Fig. 4.13. Emission spectra of Ti:Al_2O_3 measured for π and σ polarization. The dashed curve shows the gain spectrum of the same crystal [39]

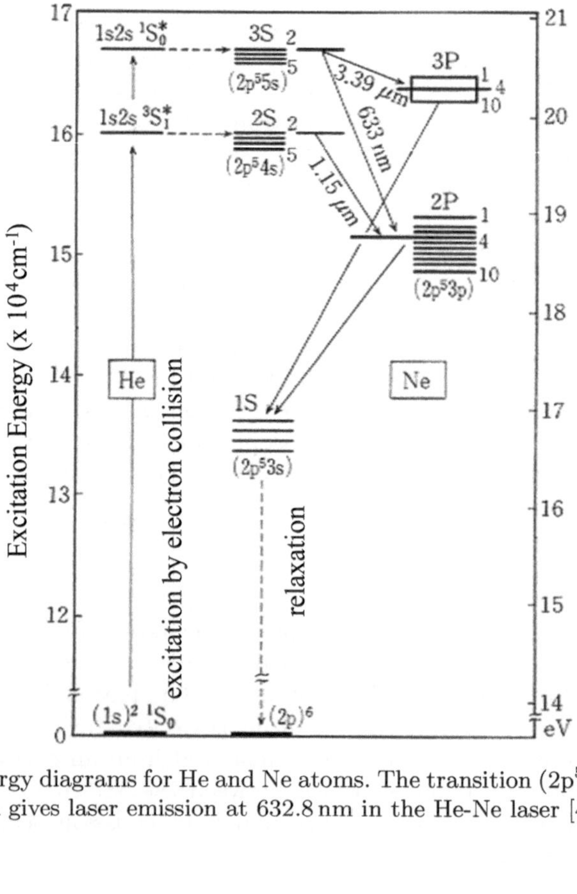

Fig. 4.14. Energy diagrams for He and Ne atoms. The transition ($2p^5 5s$) → ($2p^5 3p$) in the Ne atom gives laser emission at 632.8 nm in the He-Ne laser [40]

The carbon dioxide laser utilizes the transition between vibrational levels in the CO_2 molecule. In this case, laser emissions at 10.6 μm and 9.6 μm are obtained by the transition between levels $00^01 \rightarrow 10^00$ and $00^01 \rightarrow 02^00$, respectively. Usually, this far-infrared laser gives excellent performance with high power and high efficiency.

Solutions of organic dyes including π-electron systems often show strong fluorescence in the visible wavelength region. Because this emission has a broad gain range and there are many kinds of synthesized organic dyes available, the dye laser is the most popular tunable laser source from the near-UV to near-IR spectral region. This is a quite important property for scientific applications.

The semiconductor laser is the most important type of solid state laser in consumer electronics and optical communications. This laser has quite different characteristics from conventional solid state lasers, such as ruby, alexandrite, and Ti:sapphire lasers. Population inversion can be made by current injection across the p-n junction formed in semiconductors. Low cost, small volume, and high quality of light beams are very important properties of commercial uses.

Recently, the development of free electron lasers is in progress; these will become necessary for semiconductor lithographic processes in the near future.

4.3.4 The Road to X-ray Lasers

The generation of coherent light is relatively easy for longer wavelengths, so it has already been achieved in the infrared, visible, and near-ultraviolet regions. The difficulty of laser oscillation in the X-ray region is due to the higher transition probability of spontaneous emission described in Chap. 2. That is, the value of the constant A for the spontaneous emission rate increases with the third power of the frequency, so that the shorter the laser wavelength is, the shorter the excited state lifetime is. When the lifetime of the upper level is short, the high pumping power is required to accumulate excited states, and it requires an input power proportional to the fourth power of the angular frequency ω. In other words, high pumping power must be injected in order to overcome the strong spontaneous emission. Another problem is the difficulty of fabricating highly reflective mirrors. In the visible wavelength region, it is relatively easy to design and manufacture dielectric mirrors with nearly 100 % reflectivity by stacking two kinds of dielectrics in turn. But because the refractive indices of most materials are almost 1 in the X-ray region, it is impossible to obtain mirrors utilizing optical interference. In spite of these difficulties, great progress has been achieved in generating lasers with short wavelengths, as shown in Fig. 4.15. Here, we also cite the short-wavelength generation by using higher harmonic generations, the details of which will be introduced in Chap. 7.

In this section, we briefly describe the generation of soft X-ray laser having a wavelength of around 20 nm. In order to obtain a short-wavelength light

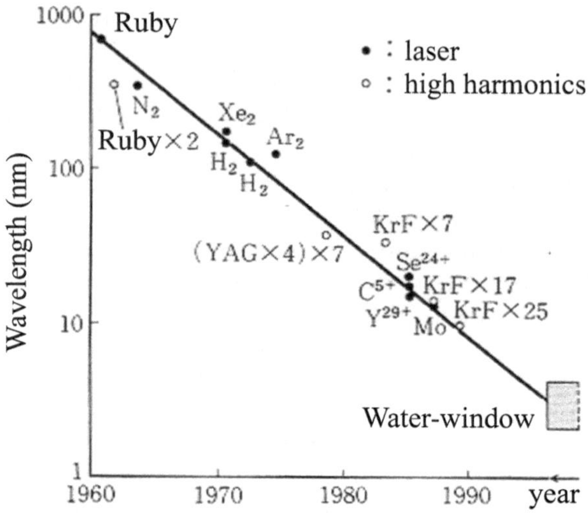

Fig. 4.15. Graphical history of the generation of short-wavelength lasers in the last 30 years since the invention of the laser. Open circles indicate the generation by higher harmonics of ruby, YAG, and KrF lasers. Others are obtained by conventional population inversion [40]

source, highly charged ions, in this case Se^{24+}, are used, and they are prepared in the form of a high-density plasma at high temperature. Se^{24+} ions have the same electronic configuration as Ne atoms, namely $(1s)^2(2s)^2(2p)^6$. Collision with electrons excites the 2p electron effectively into the 3p state instead of 3s. Then population inversion is formed between the 3s and 3p states, because the transition from 3p to 2p is dipole-forbidden, and the relaxation from 3s to 2p is quite fast, 0.5 ps. The soft X-ray laser beam was generated from an Se target deposited on a polyvinyl film with excitation by high-power pulses ($\sim$TW/cm^2) of the second harmonics of a Nd:YAG laser. X-ray beams with 20.96 nm and 20.63 nm were observed by the transition from ${}^1D_2[(1s)^2(2s)^2(2p)^5 3p]$ to ${}^1P_1[(1s)^2(2s)^2(2p)^5 3s]$, and from ${}^3P_2[(1s)^2(2s)^2(2p)^5 3p]$ to ${}^3P_1[(1s)^2(2s)^2(2p)^5 3s]$, respectively. In this case, the peak power is 24 MW, the efficiency is estimated to be 10^{-5}, the divergence angle is 11 mrad, and the pulse width is 200 ps. Also by using the Ni-like ion species Eu^{35+} and Yb^{+42}, an optical gain at 5 nm has been obtained, but laser oscillation has not yet been confirmed. For another example, a 18.2 nm X-ray laser was obtained with C^{5+}. In this case, the population inversion is formed during the recombination process of completely ionized atoms and free electrons [40].

5

Dynamics of Light

A laser is a coherent light source, which employs induced emission from excited states of materials most effectively. Interactions between the laser light and nonlinear optical materials make it possible not only to generate ultrashort light pulses but also to propagate a soliton wave in optical fibers. On the other hand, superradiance and superfluorescence take out excited state energies of materials as a coherent spontaneous emission when the material system is described as a coherent superposition of dipole moments of atoms over the system. Superradiance and superfluorescence are emitted as intense, short light pulses, the intensities of which are gigantic and proportional to N^2, the square of the atom number N, and the pulse width is proportional to $1/N$. In this chapter these two contrasting light pulses are studied. In Sect. 5.1 we discuss the generation of short light pulses from broadband lasers such as Ti-sapphire and dye lasers, using Q-switching (Sect. 5.1.1) and mode-locking (Sect. 5.1.2) techniques. We also study the mechanism of pulse compression and of soliton wave propagation in optical fibers (Sect. 5.1.3). These phenomena are closely related to the dispersion of group velocity of the light and optical Kerr effects in the medium. Over the late 1980s and early 1990s [41,42], laser intensities have increased by more than four orders of magnitude to reach enormous intensities of 10^{20} W/cm^2. The field strength at these intensities is on the order of a teravolt per centimeter (10^{12} V/cm), or a hundred times the Coulombic field binding the ground state electron in the hydrogen atom. A laser interacting with matter – solid, gas, plasma – generates higher-order harmonics of the incident beam as short as the 3 nm wavelength range, energetic ions or electrons with mega-electron-volt (10^6 eV) energies, giga-Gauss (10 K tesla) magnetic fields, and violent accelerations of $10^{21}\,g$ (g is Earth's gravity).

Figure 5.1 presents the focused intensity of lasers as a function of year [42]. It shows a rapid increase in the early 1960s, followed by a long plateau at 10^{15} W/cm^2. It took about 20 years, until the 1990s, for laser power to increase again. Note also the similarity in slopes between the early 1960s and the 1990s, and remember that it was during that period of very rapid increase in

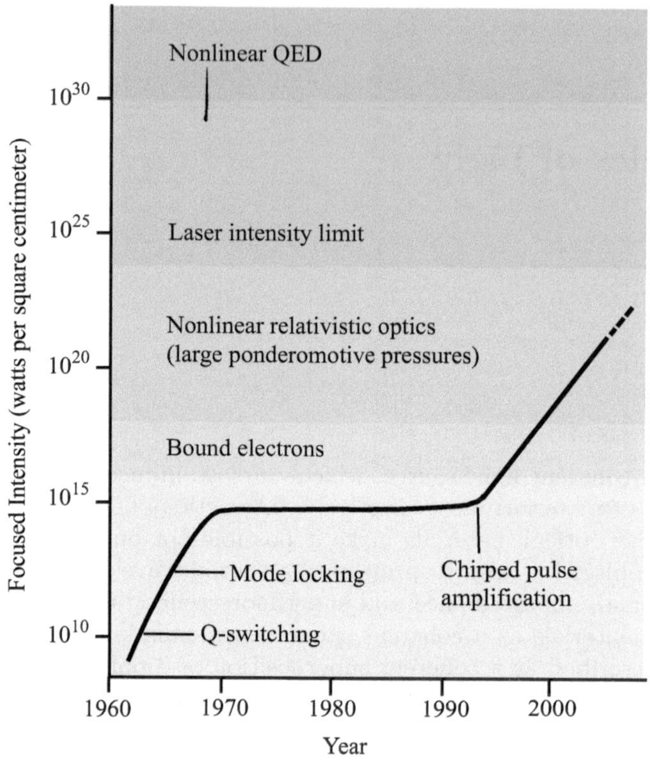

Fig. 5.1. Laser intensity vs. year to describe the development of tabletop systems, showing various breakthroughs. After an initial increase in the 1960s, the intensity levelled out for about 20 years, due to limitations caused by nonlinear effects. Over the past decade the intensity has been increasing with a slope similar to that of the 1960s, crossing into fundamentally new physical regimes. The laser intensity limit is the maximum possible stored energy (that of a complete population inversion) divided by the minimum possible pulse duration (the reciprocal of the gain bandwidth), for a beam of 1 cm^2 cross-section [42]

intensities in the 1960s that most of the nonlinear optical phenomena were discovered. In a similar way, the spectacular increase in intensity of four to five orders of magnitude that have occurred in the 1990s led to exceptional discoveries. Since their inception in 1960, lasers have evolved in peak power by a succession of leaps, each three orders of magnitude. These advances were produced each time by decreasing the pulse duration accordingly. First the lasers were free running, with durations in the 10 μs range and peak powers in the kilowatt range. In 1962, modulation of the laser cavity quality factor (Q-switching) enabled the same energy to be released on a nanosecond time scale, a thousand times shorter, to produce pulses in the megawatt range (10^6 W), as will be discussed in Sect. 5.1.1. In 1964, locking the longitudinal modes of the laser (mode locking) enabled the laser pulse duration to be reduced by another factor of a thousand, down to the picosecond level, pushing the peak power a thousand times higher, to the gigawatt (10^9 W) level. This will also be discussed in Sect. 5.1.2.

At that point, the intensities associated with the ultrashort pulses were becoming prohibitively high. At intensities of gigawatts per square centimeter (10^9 W/cm^2), the material index of refraction n becomes linearly dependent on the intensity I, varying like $n = n_0 + n_2 I$, where n_0 is the index of refraction at low intensity and n_2 the nonlinear index of refraction. The result is that, for a beam with a Gaussian radial intensity distribution, the beam on an axis sees a larger index of refraction than its surroundings. The optical elements inside the cavity thus become positive lenses that unacceptably deform the beam's wave front quality. Consequently, the only way to increase the peak power was to increase the diameter of the beam at the expense of instrument size, repletion rate and cost.

Although the pulse duration kept decreasing steadily, the intensity-dependent nonlinear effects kept the peak power about constant at the gigawatt level for a square-centimeter beam until 1985–87, when the technique of chirped pulse amplification (CPA) was demonstrated [43,44]. In CPA, the ultrashort pulse is not amplified directly, but is first stretched and then amplified, before finally being recompressed in the vacuum as will discussed in Sect. 5.1.3. The CPA reconciles two apparently conflicting needs: to have the highest fluence for efficient energy extraction, and to have minimum intensity inside the medium to avoid the undesired nonlinear effects.

In Sect. 5.1.4, we will discuss the recent development of CPA to get the most efficient ultrashort and ultrastrong laser pulses. In terms of this ultrastrong laser pulse, we can get the higher-order harmonics of the incident beam up to 3 nm wavelengths. This will be discussed in Chap. 7.

In Sect. 5.2, we summarize the important experimental results of superradiance (Sect. 5.2.1) and we extend semiclassical theory to superradiance and superfluorescence (Sect. 5.2.2). Before discussing these phenomena in detail, we make clear the difference between superradiance and superfluorescence. In general, *superradiance* means the phenomenon that many atoms (or molecules) align their electric dipole moments and cooperatively emit spontaneous radiation. If the population is completely inverted, there is no electric dipole in the system during the initial stage of the process. The electric dipole begins to grow just after spontaneous emission starts. In particular, the superradiance from a system with a fully inverted population is called *superfluorescence.* In Sect. 5.2.3 we will discuss the macroscopic manifestation of the quantum fluctuation of spontaneous emission. We also study propagation effects of superradiance. Finally, we describe phenomena observed in semiconductor microcrystals of CuCl in Sect. 5.2.4, as an example of superradiance from excitons.

5.1 Short Light Pulses and Optical Solitons

Coherency distinguishes laser light from a conventional light source. A laser is a coherent light source in the sense that its phase is spatially and temporally uniform. Using coherent laser light, we can generate an ultrashort light pulse with a pulse width as short as $2.8\,\text{fs} = 2.8 \times 10^{-15}$ s for the visible region. Since its peak power is extremely high, the electromagnetic wave has a gigantic energy within a very short interval. The generation of the ultrashort light pulse itself is interesting. The technique of the ultrashort light pulse is also important for studying relaxation processes of elementary excitations and dynamics of nonlinear optical phenomena.

5.1.1 *Q*-Switching

Q-switching is one of the most important techniques to obtain short bursts of oscillations from lasers. Consider the laser oscillation in a Fabry–Pérot resonator, consisting of external mirrors M_1 and M_2. The mirror M_1 is highly reflective, $R_1 = 1$. A laser beam outputs from the resonator through another mirror M_2 with lower reflectivity $R_2 = 1 - T < 1$. Using the transmittance T of the mirror M_2, the amplitude relaxation rate κ_T is defined as

$$\kappa_T \equiv \frac{cT}{4L} \equiv \frac{\omega}{2Q}, \tag{5.1}$$

where L is the distance between the mirrors and ω is the angular frequency of the laser light. The second equation defines the quality factor Q of the resonator. κ_T is closely related to the relaxation rate $C = (c/2L)T$ for laser modes with c the light velocity. As discussed in Sect. 4.1, laser oscillation takes place when the A coefficient, which is proportional to the pumping rate, is equal to C. The principle of Q-switching is as follows. The Q-factor of the resonator is first reduced, for example, by increasing T, so that the relaxation rate $C'(\gg C)$ is much larger than A during the pumping. The laser oscillation is thus suppressed and, simultaneously, the inverted population $\Delta N \equiv N_b - N_a$ is increased. When the Q-factor is restored abruptly to the high value, the system is now well above the oscillation threshold, $A \gg C$. Laser oscillation immediately takes place and the short and intense light pulse builds up. The simplest method of Q-switching is the rotational mirror technique: The rotation of the high-reflective mirror M_1, along the direction perpendicular to the resonator, changes the Q-value periodically. During almost the whole stage of one rotation cycle, no feedback process takes place. Within the short interval in which M_2 is nearly parallel to M_1, the resonator has a high Q-value, leading to laser oscillation. Using the rotational mirror with 400 rpm, the short light pulses with 10–100 MW peak power and with the 10–50 ns pulse width are generated from a flash-lamp-excited Nd:YAG laser and a ruby laser.

The second method is Q-switching with the use of a saturable absorber. The absorber, inserted into the laser resonator, prevents laser oscillation during the early stage of cw pumping, because the absorber acts as a lossy material. When laser oscillation starts, the absorption is abruptly saturated and the absorber becomes transparent. The sudden increase of Q leads to the generation of gigantic light pulses.

The method of Q-switching by a Pockels cell is the most popular for practical lasers. The Pockels cell is composed of an electrooptic (EO) crystal with transparent electrodes by which high voltage is applicable. Because the applied high voltage causes the change of refractive indices of the birefringent EO crystal, phase retardation of the light beam can be controlled externally. Therefore, combining with other polarization optics, the Q-factor of the cavity can be varied drastically. If we keep the low Q-factor during pumping, the energy can be accumulated in laser crystals. Then turn-on of the Q-switch induces laser oscillation, exhausting the whole energy in a short time. The giant pulses of most YAG lasers are obtained by this method.

Acoustooptic (AO) modulators are also used for Q-swtiching. Because it is possible to diffract a light beam by a grating generated in an AO crystal with ultrasound, the crystal plays the role of the rotating mirror described above. This device is usually used for lasers with high repetition rate, higher than kHz.

5.1.2 Mode Locking

The technique of mode locking enables us to make a picosecond-pulse train. Consider a Fabry–Pérot resonator that contains an amplitude modulator, such as an acoustooptical modulator, in addition to the gain medium. The wavenumber k_n and angular frequency ω_n of the resonator mode is given by

$$k_n = \frac{n\pi}{L}, \qquad \omega_n = ck_n = \frac{n\pi c}{L} \qquad (n = 1, 2, 3, \ldots), \tag{5.2}$$

respectively. The laser mode ω_n is modulated as

$$\begin{aligned} E_n \cos \omega_n t &\longrightarrow E_n(1 + M \cos \omega_{\text{mod}} t) \cos \omega_n t \\ &= E_n \cos \omega_n t + \frac{ME_n}{2} [\cos(\omega_n - \omega_{\text{mod}})t + \cos(\omega_n + \omega_{\text{mod}})t] \ , \end{aligned} \tag{5.3}$$

where M is the modulation rate. It follows from (5.3) that the amplitude modulation gives rise to sidebands $\omega_n \pm \omega_{\text{mod}}$ above and below the laser mode ω_n. The sideband and laser mode interact strongly with each other in frequency space through the third-order nonlinearity of the gain medium. When the modulation frequency ω_{mod} is just equal to the frequency spacing $\pi c/L$ between the laser modes, the laser modes are locked. Namely, all laser modes have equal frequency spacing and fixed phases with time. If mode locking is absent, the laser modes randomly oscillate and their frequency spacing is unequal, reflecting the dispersion effect of the gain medium.

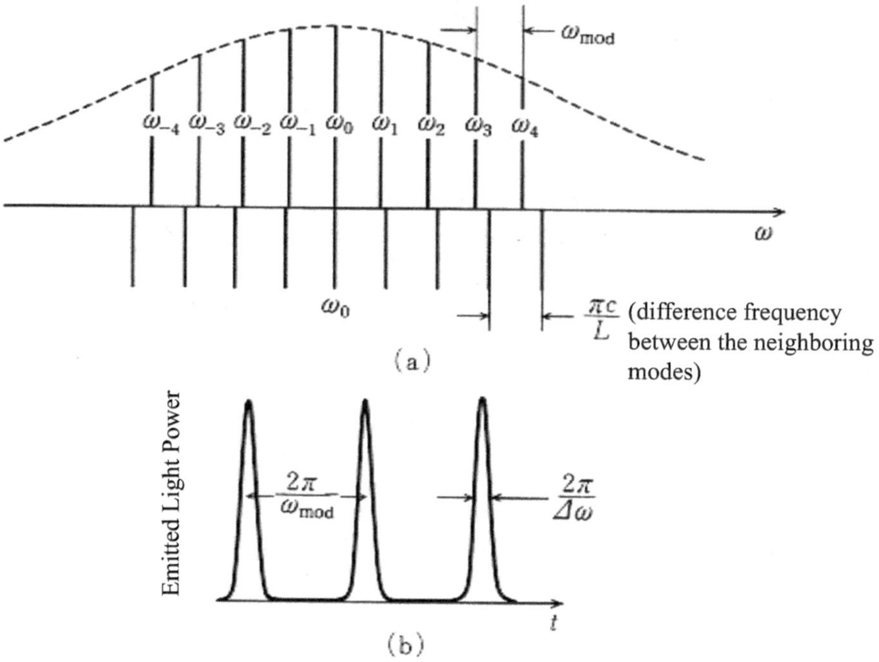

Fig. 5.2. Laser pulse generation from a mode-locked system. (**a**) Cavity modes are modulated with a frequency $\omega_{\rm mod}$ nearly equal to the frequency difference $\pi c/L$ of the neighboring modes. (**b**) Then we obtain the pulse train with pulse width $2\pi/\Delta\omega$ and periodicity $2\pi/\omega_{\rm mod}$ in the time axis. Here $\Delta\omega$ denotes the whole gain spectral width

In the presence of mode locking, the temporal profile of the total electric field $E(t)$ is approximately given by

$$
\begin{aligned}
E(t) &= \sum_n E_n \cos(\omega_0 + n\omega_{\rm mod})t \fallingdotseq \sum_{n=-N}^{N} E_0 \cos(\omega_0 + n\omega_{\rm mod})t \\
&= E_0 \frac{\sin\left[\left(N+\frac{1}{2}\right)\omega_{\rm mod}t\right]}{\sin\left(\frac{1}{2}\omega_{\rm mod}t\right)} \cos\omega_0 t, \qquad (5.4)
\end{aligned}
$$

where ω_0 and $2N+1$ are the central frequency and the number of laser modes, respectively. Here we assume for simplicity that all laser modes have equal amplitude E_0. Figure 5.2 schematically illustrates mode locking.

It is evident that the output power is emitted in the form of a train of pulses with a period of $T = 2\pi/\omega_{\rm mod}$ and the width of each pulse is $\Delta T = 2\pi/[(2N+1)\omega_{\rm mod}] = 2\pi/\Delta\omega$ with $\Delta\omega \equiv (2N+1)\omega_{\rm mod}$ the whole gain frequency region. The mode-locked dye-laser system can generate a train of picosecond pulses. Importantly, the inverse of the pulse width is proportional to the frequency width of the gain curve of the laser medium. Therefore, a Ti-sapphire crystal has the advantage of generating a short pulse, because the luminescence spectrum spreads over a wide frequency range, as shown in Fig. 4.13. Indeed, the mode-locked Ti-sapphire laser can generate a stable pulse train with a pulse width of 1–2 ps.

5.1.3 Pulse Compression

Compression of subpicosecond laser pulses with the use of optical glass fiber is a key method to generate femtosecond light pulses. The silica-glass is highly transparent in the infrared region, $\lambda \sim 1\,\mu\text{m}$. It makes it possible to confine the intense light beam to a very narrow region. As a result, the nonlinear effects in the light pulse are accumulated during propagation. In particular the dispersion of the group velocity and the self-phase-modulation effect enable us to generate either extremely short pulses or optical solitons in the fiber.

We will discuss the pulse compression process in this subsection. As already pointed out, a short light pulse is composed of a large number of electromagnetic modes. The frequency spread $\Delta\omega$ is proportional to the inverse of the pulse width ΔT, i.e., $\Delta\omega = 2\pi/\Delta T$. Consider the temporal profile of the electric field $\boldsymbol{E}(z,t) = \boldsymbol{e}E(z,t)\exp[i(kz-\omega t)]$, where $E(z,t)$ is the envelope function and $\boldsymbol{e}$ is a unit vector along $\boldsymbol{E}$. The electric field $\boldsymbol{E}$ obeys the Maxwell equation:

$$\frac{\partial^2}{\partial z^2}\boldsymbol{E}(z,t) - \frac{1}{\epsilon_0 c^2}\frac{\partial^2}{\partial t^2}\boldsymbol{D}(z,t) = \frac{1}{\epsilon_0 c^2}\frac{\partial^2}{\partial t^2}\boldsymbol{P}^{NL}(z,t). \tag{5.5}$$

Here the electric displacement $\boldsymbol{D}$ is assumed to be linear, so that $\boldsymbol{D} = \epsilon_0\boldsymbol{E} + \boldsymbol{P}^{(1)}$. The electric field $\boldsymbol{E}$, electric displacement $\boldsymbol{D}$, and nonlinear polarization $\boldsymbol{P}^{NL}$ are assumed to be along the same direction for simplicity. The Fourier transforms of $\boldsymbol{E}$ and $\boldsymbol{D}$ can be described by

$$\boldsymbol{E}(z,t) = \boldsymbol{e}\int d\eta E(z,\omega+\eta)e^{ikz-i(\omega+\eta)t} \tag{5.6}$$

and

$$\boldsymbol{D}(z,t) = \boldsymbol{e}\int d\eta\epsilon(\omega+\eta)E(z,\omega+\eta)e^{ikz-i(\omega+\eta)t}, \tag{5.7}$$

respectively, where $\epsilon(\omega+\eta)$ is a dielectric function. Using the third-order polarizability $\chi^{(3)}$, the nonlinear polarization $\boldsymbol{P}^{NL}$ approximately has the following form:

$$\frac{\partial^2}{\partial t^2}\boldsymbol{P}^{NL}(z,t) \fallingdotseq -\omega^2\chi^{(3)}\left|E(z,t)\right|^2 E(z,t)e^{i(kz-\omega t)}. \tag{5.8}$$

Combining (5.6), (5.7), and (5.8) with (5.5), and using the dispersion relation $\epsilon(\omega) = (ck/\omega)^2$, we obtain

$$
\begin{aligned}
2ik\frac{\partial}{\partial z} E(z,t) &+ \frac{1}{c^2}\int d\eta\, \eta \left[2\omega\epsilon(\omega) + \omega^2 \frac{\partial \epsilon(\omega)}{\partial \omega}\right] E(z,\omega+\eta)e^{-i\eta t} \\
&+ \frac{1}{c^2}\int d\eta\, \eta^2 \left[\omega\epsilon(\omega) + 2\omega\frac{\partial \epsilon(\omega)}{\partial \omega} + \frac{1}{2}\omega^2\frac{\partial^2 \epsilon(\omega)}{\partial \omega^2}\right] E(z,\omega+\eta)e^{-i\eta t} \\
&= -\omega^2\chi^{(3)} \left|E(z,t)\right|^2 E(z,t)e^{i(kz-\omega t)}.
\end{aligned}
\tag{5.9}
$$

Here we have neglected the second derivative with respect to z under the slowly varying approximation:

$$
\left|\frac{\partial^2}{\partial z^2}E(z,t)\right| \ll k\left|\frac{\partial}{\partial z}E(z,t)\right|.
\tag{5.10}
$$

The second and third terms of the left-hand side of (5.9) can be written, respectively, as,

$$
\frac{1}{c^2}\left[2\omega\epsilon(\omega) + \omega^2\frac{\partial\epsilon(\omega)}{\partial\omega}\right] = \frac{1}{c^2}\frac{\partial(\omega^2\epsilon)}{\partial\omega} = \frac{\partial(k^2)}{\partial\omega} = 2k\frac{1}{v_g},
\tag{5.11}
$$

$$
\begin{aligned}
\frac{1}{c^2}\left[\omega\epsilon(\omega) + 2\omega\frac{\partial\epsilon(\omega)}{\partial\omega} + \frac{1}{2}\omega^2\frac{\partial^2\epsilon(\omega)}{\partial\omega^2}\right] &= \frac{1}{2c^2}\frac{\partial^2(\omega^2\epsilon)}{\partial\omega^2} = \frac{\partial}{\partial\omega}\left[k\left(\frac{\partial k}{\partial\omega}\right)\right] \\
&= \frac{\partial}{\partial\omega}\left(\frac{k}{v_g}\right) \fallingdotseq -\frac{k}{v_g^2}\left(\frac{\partial v_g}{\partial\omega}\right),
\end{aligned}
\tag{5.12}
$$

where $v_g \equiv (\partial k/\partial\omega)^{-1}$ is the group velocity of the light pulse. Substituting (5.11) and (5.12) into (5.9) and neglecting the weak dependence of v_g and $\partial v_g/\partial\omega$ on η, we obtain the following equation that describes the propagation of the light pulse:

$$
\left(\frac{\partial}{\partial z} + \frac{1}{v_g}\frac{\partial}{\partial t}\right)E(z,t) = \frac{i\alpha}{2}\frac{\partial^2}{\partial t^2}E(z,t) + i\kappa\left|E\right|^2 E(z,t).
\tag{5.13}
$$

Here we have used the following relations,

$$
\begin{aligned}
\frac{\partial E(z,t)}{\partial t} &= -i\int d\eta\; \eta E(z,\omega+\eta)e^{-i\eta t}, \\
\frac{\partial^2 E(z,t)}{\partial t^2} &= -\int d\eta\; \eta^2 E(z,\omega+\eta)e^{-i\eta t}.
\end{aligned}
\tag{5.14}
$$

We introduce the parameters α and κ to describe the effects of group velocity dispersion and self-phase-modulation due to the optical Kerr effect, respectively:

$$
\begin{aligned}
\alpha &\equiv \frac{1}{v_g^2}\frac{\partial v_g}{\partial\omega}, \\
\kappa &\equiv \frac{1}{2\epsilon_0 c^2}\frac{\omega^2}{k}\chi^{(3)}.
\end{aligned}
\tag{5.15}
$$

Let us discuss the following three cases.

(1) $\alpha = 0$ and $\kappa = 0$. In the case that a medium is dispersionless and its nonlinearity is negligibly small, the right-hand side of (5.13) can be set to zero. Using the envelope function $E(z,t) = E_0(z)$ at $t = 0$, we obtain

$$E(z,t) = E_0(z - v_g t). \tag{5.16}$$

It is immediately evident that the light pulse maintains its line shape during propagation.

(2) $\alpha \neq 0$ and $\kappa = 0$. The light pulse is always broadened whether the medium has positive dispersion $\partial v_g/\partial\lambda > 0$ ($\partial v_g/\partial\omega < 0$: ordinary dispersion) or negative dispersion $\partial v_g/\partial\lambda < 0$ ($\partial v_g/\partial\omega > 0$: extraordinary dispersion). After passing through silica-glass optical fiber with a length of 650 m, a 10 ps light pulse with center wavelength 1.5 μm is broadened to 20 ps. As already discussed, a short light pulse is composed of different Fourier components of ω. If the medium has ordinary dispersion, that is, larger group velocity at lower frequencies, the lower frequency components of the light pulse move faster than the higher frequency components. On the other hand, the higher frequency components travel faster in an extraordinary dispersion medium. In any case, the pulse width is increased in this medium.

(3) $\alpha \neq 0$ and $\kappa \neq 0$. For $\partial v_g/\partial\omega > 0$, the pulse compression effect leads to the formation of an optical soliton. For $\partial v_g/\partial\omega < 0$, on the other hand, the shape of the light pulse changes to a rectangular one. In both cases, the self-phase-modulation effect through κ is crucial. We first discuss this self-phase-modulation effect. Neglecting the spatial variation of the intensity, $|E(z,t)|^2 \fallingdotseq |E(t)|^2$ in (5.13) and integrating it from $z = 0$ to l, we obtain

$$\begin{aligned} E(l,t)e^{-i\omega t} &\sim \exp\left[i\kappa l\,|E(t)|^2 - i\omega t\right] \\ &\sim \exp\left\{-i\left[\omega - \kappa l\left(\partial|E(t)|^2/\partial t\right)\right]t\right\}. \end{aligned} \tag{5.17}$$

We find the frequency change from ω to $\omega - \kappa l(\partial|E(t)|^2/\partial t)$. Figure 5.3 illustrates the t-dependence of the intensity and frequency modulation for the case of $\kappa > 0$. In the front side of the light pulse ($t < 0$), the frequency shows a redshift, whereas it exhibits a blueshift in the back side ($t > 0$).

In the extraordinary dispersion medium, the group velocity v_g of the redshifted light is smaller than that of the blueshifted light. Therefore, the front side of the pulse consisting of redshifted light travels slowly in the medium, while the blueshifted backside moves faster. As a result, the light pulse is compressed from both sides. An optical fiber made of silica-glass has extraordinary dispersion for $\lambda > 1.3$ μm, whereas it has ordinary dispersion for $\lambda < 1.3$ μm. When a light pulse with center wavelength 1.55 μm is made to enter this optical fiber, the light pulse encounters the compressible effect and propagates as a soliton. The characteristic of the soliton depends on the intensity of the light pulse. Figure 5.4 demonstrates the results of a numerical calculation for an $N = 3$ soliton [45]. The line shape of the light pulse is very

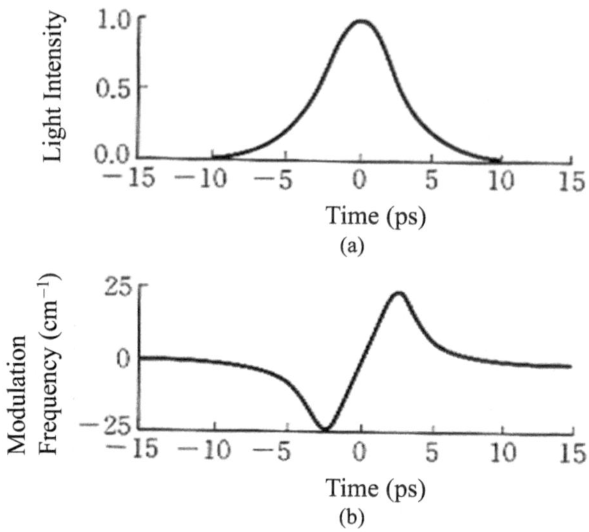

Fig. 5.3. (**a**) Profile of a laser pulse and (**b**) its self-phase-modulation. Negative time corresponds to the front side of the pulse. The material constants of the glass fiber are used

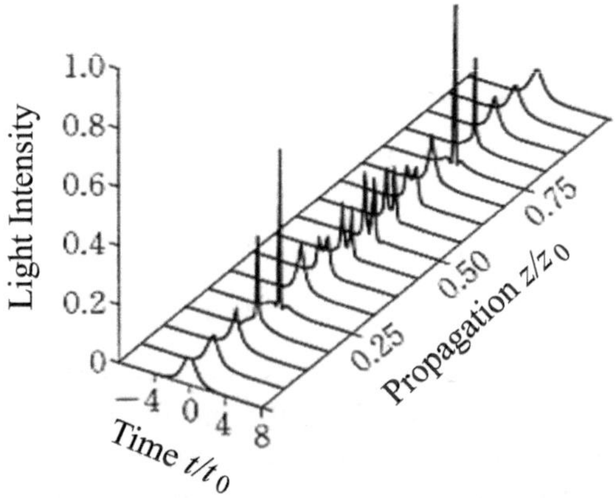

Fig. 5.4. Pulse shape of the carrier wavelength 1.55 μm (>1.3 μm) in the anomalous dispersion region is modified into a soliton as the pulse propagates in the glass fiber. The normalized distance z_0 is 700 m and the normalized time is $t_0 \equiv \phi_0^{-1}$ (see text) [45]

sharp at $z/z_0 = 1/4$ and splits into two peaks at $z/z_0 = 1/2$. The sharp peak emerges again at $z/z_0 = 3/4$ and is restored to the initial line at $z/z_0 = 1$. This periodic change of soliton propagation will be discussed in Sect. 5.1.4.

If the light pulse is in an ordinary dispersion medium, for example, visible light is sent into a silica-glass fiber, the light pulse exhibits different evolution

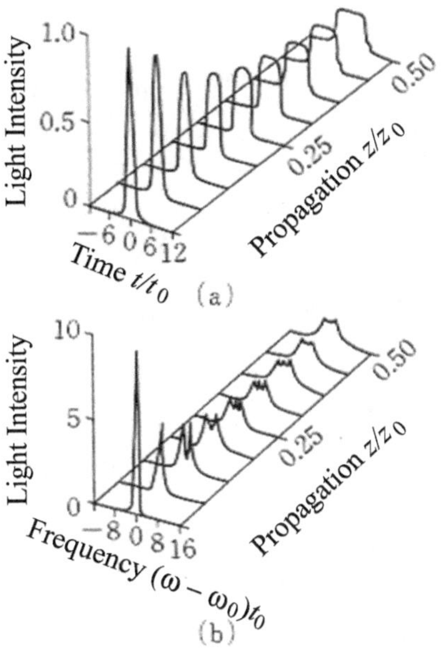

Fig. 5.5. The pulse shape is modified into a rectangular wave as a function of both (**a**) time and (**b**) frequency as it propagates in a glass fiber in the normal dispersion region [46]

[46]. Figure 5.5 shows numerical calculations for the temporal profile and the spectrum of the light pulse as a function of propagation distance in the optical fiber. In an ordinary dispersion medium the group velocity v_g of the redshifted light is higher that the blueshifted light. Therefore, the front side of the light pulse, showing the redshift due to the self-phase-modulation effect, moves faster, while the blueshifted backside propagates slowly. As a result, the pulse shape is gradually suppressed and changed to a rectangular shape at $z/z_0 = 1/2$.

We describe the principle of pulse compression with the use of a diffraction grating that is employed instead of an extraordinary dispersion medium. Figure 5.6 illustrates the optical setup for the pulse compression. Figure 5.7 displays the autocorrelation traces of the incident and compressed pulses [47]. The light pulse from a mode-locked dye laser with peak power of 2 kW and pulse width of 5.9 ps is sent into the optical fiber which is 3 m long. Since the wavelength of the incident laser is smaller than 1.3 μm, it is in the ordinary dispersion region. Accordingly, the line shape of the light pulse is changed to the rectangular shape with a pulse width of 10 ps. The output light is spatially dispersed by the diffraction grating and reflected by the prism. The dispersed light is again incident on the diffraction grating and, then, is restored. Here the redshifted light, the front side of the rectangular pulse, travels over a longer path than that for the blueshifted backside. As a result, the 10 ps pulse

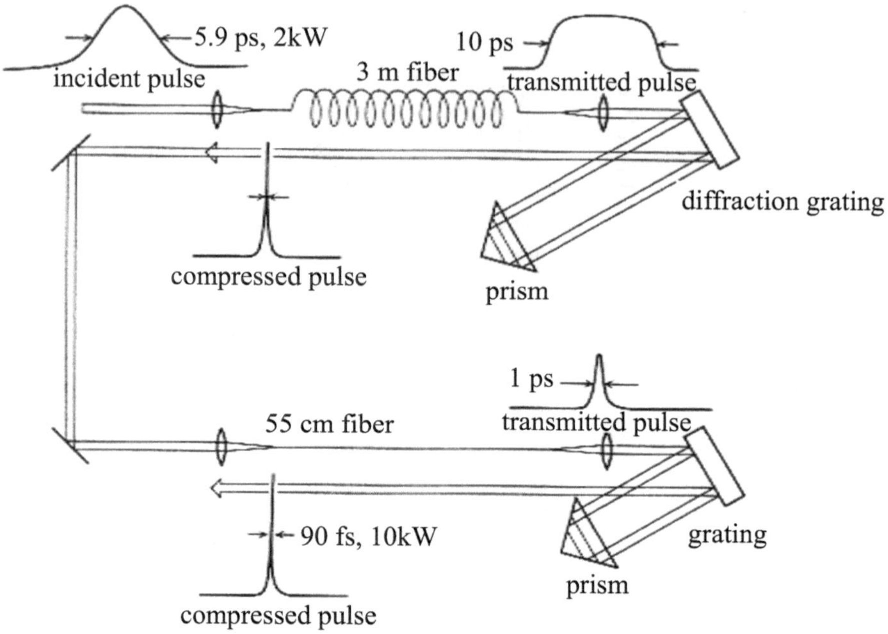

Fig. 5.6. Light pulse compression using two sets of optical fiber and prism-diffraction grating pairs [47]

is compressed to a 200 fs pulse. The second optical fiber, 55 cm long, makes a 1 ps rectangular pulse. The second grating-prism coupling compresses it to a 90 fs pulse. Recently, an ultrashort light pulse of 6 fs has been successfully generated. During 6 fs the electromagnetic wave ($\lambda = 900\,\text{nm}$) oscillates over only two cycles. When the higher-order dispersion is corrected, a pulse width as short as 2.8 fs was obtained [48].

5.1.4 Optical Solitons

We will discuss the propagation of solitons in the optical fiber in detail. Using the pulse width ΔT, which is self-consistently determined, we define dimensionless parameters for t, z and $E(z,t)$ as

$$\tau \equiv \frac{1}{\Delta T}\left(t - \frac{z}{v_g}\right), \quad \xi \equiv \left|\frac{\partial v_g^{-1}}{\partial \omega}\right| \frac{z}{(\Delta T)^2}, \quad \phi \equiv \Delta T \left|\frac{\kappa}{\partial v_g^{-1}/\partial \omega}\right|^{1/2} E(z,t), \tag{5.18}$$

respectively. Substituting these parameters into (5.13), we obtain the differential equation

$$i\frac{\partial \phi}{\partial \xi} + \frac{1}{2}\frac{\partial^2 \phi}{\partial \tau^2} + |\phi|^2 \phi = 0. \tag{5.19}$$

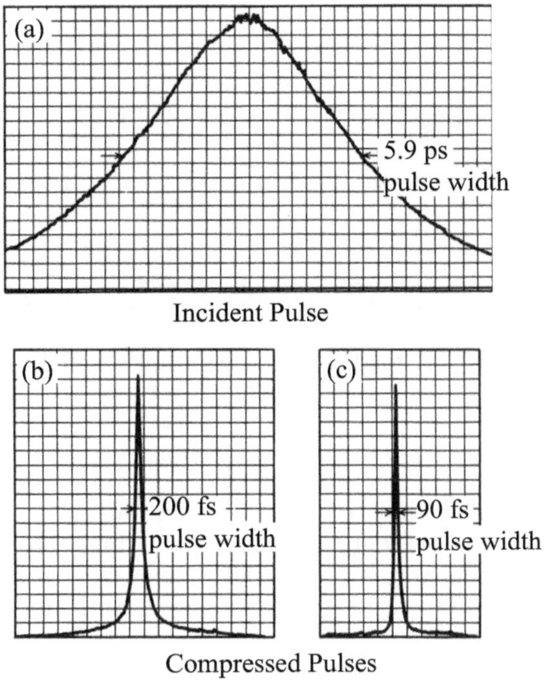

Fig. 5.7. Pulse shape of (**a**) incident pulse, (**b**) compressed pulse in the first stage, and (**c**) that in the second stage [47]

The first, second, and third terms describe the wave propagation, dispersion effect of the group velocity, and the self-phase-modulation effect, respectively. Here we consider an extraordinary dispersion medium. Note that the light pulse in an ordinary dispersion medium yields a negative sign for the second term in (5.19).

The one-soliton solution can be obtained by using separation of variables as follows:

$$E(z,t) \propto \phi(\xi,\tau) = \phi_0 \mathrm{sech}(\phi_0 \tau) e^{i\phi_0^2 \xi/2}. \tag{5.20}$$

Assume that the solution has the form $\phi(\xi,\tau) = \psi(\xi)\phi(\tau)$ with $|\psi(\xi)|^2 = 1$. Substituting $\psi(\xi)\phi(\tau)$ into (5.19), we obtain

$$-\frac{i}{\psi(\xi)}\frac{\partial\psi(\xi)}{\partial\xi} = \frac{1}{2\phi(\tau)}\frac{\partial^2\phi(\tau)}{\partial\tau^2} + \phi(\tau)^2 = \frac{1}{2}\phi_0^2, \tag{5.21}$$

where the constant ϕ_0 is independent of ξ and τ. Integrating (5.21) with respect to ξ, we obtain

$$\psi(\xi) = \exp\left(i\frac{1}{2}\phi_0^2\xi\right), \tag{5.22}$$

which satisfies the condition $|\phi(\xi,\tau)|^2 = \phi(\tau)^2$ given before. As for τ, we obtain

$$\frac{1}{2}\frac{d^2\phi(\tau)}{d\tau^2} + \phi(\tau)^3 - \frac{1}{2}\phi_0^2\phi(\tau) = 0. \tag{5.23}$$

Using the boundary condition $d\phi/d\tau = \phi = 0$ at $\tau = \pm\infty$, we integrate this equation and obtain

$$\frac{d\phi}{\phi\sqrt{1-(\phi/\phi_0)^2}} = \phi_0 \, d\tau. \tag{5.24}$$

We again integrate (5.24) for the initial condition $\phi(0) = \phi_0$ at $\tau = 0$ and obtain

$$\phi(\tau) = \phi_0 \mathrm{sech}(\phi_0\tau). \tag{5.25}$$

We confirm that (5.20) is indeed one of the solutions.

It is evident that the soliton wave has a hyperbolic envelope and its velocity is v_g. It follows from (5.25) that (*normalized amplitude* ϕ_0) $\times$ (*normalized pulse width* ϕ_0^{-1}) is always a constant. More precisely, there is a conservation law,

$$\int_{-\infty}^{\infty} \phi(\tau)d\tau = \int_{-\infty}^{\infty} d\tau\phi_0\mathrm{sech}(\phi_0\tau) = \pi. \tag{5.26}$$

Consequently, the pulse width $\Delta T = \phi_0^{-1}$ is self-consistently determined. Another important feature is the periodic variation of the wave form. If an input pulse is $\phi(\tau) = a\phi_0\mathrm{sech}(\phi_0\tau)$ at $\xi = 0$, the light pulse is propagated as

a fundamental soliton	for	$0.5 \leqq a < 1.5,$
an N soliton	for	$N - \frac{1}{2} \leqq a < N + \frac{1}{2},$

where N is the number of solitons.

Figure 5.8 illustrates the profile of the soliton wave calculated for $N = 1, 2, 3$. The $N = 1$ soliton exhibits normal propagation. In contrast, the profiles of $N = 2, 3$ solitons vary with z, as already shown for the $N = 3$ soliton in Fig. 5.4 [49]. The transition from $N = 1$ to higher N has been experimentally observed. Figure 5.9(a) illustrate the spectrum and autocorrelation trace of the incident pulse. Figures 5.9(b)–(f) show the autocorrelation traces of the soliton wave at $z = z_0/2$, half of the $N = 1$ soliton cycle, as a function of the incident laser power P. When the incident laser is very weak, (b) the pulse width of the $N = 1$ soliton is 7.2 ps, which is the same as the incident laser. At $P = 1.2$ W (c) the profile is of the $N = 1$ soliton, but the pulse width is slightly narrower than the incident pulse. At $P = 5.0$ W (d) the autocorrelation has a sharp peak that accompanies small sidebands, indicating the $N =$

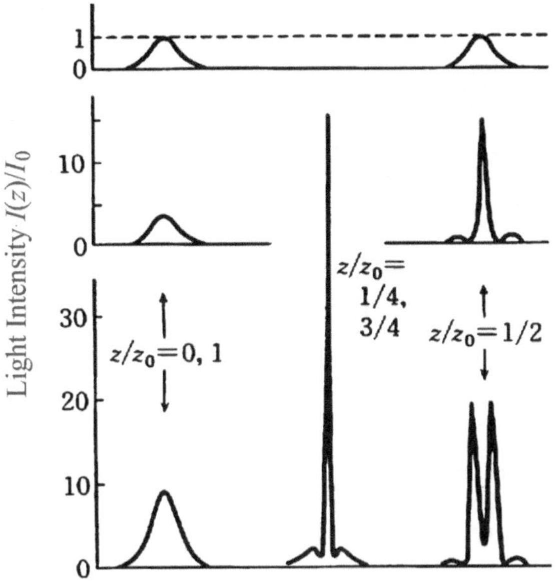

Fig. 5.8. Computer generated solutions of (5.23). *Top*: the fundamental soliton; *middle*: the $N = 2$ soliton; *bottom*: the $N = 3$ soliton. These solutions agree well with the observed ones in Fig. 5.9 [49]

2 soliton, shown also in the middle panel of Fig. 5.8. At $P = 11.4\,\text{W}$ (e) the trace shows three peaks and the intensity ratio is 1:2:1. This implies that the temporal profile has two intense peaks, shown in the lower panel of Fig. 5.5, that is, the $N = 3$ soliton. At $P = 22.5\,\text{W}$ (f) the autocorrelation exhibits five peaks, which is expected from the $N = 4$ soliton showing three sharp peaks.

According to multisoliton theory, there is a relation between the soliton cycle z_0 and the pulse width ΔT:

$$z_0 = \frac{\pi\,(\Delta T)^2}{2\,|\alpha|^2}. \tag{5.27}$$

Here α describes the dispersion of the group velocity v_g in the optical fiber,

$$|\alpha| = \frac{1}{v_g^2}\left|\frac{\partial v_g}{\partial \omega}\right|. \tag{5.28}$$

The power P_N of the N soliton is given by

$$P_N = N^2 P_0. \tag{5.29}$$

The power P_0 of the $N = 1$ soliton is $P_0 = I_0/A_{\text{eff}}$, where A_{eff} is the effective cross-section of the optical fiber. The intensity I_0 of the incident laser is given by

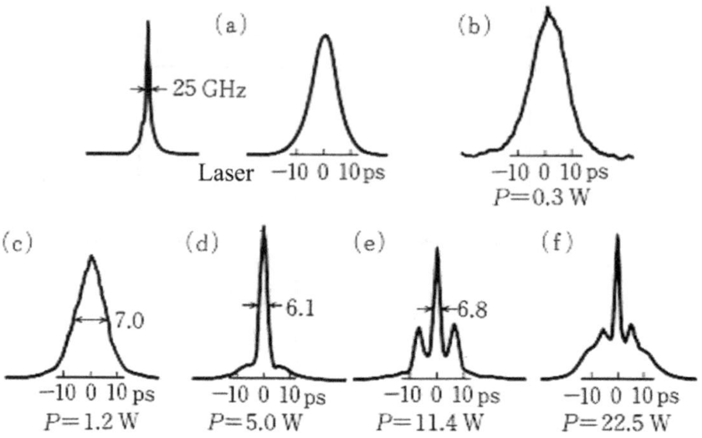

Fig. 5.9. (**a**) Incident laser spectrum and temporal evolution of the initial shape with the half-width 7.2 ps. Autocorrelation function of the soliton signals at $z_0/2$ for (**b**) weak laser pulse, (**c**) $N = 1$ soliton, (**d**) $N = 2$ soliton, (**e**) $N = 3$ soliton and (**f**) $N = 4$ soliton [49]

$$I_0 = \frac{\epsilon_0 n_0 c \lambda}{4 z_0 n_2}. \tag{5.30}$$

Using the linear and nonlinear refractive indices $n_0 = 1.45$ and $n_2 = 1.2 \times 10^{-22}\,\mathrm{m}^2/V^2$ of the optical fiber, $|\alpha| = 16\lambda^2/2\pi c$ (ps/nm·km) at $\lambda = 1.55\,\mu\mathrm{m}$. For the pulse width $\Delta T = 4\,\mathrm{ps}$, the soliton cycle $z_0 = 1260\,\mathrm{m}$ and $I_0 = 1.0 \times 10^{10}\,\mathrm{W/m}^2$. Using the geometrical cross-section A_{geo}, we obtain the effective cross-section $A_{\mathrm{eff}} \fallingdotseq 1.5\,A_{\mathrm{geo}} \sim 1.0 \times 10^{-6}\,\mathrm{cm}^2$. Thus, the power is $P_0 \fallingdotseq 1.0 \times 10^6\,\mathrm{W/cm}^2 \times 10^6\,\mathrm{cm}^2 = 1.0\,\mathrm{W}$. On the other hand, analyzing the data shown in Fig. 5.9 and using the relation $P_0 = P_N/N^2$, the value of P_0 is estimated to be 1.2 W. This agrees excellently with the above theoretical value.

5.1.5 Chirped Pulse Amplification

Chirped pulse amplification (CPA) had a dramatic impact in short-pulse amplification. First, one could use superior (by a factor of 1000) energy storage media such as Nd:glass, Cr:alexandrite, Ti:sapphire and Cr:$\mathrm{LiSrAlF_6}$ instead of dye and excimer. So a CPA laser system, using these good energy storage media, could produce a peak power 10^3–10^4 times higher than dye or excimer systems of equivalent size. Second, CPA could be easily adapted for use with the very large scale, and huge sized lasers have already been built for the purpose of laser fusion.

CPA involves impressive manipulations as Fig. 5.10 shows: stretching by 10^3–10^5, amplification by 10^{11} (from nanojoules to tens of joules) and recompression by 10^3–10^5. Here two main hurdles had to be overcome: the accommodation of the large stretching/compression ratio, and the amplification of large pulse spectra. The first part of the CPA system used the positive group velocity dispersion (in which the redshifted light goes faster than the blueshifted light) of a single-mode fiber to temporally spread the frequency distribution of the ultrashort pulses [43]. After passing through the fiber, the pulse is stretched with the red component first, followed by the blue. It is then amplified to the desired level and recompressed by a pair of parallel diffraction gratings. Pulse compression using this fiber and a single diffraction grating together with a prism was shown in Sect. 5.1.3. A grating pair exhibits a negative group velocity dispersion in which blue light goes faster than red light as shown in the bottom part of Fig. 5.10b. This is called the Treacy compressor. However, the fiber stretcher and diffraction grating compressor did not have their dispersive characteristics exactly compensated, so that the recompression was not perfect, leading to temporal wings in the pulse, and hence limiting the stretching/compression ratio to about 100.

Let us consider a telescope of magnification 1 placed between two antiparallel gratings as shown at the top of Fig. 5.10(b). It was found that this device had the exact same dispersion function as the Treacy compressor, but with the opposite sign [50]. In other words, it was the Treacy compression's perfect conjugate, with the important consequence that any arbitrarily short pulse in principle could be stretched to any pulse duration and then recompressed to its original shape. With the discovery of the matched stretcher–compressor, a very high first hurdle was cleared.

From the uncertainty principle, a pulse with a Gaussian envelope will have a minimum-duration-bandwidth product $\Delta\nu \cdot \tau$ of 0.4, where $\Delta\nu$ is the pulse bandwidth and τ is the pulse duration (full width at half maximum). A 10 fs pulse will have a bandwidth of 80 nm. All the spectral components must be amplified equally over many orders of magnitudes to avoid the risk of narrowing the pulse's bandwidth and lengthening its duration after compression. For ultrashort pulses, this limitation was removed with the invention of ultrabroadband amplification media such as Ti:sapphire. Ti:sapphire has a gain bandwidth that can theoretically support the amplification of pulses of less than 5 fs in duration. The second hurdle was also cleared. It also has excellent thermal properties, making amplification possible at a high repetition rate, 10–1000 Hz, an improvement of two-to-three orders of magnitude over the systems based on dye or excimer. A higher repetition rate at constant peak power leads directly to higher experimental utility. With 10–1000 Hz Ti:sapphire CPA systems, it is possible to apply signal-averaging techniques to investigate interactions between high-field lasers and matter. This capability is used fully in experiments involving signal-to-noise ratios, where low signal averaging is necessary.

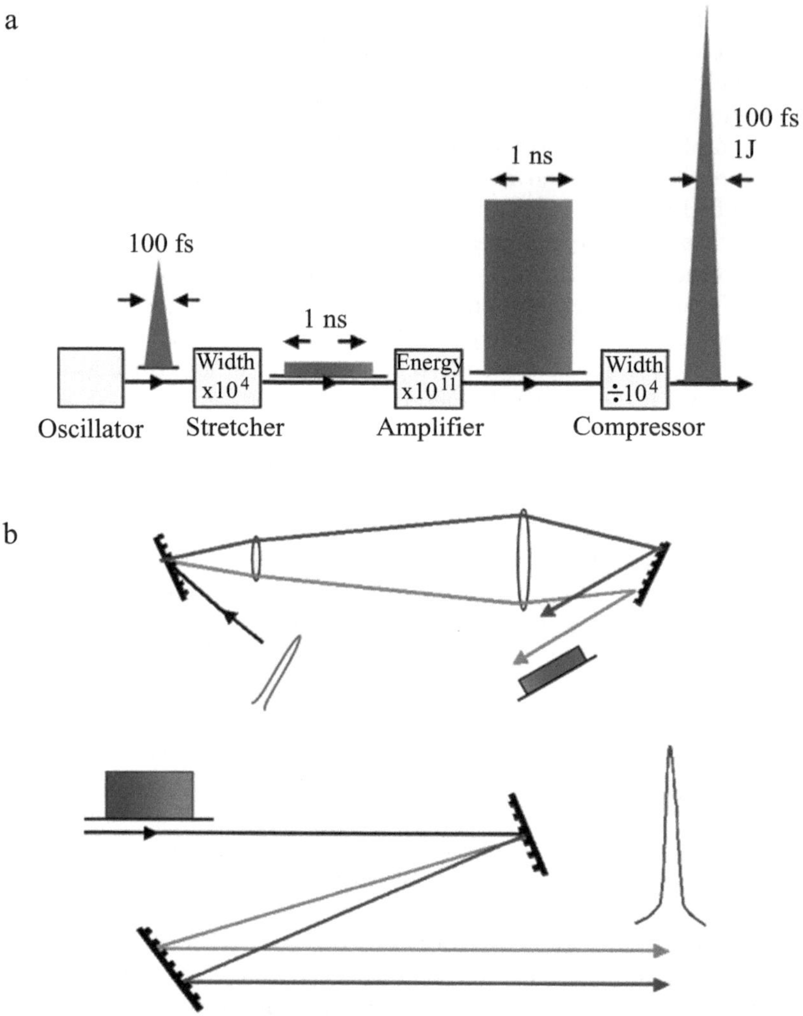

Fig. 5.10. Chirped pulse amplification (CPA). (**a**) The CPA concept. An oscillator produces a short pulse, which is then stretched by a factor of 10^3–10^5, from the femtosecond to the nanosecond regime, reducing its intensity accordingly. The intensity is now low enough that the pulse can be amplified and the stored energy can be safely extracted from the amplifier without fear of beam distortions and damage in optics. After extraction, the pulse is recompressed, ideally to its initial width. (**b**) The matched stretcher and compressor. The stretcher (top) is composed of a telescope of magnification 1 between two antiparallel diffraction gratings. Note that in this configuration the redshifted light has a shorter path than the blueshifted one. Conversely, the compressor (bottom) is composed of a pair of parallel gratings in which the optical path length for the blue is shorter than that for the red. The gratings are matched over all orders [42]

5.2 Superradiance

When many atoms (or molecules) have electric dipole moments along the same direction and with the same phase, these atoms (or molecules) emit their energies as a gigantic and ultrashort light pulse. Suppose that N atoms are in a volume V so that the density is $n = N/V$ and that each atom has an electric dipole moment μ. Consider that the macroscopic polarization density $\boldsymbol{P}$ is polarized along the z-direction and propagating along the x-direction. The polarization density $\boldsymbol{P}$ is given with $k \equiv \boldsymbol{k} \cdot \hat{\boldsymbol{x}}$ by

$$\boldsymbol{P} = n\mu\hat{\boldsymbol{z}}e^{-i(\omega t - kx)}, \tag{5.31}$$

where $\hat{\boldsymbol{x}}$ and $\hat{\boldsymbol{z}}$ are unit vectors along the x- and z-directions, respectively.

Dicke [51] predicted that the intensity $I(\boldsymbol{k}')$ of the electromagnetic wave emitted along the $\boldsymbol{k}'$-direction is given by

$$I(\boldsymbol{k}') = I_0(\boldsymbol{k}')\frac{N^2}{4}\left|[e^{i(\boldsymbol{k}-\boldsymbol{k}')\cdot\boldsymbol{r}}]_{av}\right|^2 + O(N), \tag{5.32}$$

where $[\cdots]_{av}$ means an average over a volume V. $I_0(\boldsymbol{k}')$ is the radiative intensity from one atom along the $\boldsymbol{k}'$-direction, $\boldsymbol{k} = (k, 0, 0)$ is the wavevector of the excitation, and $O(N)$ means a term proportional to N. When the magnitude $|\boldsymbol{k} - \boldsymbol{k}\prime|$ is small enough, $[\cdots]_{av}$ is the order of unity and, therefore, the intensity $I(\boldsymbol{k}')$ is proportional to N^2. When atoms are filled into a cylindrical container with a cross-section A, an intense light pulse with a wavelength of λ is emitted into a cone with solid angle λ^2/A along the cylinder axis. The total emission is of order $N\lambda^2/A$, and the peak intensity is proportional to N^2 and it is predicted that the pulse width is proportional to $1/N$. As a result, the peak intensity should reach the order of 10^{10} stronger than that of conventional spontaneous emission and the pulse width should become very short. Dicke named this phenomenon superradiance.

In this section we first present experimental results for superradiance in Sect. 5.2.1. In Sect. 5.2.2 we compare these results to theoretical predictions based on the semiclassical theory for the electromagnetic field. We discuss the effects of quantum fluctuations that are characteristics of spontaneous emission and discuss propagation effects of ultrashort light pulses in Sect. 5.2.3. In Sect. 5.2.4 we describe superradiance from excitons in solids when a single exciton has a mesoscopic dipole moment of transition.

5.2.1 Experiments of Superradiance

Feld and coworkers [52] reported for the first time the observation of superradiance in 1973, about 20 years after Dicke [51] predicted the phenomenon. Figure 5.11 shows a schematic illustration of superradiance, originating from a dipole transition between rotational levels of the HF molecule. A HF gas of 1–20 mtorr is held in a cylindrical container 12–28 mm diameter and 30–100 cm

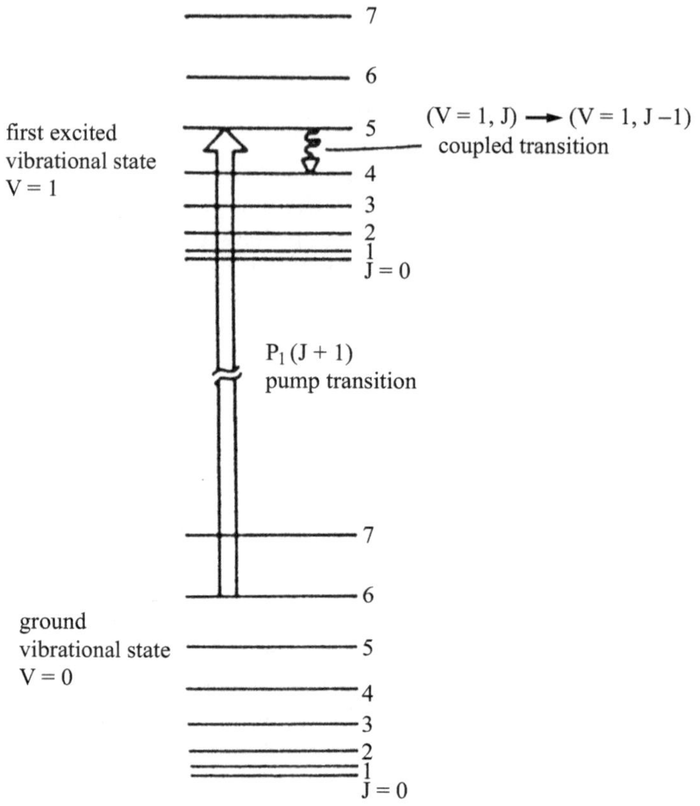

Fig. 5.11. Energy levels of the HF molecule and its transitions. The superrradiant output pulse occurs at the coupled transition

long. The HF molecule is excited from the $(v, J) = (0, 6)$ molecular state to the $(v, J) = (1, 5)$ state by an HF laser, where v and J mean vibrational and rotational quantum numbers, respectively. Since the transition between $(1, J)$ and $(1, J \pm 1)$ is dipole allowed, the population inversion is completed between the $(1, 5)$ and $(1, 4)$ states. Further, the transition from the $(1, 5)$ state to the $(1, 4)$ state has a maximum dipole moment. Figure 5.12(a) shows the intensity profile of the excitation laser pulse. The pulse width is 100 ns ($= 10^{-7}$ s). If each molecule independently emits luminescence, it should be characterized by the radiative lifetime of 1 s, as illustrated in Fig. 5.12(b), where $I_0(t)$ describes the intensity profile of the emission. However, the observed emission exhibits a sharp pulse with a peak height of $10^{10} \times I_0(t)$ and decays much more rapidly (Fig. 5.12c).

The facts above demonstrate that the emission is not incoherent spontaneous radiation. Furthermore, we can rule out the possibility that the emission results from amplification of the spontaneous emission, as was expected from

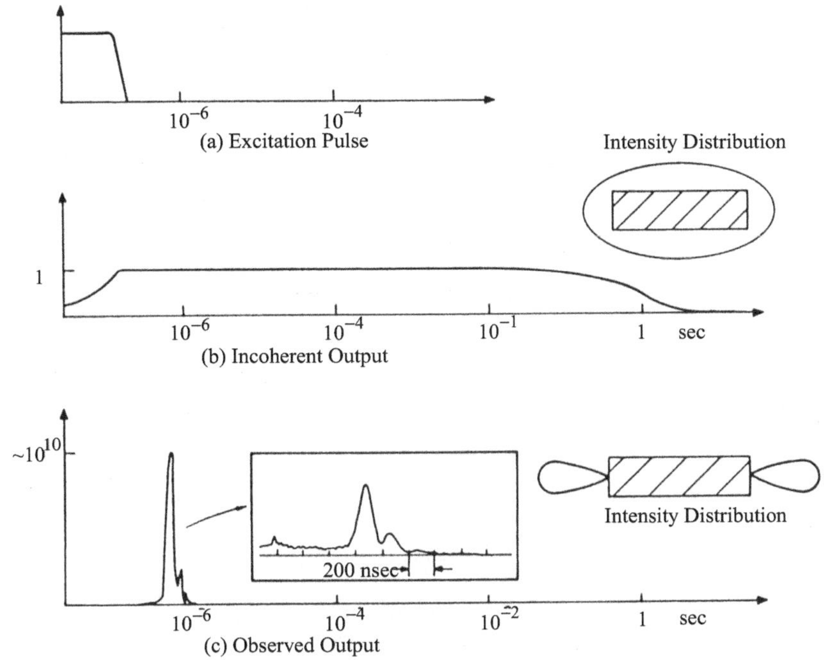

Fig. 5.12. Comparison of observed superradiant output and incoherent spontaneous emission. Time is plotted on a logarithmic scale. (**a**) Pump laser pulse. (**b**) Output expected from incoherent spontaneous emission, exhibiting exponential decay and an isotropic radiation pattern. (**c**) Observed output, exhibiting ringing, a highly directional radiation pattern, and a peak intensity of 10^{10} times that of (b). The inset shows the time evolution of the same pulse with a linear time scale [55]

the temporal profile. Namely, if it came from the amplification, the event should be finished within 10^{-8} s, because it takes less than 10^{-8} s when the light passes though the 100 cm-long container. However, the emission is observed to be delayed by about 10^{-6} s and its pulse width is 2×10^{-7} s, as can be seen in Fig. 5.12(c). The possibility of laser generation is also ruled out. As discussed in Chap. 4, the laser power should be proportional to the number in the inverted population, and so proportional to the HF pressure. However, Fig. 5.13 demonstrates that the intensity is proportional to the square of the pressure. From theoretical considerations discussed later, it is concluded that the emission arises from a coherent spontaneous process due to the $(v = 1, J = 3) \rightarrow (v = 1, J = 2)$ transition of the HF molecule, pumped into the $(v = 1, J = 3)$ level by the $P_1(4)$ laser line shown in Fig. 5.11.

In addition to these three features, the following interesting phenomena were also found. Figure 5.14(a) shows the superradiance due to the $(v = 1, J = 3)$ to $(v = 1, J = 2)$ transition of the HF molecule observed at 4.5 mtorr. In these experiments, both the $P_1(4)$ and $R_1(2)$ laser lines were

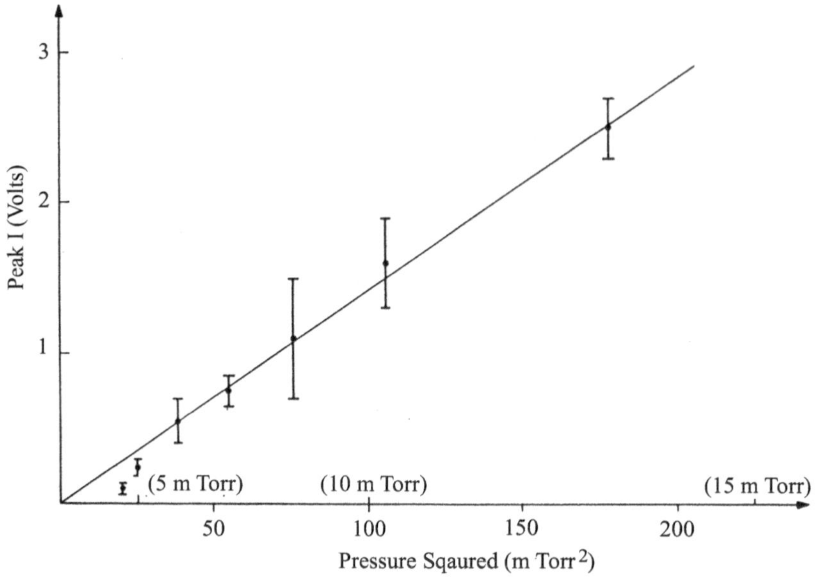

Fig. 5.13. Peak intensity of the superradiant pulse at 84 µm ($J = 3 \rightarrow 2$), pumped by the $P_1(4)$ laser line, as a function of the square of the HF pressure in the sample cell [55]

used as the pump pulse. The $R_1(J-1)$ laser line is induced between the transition between ($v = 0$, $J - 1$) and ($v = 1$, J) shown in Fig. 5.11. The time delay is 0.6 µs and the pulse width is 0.1 µs. When the pressure is lowered to 2.1 mtorr, the peak intensity is decreased and the time delay and pulse width are increased to 1.5 µs and 0.7 µs, respectively. Moreover, the time delay is randomly distributed. As will be discussed in Sect. 5.2.2, this is due to the quantum fluctuation of the initial emission. Figures 5.14(c) and (d) demonstrate the dependence on pump intensity. With the decrease of the pump intensity, the intensity decreases and the time delay increases.

Another important observation is superfluorescence from Cs atoms, reported by Gibbs and coworkers [53]. The valence 6s electron of the Cs atom is excited to the 7p state by a dye laser with a wavelength of 0.455 µm and the emission due to the 7p to 7s transition is monitored at $\lambda = 2.9$ µm as shown in Fig. 5.15. Here we call this phenomenon superfluorescence, in order to distinguish it from superradiance. If the electron population is fully inverted, the Bloch vector is metastable at the very beginning and, therefore, the emission can be induced by fluctuation of the electromagnetic field due to the initial spontaneous emission. Namely, the coherent spontaneous emission, which accompanies the time delay and is originated from a system of fully inverted population, is called superfluorescence. In this mean, *superradiance* from the HF molecule system is also superfluorescence. On the other hand,

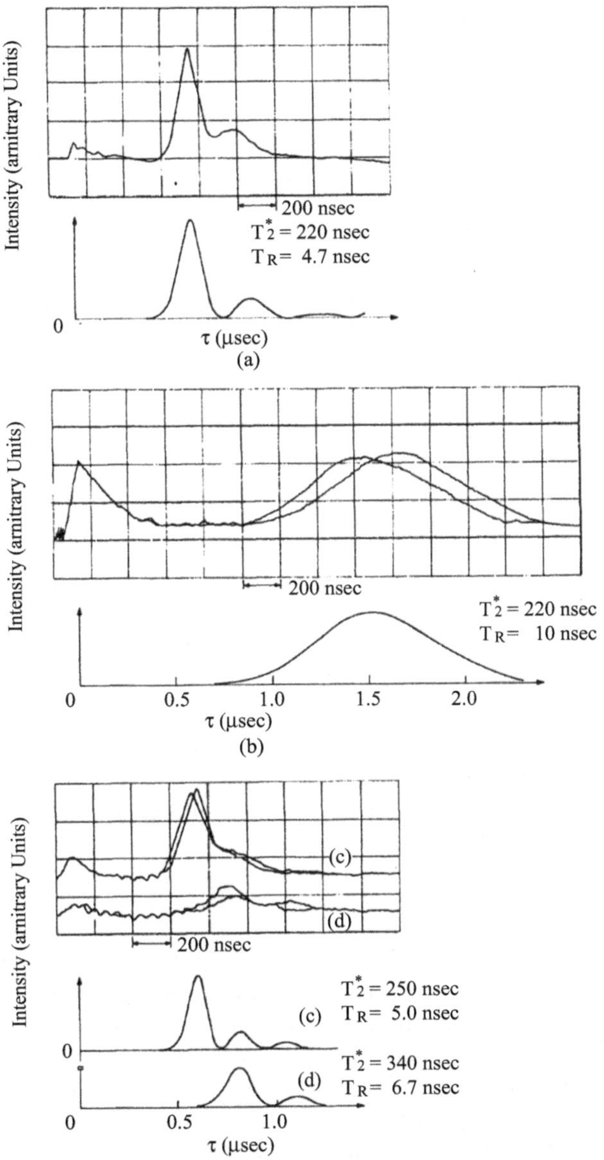

Fig. 5.14. Oscilloscope traces of superradiant pulses and computer fits. (**a**) $J = 3 \rightarrow 2$ transition at 84 μm pumped by $P_1(4)$ laser line. $I = 2.2\,\mathrm{kW/cm^2}$, $p = 4.5\,\mathrm{mtorr}$, $\kappa L = 2.5$ ($L = 100\,\mathrm{cm}$), giving $T_2^* = 220\,\mathrm{nsec}$, $T_R = 4.7\,\mathrm{nsec}$. (**b**) Same as (a) except that $p = 2.1\,\mathrm{mtorr}$, giving $T_2^* = 220\,\mathrm{nsec}$, $T_R = 10\,\mathrm{nsec}$. Note increased delay and broadening of pulse. (**c**) Same transition as (a), but pumped by $R_1(2)$ laser line. $I = 1.7\,\mathrm{kW/cm^2}$, $P = 1.2\,\mathrm{mtorr}$, $\kappa L = 3.5$ ($L = 100\,\mathrm{cm}$) giving $T_2^* = 250\,\mathrm{nsec}$, $T_R = 5.0\,\mathrm{nsec}$. (**d**) Same as (c) except $I = 0.95\,\mathrm{kW/cm^2}$, giving $T_2^* = 340\,\mathrm{nsec}$, $T_R = 6.7\,\mathrm{nsec}$. The same intensity scale is used in fitting curves (a) and (b), and (c) and (d). Note no reproducibility of the oscilloscope traces in double exposure [52]

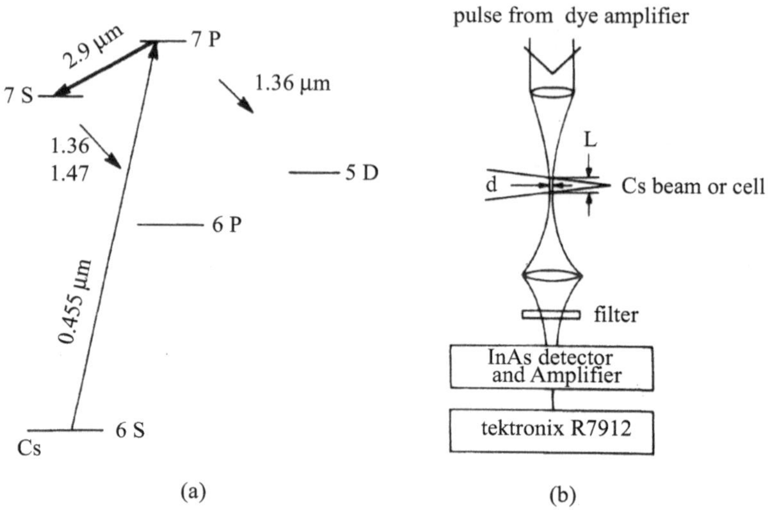

Fig. 5.15. (**a**) Simplified level scheme of Cs and (**b**) diagram of the experimental apparatus [53]

once an electric dipole appears in the superfluorescence process, it accompanies superradiance.

Figure 5.15(b) illustrates the setup of the Gibbs experiment and Fig. 5.16 shows the result. We immediately see that the light pulse has a time delay of 10 ns and its line shape is well fitted by $\mathrm{sech}^2(t)$. Figure 5.17 shows the light pulses at different Cs concentrations. The time delay gradually increases with decreasing Cs concentration.

In order to observe superfluorescence clearly, the following three conditions are required.

(1) The Cs atoms are in a fully inverted population state, because superfluorescence is grown from the fluctuation of the electromagnetic field. Thus, one has to make the population inversion within a short interval, compared to the time delay τ_D of superfluorescence.
(2) $\tau_E = L/c$, the time constant required for the electromagnetic wave to escape from an atom system, must be much shorter than both the longitudinal relaxation time T_1 and the transverse relaxation time T_2. Otherwise, the electromagnetic wave feeds back to the atomic system. As a result, induced emission occurs and suppresses the superfluorescence originated from spontaneous emission. It is evident that this condition is opposite to laser oscillation.
(3) The time constant τ_E, pulse width τ_R, and time delay τ_D must obey the conditions

$$\tau_E < \tau_R < \tau_D < T_1,\, T_2. \tag{5.33}$$

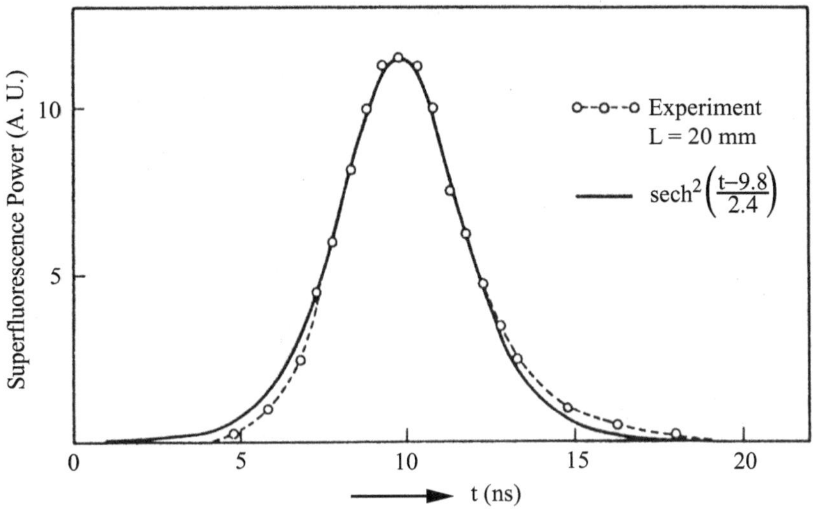

Fig. 5.16. Example of the very symmetrical pulses that have often been observed in Cs gas [53]

In the Gibbs experiment these conditions are actually satisfied:

$$\tau_E = 0.067\,\text{ns} < \tau_R = 5\,\text{ns} < \tau_D = 10\,\text{ns} < T_1 = 70\,\text{ns}, \;\; T_2 = 80\,\text{ns}. \tag{5.34}$$

5.2.2 Theory of Superradiance

We will theoretically discuss superradiance and superfluorescence from two-level atoms or molecules. Suppose each atom has the energy level $\hbar\omega_a$ and $\hbar\omega_b$, where a and b denote the electron ground state and electron excited state, respectively. Using the electron annihilation operators (a_i, b_i) and electron creation operators $(a_i^\dagger, b_i^\dagger)$ on the ith atom, the Hamiltonian $\mathcal{H}_A + \mathcal{H}_{AR}$ is given by

$$\mathcal{H}_A = \hbar \sum_i (\omega_a a_i^\dagger a_i + \omega_b b_i^\dagger b_i), \tag{5.35}$$

$$\mathcal{H}_{AR} = \hbar g \sum_i (a_i^\dagger b_i B^\dagger e^{i\omega t} + b_i^\dagger a_i B e^{-i\omega t}), \tag{5.36}$$

where B and $B^\dagger$ are annihilation and creation operators of the radiation mode ω. Here we assume the electric dipole interaction between the electron system and the radiation field, and neglect the antiresonant terms. The coupling constant g is given by

$$\hbar g \equiv \mu \sqrt{\frac{\hbar\omega}{2\epsilon_0 V}}, \tag{5.37}$$

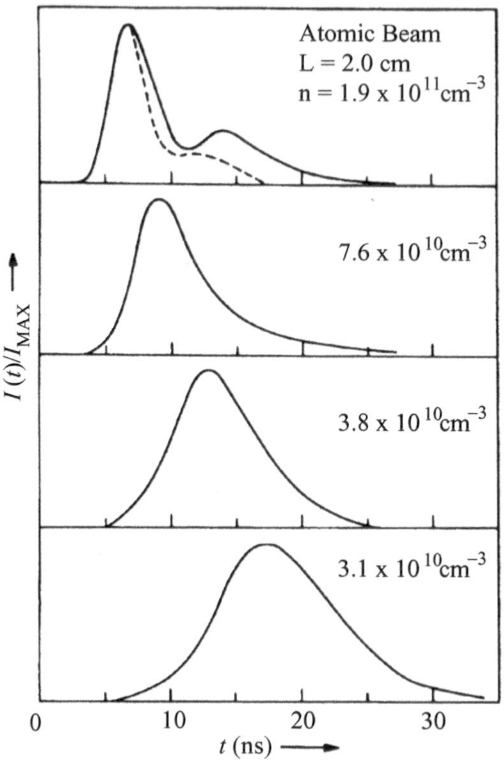

Fig. 5.17. Normalized single-shot pulse shapes for several Cs densities n; Fresnel number $F \approx 1$. Uncertainties in the values of n are estimated to be (+60, −40)% [53]

where μ is the electric dipole moment. We introduce the following *spin* operators:

$$s_i^+ e^{i\omega t} \equiv b_i^\dagger a_i, \quad s_i^- e^{-i\omega t} \equiv a_i^\dagger b_i, \quad s_i^z \equiv \frac{1}{2}(b_i^\dagger b_i - a_i^\dagger a_i). \tag{5.38}$$

The equations of motion for these operators are

$$\frac{\partial}{\partial t} s_i^+ = -i\Delta s_i^+ - 2igB^\dagger s_i^z, \tag{5.39}$$

$$\frac{\partial}{\partial t} s_i^- = i\Delta s_i^+ + 2igBs_i^z, \tag{5.40}$$

$$\frac{\partial}{\partial t} s_i^z = ig(B^\dagger s_i - Bs_i^+), \tag{5.41}$$

where $\Delta \equiv \omega - \omega_{ba}$. The radiation field always follows the motion of the atomic system, i.e., it obeys the adiabatic approximation, because the lifetime τ_E is shorter than other time constants. Simultaneously, (5.39)–(5.41) describe a phenomenon that occurs during a time interval much shorter than the relaxation time T_1 and T_2. Accordingly, we can neglect the relaxation terms. We will consider the following two cases.

Case 1 :L (length of the system) < λ (wavelength of the light)

We introduce the polarization operators S^+ and S^- of the total electron system and the inverted population operator S^z as

$$S^+ \equiv \sum_i s_i^+, \qquad S^- \equiv \sum_i s_i^-, \qquad S^z \equiv \sum_i s_i^z, \tag{5.42}$$

respectively. Equations (5.39)–(5.41) then become

$$\frac{\partial}{\partial t}S^+ = -i\Delta S^+ - 2igB^\dagger S^z, \tag{5.43}$$

$$\frac{\partial}{\partial t}S^- = i\Delta S^- + 2igBS^z, \tag{5.44}$$

$$\frac{\partial}{\partial t}S^z = ig(B^\dagger S^- - BS^+). \tag{5.45}$$

Here we have neglected the effects of relaxation $1/T_1$ and $1/T_2$, as was expected from (5.33). Namely, the values of τ_R and τ_D, which are obtained from (5.39), (5.40), and (5.41), should be much smaller than T_1 and T_2.

From (5.43)–(5.45), we have the conservation relation

$$S^-S^+ + S^+S^- + 2(S^z)^2 = \text{const.} \tag{5.46}$$

Introducing $S^- \equiv S^x - iS^y$ and $S^+ \equiv S^x + iS^y$, we obtain

$$(S^x)^2 + (S^y)^2 + (S^z)^2 \equiv \boldsymbol{S}^2 = S(S+1). \tag{5.47}$$

This equation gives eigenvalues of spin $\boldsymbol{S}$ for the total atomic system, because $\boldsymbol{S}^2$ obeys a commutation relation

$$\left[\mathcal{H}_A + \mathcal{H}_{AR}, \boldsymbol{S}^2\right] = 0, \tag{5.48}$$

namely, $\boldsymbol{S}^2$ is a constant of the motion. Similarly, since S^z has a commutation relation

$$[\mathcal{H}_A, S^z] = 0, \tag{5.49}$$

S^z is also a constant of the motion for the Hamiltonian $\mathcal{H}_A$. By analogy with spin systems, the quantum number m of S^z and the quantum number S are restricted as

$$|m| \leqq S \leqq \frac{1}{2}N, \tag{5.50}$$

where N is the number of atoms. The quantum number $m \equiv (1/2)(N_b - N_a)$ displays the magnitude of the population inversion. The cooperation number S describes the magnitude of the Bloch vector.

Case 2: L (length of the system) > λ (wavelength of the light)

In this case we take $\Delta \equiv \omega - \omega_{ba} = 0$ for simplicity and assume $B^\dagger$ and B are classical values. Taking the average over ΔV for $\lambda < (\Delta V)^{1/3} < L$, we introduce the polarization density S^+ and inverted population density S^z as

$$\frac{1}{\Delta V} \sum_{i \in \Delta V(x)} s_i^+ \exp(-ikx_i) = S^+(x,t) \exp(-ikx), \tag{5.51}$$

$$\frac{1}{\Delta V} \sum_{i \in \Delta V(x)} s_i^z = S^z(x,t). \tag{5.52}$$

Then the equations of motion are

$$\frac{\partial}{\partial t} S^+(x,t) = -2igB^\dagger(x,t) S^z(x,t), \tag{5.53}$$

$$\frac{\partial}{\partial t} S^z(x,t) = ig\left[B^\dagger(x,t) S^-(x,t) - B(x,t) S^+(x,t)\right]. \tag{5.54}$$

Similarly, we quantize the electric field $E(x,t)$ in a volume ΔV and obtain

$$E(x,t) = i\sqrt{\frac{\hbar\omega}{2\epsilon_0 \Delta V}}\left[B(x,t)e^{-i(\omega t - kx)} - B^\dagger(x,t)e^{i(\omega t - kx)}\right]. \tag{5.55}$$

Therefore, $\hbar g \equiv \mu(\hbar\omega/2\epsilon_0\Delta V)^{1/2}$. Using the operators S^+ and S^-, we can rewrite the electric polarization density $P(x,t)$ as

$$P(x,t) = i\mu\left[S^+(x,t)e^{i(\omega t - kx)} - S^-(x,t)e^{-i(\omega t - kx)}\right]. \tag{5.56}$$

Substituting (5.55) and (5.56) into the Maxwell equation

$$\nabla^2 E - \frac{1}{c^2}\frac{\partial^2}{\partial t^2} E = \frac{1}{\epsilon_0 c^2}\frac{\partial^2}{\partial t^2} P, \tag{5.57}$$

and using the dispersion relation $ck = \omega$, we obtain

$$\left(\frac{\partial}{\partial t} + c\frac{\partial}{\partial x} + \kappa\right) B^\dagger(x,t) = ig\Delta V S^+(x,t). \tag{5.58}$$

Here we have used the rotating wave approximation, and slowly varying envelope approximation

$$\left|\omega\frac{\partial}{\partial t} B^\dagger\right| \gg \left|\frac{\partial^2}{\partial t^2} B^\dagger\right|, \qquad \left|k\frac{\partial}{\partial x} B^\dagger\right| \gg \left|\frac{\partial^2}{\partial x^2} B^\dagger\right|. \tag{5.59}$$

We introduce empirically $\kappa \equiv c/L$, the decay rate of the radiation wave that escapes from the atomic system. For the maximum cooperation number $S = n/2 (\gg 1)$ we have

$$\left|S^{+}\right|^{2}+\left|S^{z}\right|^{2} \fallingdotseq \left(\frac{n}{2}\right)^{2}, \tag{5.60}$$

where n is the number of atoms per unit volume. Introducing the polar angle $\Phi(x,t)$ of the Bloch vector (S^x, S^y, S^z), we obtain

$$S^{z}(x,t) = -\frac{n}{2}\cos\Phi(x,t), \tag{5.61}$$

$$S^{-}(x,t) = S^{+}(x,t) = \frac{n}{2}\sin\Phi(x,t). \tag{5.62}$$

Here we take $\Phi = 0$ for the ground state of the electron system. Substituting these equations into (5.53), we obtain

$$B^{\dagger} = -\frac{i}{2g}\frac{\partial\Phi}{\partial t} = -B. \tag{5.63}$$

Substituting (5.62) and (5.63) into (5.58) gives

$$-\frac{i}{2g}\left(\frac{\partial^2}{\partial t^2} + c\frac{\partial^2}{\partial x\partial t} + \kappa\frac{\partial}{\partial t}\right)\Phi = ig\Delta V\frac{n}{2}\sin\Phi. \tag{5.64}$$

In order to rewrite this differential equation in dimensionless form, we introduce τ and ξ defined as

$$\frac{ng^2\Delta V}{\kappa} = \frac{N\mu^2\omega}{2\epsilon_0\hbar\kappa V} \equiv \frac{1}{\tau}, \qquad c\tau \equiv \xi. \tag{5.65}$$

Using τ and ξ, we obtain

$$\frac{\partial}{\partial(t/\tau)}\Phi + \sin\Phi = -\frac{1}{\kappa\tau}\left[\frac{\partial^2\Phi}{\partial(t/\tau)^2} + \frac{\partial^2\Phi}{\partial(t/\tau)\partial(x/\xi)}\right]. \tag{5.66}$$

Now we turn back to *Case 1*. If we have the condition

$$\frac{1}{\kappa\tau} = \frac{n\mu^2\omega}{2\epsilon_0\hbar\kappa^2} \equiv \left(\frac{L}{L_c}\right)^2 \ll 1, \tag{5.67}$$

we can neglect the contribution of the right-hand side of (5.66). Then Φ depends only on t:

$$\frac{\partial\Phi}{\partial t} = -\frac{1}{\tau}\sin\Phi. \tag{5.68}$$

Namely, if the length of the system is smaller than the cooperation length $L_c \equiv c\sqrt{2\epsilon_0\hbar/n\mu^2\omega}$, the superradiance and superfluorescence can be described by (5.68). If not, the right-hand side of (5.66) leads to the propagation effect. This effect will be discussed in Sect. 5.2.3. By solving (5.68) for an initial condition $\Phi(0)$ at $t = 0$, we obtain

$$\log \left| \frac{\tan \left[\frac{1}{2}\Phi(t)\right]}{\tan \left[\frac{1}{2}\Phi(0)\right]} \right| = -\frac{t}{\tau}. \tag{5.69}$$

That is

$$\tan \left[\frac{1}{2}\Phi(t)\right] = \exp \left[-\frac{1}{\tau}(t - t_{\max})\right], \tag{5.70}$$

where

$$\tan \left[\frac{1}{2}\Phi(0)\right] = \exp \left(\frac{t_{\max}}{\tau}\right). \tag{5.71}$$

Note that (5.70) can be written as

$$\sin \Phi(t) = \operatorname{sech} \left(\frac{t - t_{\max}}{\tau}\right). \tag{5.72}$$

Using the above results, we obtain the t-dependence of the radiation intensity $I(t)$:

$$I(t) = 2\kappa\hbar\omega B^{\dagger}(t)B(t)\frac{V}{\Delta V} \tag{5.73}$$

$$= 2\kappa\hbar\omega\frac{1}{4g^2}\left(\frac{\partial\Phi}{\partial t}\right)^2 \frac{V}{\Delta V} \tag{5.74}$$

$$= \hbar\omega\left(\frac{\kappa}{2g^2}\right)\frac{1}{\tau^2}\left(\sin^2 \Phi\right)\frac{V}{\Delta V} \tag{5.75}$$

$$= \frac{\epsilon_0\hbar^2\kappa V}{\mu^2\tau^2}\operatorname{sech}^2\left(\frac{t - t_{\max}}{\tau}\right). \tag{5.76}$$

Let us summarize the predictions of the semiclassical theory and compare it to the experimental results shown in Sect. 5.2.1.

(1) The temporal profile of the light pulse is described by (5.76). From (5.65) the pulse width is given by

$$\tau = \frac{2\epsilon_0\hbar\kappa}{\omega n\mu^2} = \frac{4\epsilon_0\tau_0}{3\pi n\lambda^2 L}, \tag{5.77}$$

where τ_0 is the lifetime of the spontaneous emission from an isolated two-level atom. It follows from (5.77) that the pulse width becomes shorter in inverse proportion to the atom density n, i.e., the number of atoms in a volume $\lambda^2 L$. This well accounts for the dependences of the pulse width on the HF pressure (Figs. 5.14a and b) and on the Cs atom density (Fig. 5.17).

(2) Using (5.65) and (5.76), the peak intensity I_0 is given by

$$I_0 = \frac{\hbar\omega\kappa V}{2g^2\tau^2 \Delta V} = \frac{\omega^2\mu^2 n^2}{4\varepsilon_0 \kappa} V. \tag{5.78}$$

This equation explains the experimental result that I_0 is proportional to the square of the HF pressure (Fig. 5.13).

(3) Equation (5.76) well describes the observed profile of the pulse (Fig. 5.16).

5.2.3 Quantum and Propagation Effects on Superradiance

The remaining problems, which cannot be explained by the semiclassical theory, are associated with the quantum effect and propagation effect on superradiance. The former is observed as the fluctuation and n-dependence of the time delay, as shown in Fig. 5.14(b). From (5.71) the time delay $t_{\max}$is given by

$$t_{\max} = \tau \log \left| \tan \left[\frac{\Phi(0)}{2} \right] \right| . \tag{5.79}$$

$t_{\max} = 0$ for $\Phi(0) = \pi/2$ and $t_{\max} = \infty$ for $\Phi(0) = \pi$. When the system is in a fully population-inverted state, the Bloch vector of (S^x, S^y, S^z) is along the polar axis so that $\Phi(0) = \pi$. Therefore, the time delay $t_{\max}$ is infinite, indicating no superradiance. The superradiance from the actual system is initiated by spontaneous emission of one atom or one molecule. The time delay and its fluctuation depend on when and which atom or molecule triggers the process. In other words, these phenomena demonstrate the macroscopic manifestation of quantum fluctuations. To describe these, we have to take account of the quantum fluctuation of the photon, namely the fluctuation of the vacuum. Here we introduce the effective initial angle θ_0 of the Bloch vector to describe the fluctuation effect of the radiation field:

$$\theta_0 \equiv \sqrt{\langle \Delta\Phi(x, t=0)^2 \rangle}, \tag{5.80}$$

where $\Delta\Phi(x, t=0) = \Phi(x, t=0) - \pi$ is the angle between the polar axis and the direction of the Bloch vector. A complicated quantum-mechanical calculation yields $\theta_0 \fallingdotseq \sqrt{2/N}$, where N is the total number of atoms or molecules involved in the process. Here we skip this calculation. Instead, we will show that the angle θ_0 is directly obtained by experiment.

Suppose that, immediately after excitation, light creates a full population inversion; a weak light pulse with the same frequency illuminates this system. When the pulse area $\theta \equiv \int \mu E(t) dt/\hbar$ of the weak light pulse is smaller than θ_0, the effect of the light pulse is negligibly small and, thus, the fluctuation of the vacuum for the radiation field triggers superfluorescence. On the other hand, for $\theta > \theta_0$, superfluorescence is started by the incident light pulse. In the latter case the average value τ_D of the time delay of superfluorescence is

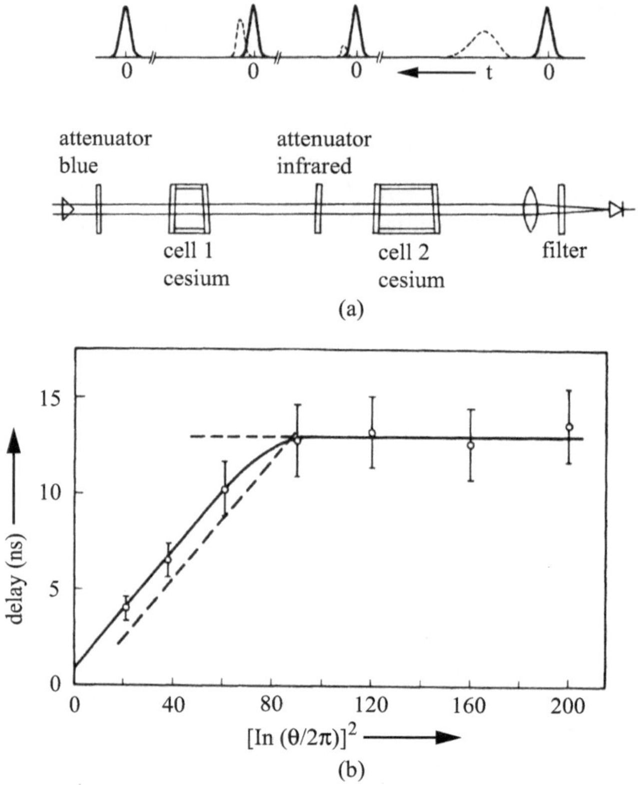

Fig. 5.18. (**a**) Setup for the measurement of the effective initial tipping angle. (**b**) Delay time τ_D of the superfluorescence output pulse in Cs vs. $[\text{In}(\theta/2\pi)]^2$. The dashed line is used to correct for the delay of the injection pulse with respect to the pump pulse [54]

greatly shortened. Thus, we can estimate θ_0 by measuring the θ-dependence of τ_D.

Figure 5.18(a) illustrates the setup for the above experiment. Cs atoms are filled into two glass containers. An ultraviolet laser excites the Cs atoms and, thus, full population inversion occurs between the 7p and 7s states. The superradiance arises from the 7p $\rightarrow$ 7s transition. The average time delays of superradiance from cell 1 and cell 2 are 1.5 ns and 13 ns, respectively. This difference is due to the different density of the Cs atoms between them. Namely, since the density of the Cs atoms is larger for cell 1, the characteristic time τ is smaller, leading to a small pulse width τ_D and a short time delay τ_R (see (5.65)). In this experiment the light pulse from cell 1 initiates superradiance of Cs atoms in cell 2. The absorption filter is set between the cells to change the pulse area θ of the initiating pulse. Figure 5.18(b) shows a plot of τ_D vs. $[\ln(\theta/2\pi)]^2$. We find that τ_D strikingly changes its slope at $[\ln(\theta/2\pi)]^2 = 90$.

For $[\ln(\theta/2\pi)]^2 < 90$, τ_D decreases with increasing θ up to 2π. This suggests that superradiance is affected by the light pulse area from cell 1. For $[\ln(\theta/2\pi)]^2 > 90$, on the other hand, the time delay is constant, τ_D=13ns, demonstrating that the superradiance is triggered by spontaneous emission in cell 2. The effective angle θ_0 is estimated to be $\theta_0 = 5 \times 10^{-4}$. This value agrees basically with the theoretical value $\theta_0 \fallingdotseq \sqrt{2/N} = 1 \times 10^{-4}$.

Another problem is related to the propagation effects of superradiance. As can be seen in Fig. 5.19, the radiation intensity has an oscillating character at higher density n. Needless to say, the semiclassical theory cannot account for this phenomenon. In the previous subsection we assumed $1/\kappa\tau \ll 1$ to obtain (5.68). For larger n, however, the contribution of the right-hand side of (5.66), which describes the propagation effect of the radiation, becomes important. Similarly, when the length L of the total system is larger than the cooperation length L_c, we have to take into account the right-hand side of (5.66).

The initial stage of superfluorescence is triggered by spontaneous emission from one excited atom or molecule. The weak radiation field due to the spontaneous emission travels in the medium and induces a weak macroscopic dipole, which enlarges the radiation field further. Repeating this process, the electric dipole grows spatially and temporally, and emanates as the light from the endplate. Figure 5.19 shows this propagation effect. For $L > L_c$ the light pulse is periodically emitted from the system and its intensity decays in time.

5.2.4 Superradiance from Excitons

We have studied the superradiance from two-level atoms or molecules. In this process the dipole moments of the two-level atoms or molecules align their phase through virtual interaction with the radiation field. As a result, the dipole moments of $N/2$ excitations are oriented and a macroscopic dipole moment is formed at $\Phi = \pi/2$. This macroscopic dipole leads to superradiance. In other words, the electromagnetic interaction between two levels results in the $S = N/2$ state. In this subsection we show superradiance from excitons in solids.

Since atoms and/or molecules are periodically arranged in crystals, one excitation in an atom/molecule propagates among all atoms/molecules. This is called an *exciton.* The lowest exciton state can be described by a linear combination of the excited state of the constituted atoms/molecules:

$$\Psi_{\boldsymbol{k}} = \frac{1}{\sqrt{N}} \sum_{i=1} e^{i\boldsymbol{k}\cdot\boldsymbol{r}_i} (b_i^\dagger a_i)|g\rangle, \tag{5.81}$$

where a_i and b_i are electron annihilation operators of the conduction and valence bands, respectively. These bands are based on the Wannier function centered on the ith atom. N is the number of constituent atoms/molecules and $\Psi_{\boldsymbol{k}}$ is normalized in the crystal. The wavefunction (5.81) means that the excitation of an atom/molecule propagates from site to site with wavevector

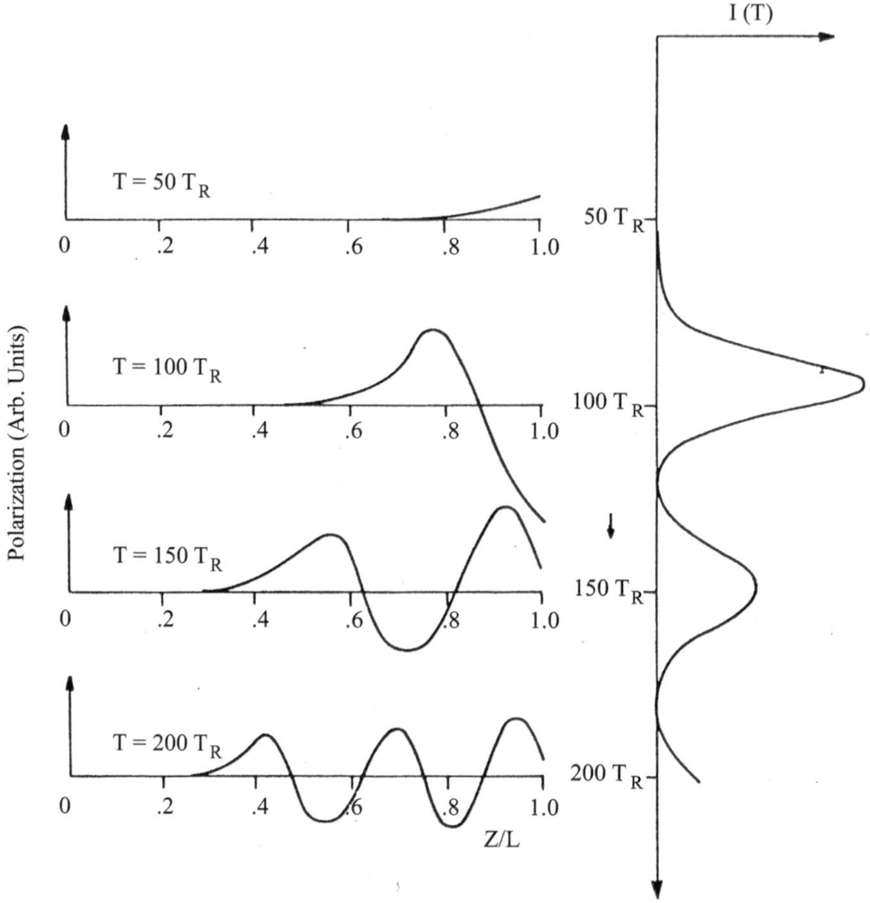

Fig. 5.19. Stretch of the buildup of polarization in the medium, showing the polarization as a function of z at $t = 50\,T_R$, $100\,T_R$, $150\,T_R$, and $200\,T_R$. The corresponding output intensity pattern is shown at the right [55]

$\boldsymbol{k}$. This type of excitations is called a *Frenkel exciton*. The electric dipole moment of the transition between the ground state $|g\rangle$ of the crystal and the one Frenkel exciton state $\Psi_{\boldsymbol{k}}$ is

$$\left\langle \Psi_{\boldsymbol{k}} \left| \sum_{i=1} (-e\boldsymbol{r}_i e^{i\boldsymbol{k}\cdot\boldsymbol{r}_i}) b_i^\dagger a_i \right| g \right\rangle = \sqrt{N} \boldsymbol{\mu} \delta_{\boldsymbol{k}\boldsymbol{K}}. \tag{5.82}$$

From (5.82) it follows that the dipole moment is $\sqrt{N}$ times larger than the dipole moment $\boldsymbol{\mu}$ of one atom/molecule. When the wavevector $\boldsymbol{k}$ of the exciton is just the same as the wavevector $\boldsymbol{K}$ of electromagnetic wave, the electric dipole transition is allowed.

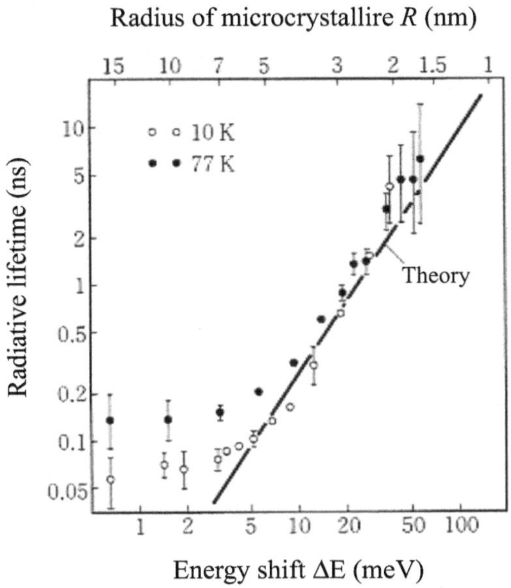

Fig. 5.20. The radiative lifetime of the exciton (ns) depends on the radius R (nm) of the microcrystallite of CuCl, which determines the energy shift ΔE(meV) of the emitted light [58]. The observed lifetime is compared with the theory [56, 57]

In semiconductor crystals, an electron in the conduction band and a hole in the valence band form a different type of exciton, a *Wannier exciton.* The electric dipole moment of the Wannier exciton is given by $\sqrt{Nu^3/\pi a_B^3}\boldsymbol{\mu}_{cv}\delta_{\boldsymbol{kK}}$, where u^3 is the unit cell volume and $\boldsymbol{\mu}_{cv}$ is the transition moment between the valence and conduction bands; a_B is the exciton Bohr radius, which measures the average distance between the electron and hole in the exciton. Since an electron–hole pair of the Wannier exciton spreads over several lattice sites, the dipole moment of the Wannier exciton is $\sqrt{u^3/\pi a_B^3}$ times smaller than that of the Frenkel exciton. However, it still has the macrascopic enhancement factor $\sqrt{N}$.

Since the exciton is scattered by lattice imperfections and lattice vibrations, the exciton has a finite coherence length L_{ex}. As a result, N is restricted to the number of atoms/molecules in a volume L_{ex}^3. If we greatly reduce the crystal volume, the crystal size determines the coherent length. In addition, the exciton is confined in the spherical microcrystallite and, therefore, the motion of the center of gravity is quantized. In this case the electric dipole moment is given by

$$\boldsymbol{P}_n = \frac{2\sqrt{2}}{\pi}\left(\frac{R}{a_B}\right)^3 \frac{1}{n}\boldsymbol{\mu}_{cv}, \tag{5.83}$$

where n is a principal quantum number for the center-of-mass motion and R is the radius of the spherical microcrystallite. It is evident that the transition to the lowest energy level $n = 1$ has a maximum dipole moment. Thus we expect that this lowest energy exciton emits fast spontaneous radiation as superradiance [56, 57]. Since this exciton has a mesoscopic dipole moment, the inverse of the radiative lifetime, $1/T_1$, is increased mesoscopically by $64\pi(R/a_B)^3$ compared to $4\mu_{cv}^2/3\hbar\lambda^3$ of the interband transition:

$$2\gamma \equiv \frac{1}{T_1} = 64\pi \left(\frac{R}{a_B}\right)^3 \frac{4\mu_{cv}^2}{3\hbar\lambda^3}, \tag{5.84}$$

where λ is the wavelength of the excitation. Although this superradiance comes from the one-exciton state, the excited states of all constituent atoms/molecules cooperatively emit radiation. Namely, the emission originates from the maximum cooperation-number state.

Itoh et al. [58] observed the superradiance from CuCl microcrystals with $R = 1.7\,\mathrm{nm} \sim 10\,\mathrm{nm}$, embedded in NaCl crystals. The crystal size was controlled by adjusting the annealing temperature, the annealing time interval, and the quenching speed. Figure 5.20 shows the radiative lifetime T_1 as a function of the energy shift ΔE of the emission peak, which is related to the mean radius R through the relation $\Delta E = \hbar^2\pi^2/2MR^2$. Here M is the effective mass for center-of-mass motion. For $1.7\,\mathrm{nm} < R < 80\,\mathrm{nm}$ the result agrees excellently with the theoretical prediction (5.84). Nakamura et al. [59] studied a CuCl microcrystallite embedded in a glass matrix, and obtained a similar R-dependence of the radiative decay as with the microcrystallites of CuCl.

6

Nonlinear Optical Responses I

A laser is an intense, monochromatic, and coherent light source. By making the best use of these characteristics, we have become able to easily observe the nonlinear optical responses of atoms, molecules, and solids such as higher-harmonic generation, four-wave mixing, parametric oscillation, multiphoton absorption, induced-Raman scattering, etc. In general, the interaction between the radiation field and matter is relatively weak so that the effects of this interaction can be described in terms of perturbational methods. These nonlinear optical responses are discussed in this chapter. However, we have such strong power in ultrashort laser pulses, as discussed in Chap. 5, that the perturbative treatment is not justified in describing the nonlinear optical phenomena under these laser fields. These phenomena will be discussed in Chap. 7. On the other hand, nonlinear optical responses which are discussed in this chapter can be classified according to how many times this interaction works in each nonlinear optical response. The lowest-order nonlinear optical responses are sum-frequency and second-harmonic generation (SHG). These will be discussed in Sect. 6.1. This sum-frequency and higher harmonic generation in solids is useful to understand the electronic structure as well as the microscopic optical processes of solids but also is important from an engineering point of view. First of all, laser light generation with shorter wavelengths is the more difficult, as already discussed in Sect. 4.3.4. Because of this, we are now producing coherent ultraviolet light as higher harmonics in crystals by using Nd:glass lasers or YAG lasers as a strong fundamental source. Secondly, we obtain the light source in the visible region by converting the infrared ($\approx 1\,\mu$m) laser light of cheap and stable semiconductor lasers into the second harmonics.

Optical parametric oscillation which is used to generate squeezed light (see Chap. 2) is also a second-order nonlinear optical response. This will be understood as just an optical process reversal to sum-frequency generation. This will be discussed in Sect. 6.1.4.

We have very colorful nonlinear optical phenomena from third-order optical processes. In Sect. 6.2, four-wave mixing will be discussed. This includes CARS (coherent anti-Stokes Raman scattering), the generation of phase-conjugated waves, and optical bistability. All these phenomena are described by the third-order susceptibility $\chi^{(3)}$. We will show in Sect. 6.3 that this $\chi^{(3)}$ will be extremely enhanced when the exciton, i.e., a kind of collective elementary excitation in crystals, is resonantly pumped. As discussed in Sect. 5.2.4, this exciton in quantum dots (microcrystallites) or quantum wells can radiatively decay very rapidly by a superradiant process. As a result of these two characteristics, the exciton in these confined systems possibly satisfies both requirements of large $\chi^{(3)}$ and rapid switching.

In Sect. 6.4, we introduce, as an example of a nonlinear dissipative optical process, two-photon absorption spectroscopy. From this spectroscopy, we will be able to obtain useful information about the electronic structure of materials, which is complementary to one-photon spectroscopy. As an example of the application of SHG, we will discuss how to determine the sign of order parameters such as ferroelectric polarization or sublattice magnetization of ferroelectric antiferromagnets by using the inference effects of the SHG signals. This will be shown in Sect. 6.5

6.1 Generation of Sum-Frequency and Second Harmonics

Laser light has a well-defined frequency and wavevector so that it is temporally and spatially coherent, and has a large amplitude of the radiation field. As a result, nonlinear optical phenomena are easily induced by pumping materials by this laser light, and are also easily observable. The generation of the sum-frequency and second harmonics is the lowest order nonlinear optical phenomenon so that it is also important for applications to engineering.

6.1.1 Principle of Higher-Harmonic Generation

Consider irradiating an insulating crystal by radiation fields $\boldsymbol{E}_1 = \boldsymbol{E}(\omega_1) \exp [i(\boldsymbol{k}_1 \cdot \boldsymbol{r} - \omega_1 t)]$ and $\boldsymbol{E}_2 = \boldsymbol{E}(\omega_2) \exp[i(\boldsymbol{k}_2 \cdot \boldsymbol{r} - \omega_2 t)]$. These two fields induce not only linear polarizations $\boldsymbol{P}^{(1)}(\omega_1) = \chi(\omega_1)\boldsymbol{E}(\omega_1)$ and $\boldsymbol{P}^{(1)}(\omega_2) = \chi(\omega_2)\boldsymbol{E}(\omega_2)$ but also second harmonics oscillating at $2\omega_1$ and $2\omega_2$, and the sum-frequency at $\omega = \omega_1 + \omega_2$. The induced electric dipole moment with the sum-frequency ω is written as

$$\begin{aligned}\boldsymbol{P}_{\omega}^{(2)}(\boldsymbol{r}, t) &= \chi^{(2)}(\omega = \omega_1 + \omega_2) : \boldsymbol{E}_1 \boldsymbol{E}_2 \\ &= \chi^{(2)} : \boldsymbol{E}(\omega_1) \boldsymbol{E}(\omega_2) \exp\left[i(\boldsymbol{k}_1 + \boldsymbol{k}_2) \cdot \boldsymbol{r} - i(\omega_1 + \omega_2) t\right] . \quad (6.1)\end{aligned}$$

Here χ and $\chi^{(2)}$ are linear and second-order polarizabilities, and these second- and third-rank tensors are denoted by χ_{ij} and $\chi_{ijk}^{(2)}$, respectively. In (6.1),

$\boldsymbol{P}^{(2)} = \chi^{(2)} : \boldsymbol{E}_1\boldsymbol{E}_2$ is an abbreviation for a vector whose i-component is $P_i^{(2)} = \chi_{ijk}^{(2)} E_{1j} E_{2k}$. The microscopic description of the second-order polarizability will be derived in Sect. 6.1.2. With this second-order polarization of (6.1) as a source term, the radiation field with angular frequency ω is produced. This process is described by the following Maxwell equation with the source term (6.1) on the right-hand side:

$$\nabla^2 \boldsymbol{E}_\omega + \frac{\omega^2}{c^2} \epsilon(\omega) \boldsymbol{E}_\omega = -\frac{\omega^2}{\epsilon_0 c^2} \boldsymbol{P}_\omega^{(2)} . \tag{6.2}$$

Here $\epsilon(\omega) \equiv 1 + \chi(\omega)/\epsilon_0$ is a linear dielectric function of the insulator. Let us choose the propagating direction of the second-harmonic signal in z-axis and denote the polarization direction as a unit vector $\boldsymbol{e}_\omega$, i.e.,

$$\boldsymbol{E}_\omega = \boldsymbol{e}_\omega E(z) \exp\left[i(kz - \omega t)\right] . \tag{6.3}$$

Then the signal amplitude $E(z)$ may gradually increase in the z-direction. Therefore we accept the slowly varying envelope approximation, that is, we may neglect $d^2E(z)/dz^2$ in comparison to $kdE(z)/dz$. As a result, we have

$$2ik\frac{d}{dz}E(z) = -\frac{\omega^2}{\epsilon_0 c^2}\chi^{(2)} : \boldsymbol{E}(\omega_1)\boldsymbol{E}(\omega_2)\exp(i\Delta kz) . \tag{6.4}$$

Here the dispersion relation $\epsilon(\omega) = (ck/\omega)^2$ was used and note that only the component parallel to $\boldsymbol{e}_\omega$ is chosen for the right-hand side of (6.4). When both $\boldsymbol{E}_1$ and $\boldsymbol{E}_2$ propagate in the z-direction, $\Delta k = k_1 + k_2 - k$ in (6.4). Integrating (6.4) from $z = 0$ to $z = l$ with the boundary condition $E(z = 0) = 0$, the signal intensity $I_\omega(l)$ is given by

$$\begin{aligned} I_\omega(l) &= \frac{c\epsilon_0\sqrt{\epsilon(\omega)}}{2}|E(l)|^2 \\ &= \frac{\omega^2}{8c\epsilon_0\sqrt{\epsilon(\omega)}}\left|\chi^{(2)} : \boldsymbol{E}(\omega_1)\boldsymbol{E}(\omega_2)\right|^2\left\{\frac{2\sin(\Delta kl/2)}{\Delta k}\right\}^2 . \end{aligned} \tag{6.5}$$

The thickness dependence of the SHG signal $I_\omega(l)$ was observed by Maker et al. [60] as shown in Fig. 6.1(a). The oscillation shown in the figure is called a Maker fringe. When the incident light hits the crystal surface at an angle θ, as shown in Fig. 6.1(b), the light-path length l inside the crystal with thickness d varies as $l = d/\cos\theta$ against the incident angle θ. Here a quartz crystal with thickness $d = 0.787$ mm was used and this crystal was rotated around the crystal c-axis which is parallel to the crystal surface so that the phase mismatch Δk was independent of the angle θ. Red light from a ruby laser with wavelength 0.694 µm is sent to the quartz crystal and the blue light intensity of the SHG signal was observed as a function of θ as shown in Fig. 6.1(a).

From (6.5) and the experiment shown in Fig. 6.1, it has becomes clear that two conditions must be fulfilled to obtain a strong SHG signal: (1) the

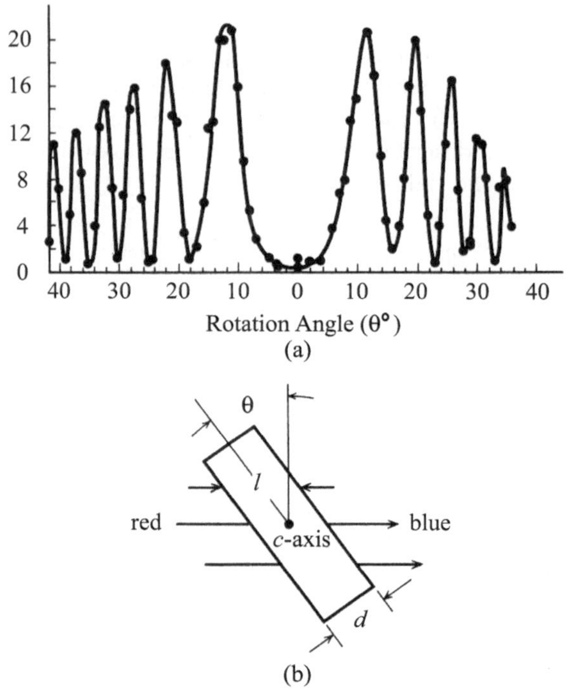

Fig. 6.1. (**a**) The second-harmonic intensity against the angle θ of the incident light, when a quartz crystal with thickness 0.787 mm is irradiated by a ruby laser [60]. (**b**) The crystal surfaces of a quartz crystal are chosen to be parallel to its c-axis so that the path length of the incident light $l = d/\cos\theta$ is changed as a function of the rotation angle θ. The polarization of both the fundamental light of ruby and the blue second-harmonics is chosen to be parallel to the c-axis

second-order polarizability $\chi^{(2)}$ must be large; and (2) the phase-matching condition $\Delta k = k_1 + k_2 - k = 0$ must be satisfied. In Sect. 6.1.2, we will give a microscopic description of the second-order polarizability $\chi^{(2)}(\omega = \omega_1 + \omega_2) \equiv \chi^{(2)}(\omega; \omega_1, \omega_2)$ and discuss the first condition mentioned above. In Sect. 6.1.3, we will discuss how to satisfy the phase-matching condition.

6.1.2 Second-Order Polarizability $\chi^{(2)}$

In this subsection, we show the procedure to derive $\chi^{(2)}(\omega; \omega_1, \omega_2)$:

(1) The interaction of the material with the radiation fields $\boldsymbol{E}_1$ and $\boldsymbol{E}_2$ is expressed in terms of the electric dipole $\boldsymbol{P}$ of the material as

$$\mathcal{H}' = -\boldsymbol{P} \cdot (\boldsymbol{E}_1 + \boldsymbol{E}_2) . \tag{6.6}$$

We derive the density matrix of the material $\varrho(t)$ to second-order in $\mathcal{H}'$, and denote this as $\varrho^{(2)}(t)$.

(2) The k-component $E_k(\omega_1)$ of a vector $\boldsymbol{E}(\omega_1)$ and the l-component $E_l(\omega_2)$ of $\boldsymbol{E}(\omega_2)$ induce the second-order polarization $\boldsymbol{P}_\omega^{(2)}$, the jth component of which is denoted by $P_j^{(2)}(\omega)$, as follows:

$$\begin{aligned} P_j^{(2)}(\omega) &= \mathrm{Tr} P_j \varrho^{(2)}(\omega) \\ &\equiv \chi_{jkl}^{(2)}(\omega;\omega_1,\omega_2) E_k(\omega_1) E_l(\omega_2) \,. \end{aligned} \tag{6.7}$$

The electric polarization in the visible region comes mainly from the electronic excitations of atoms, molecules, or solids. Therefore we derive the equations of motion for $\varrho(t)$ under the Hamiltonian $\mathcal{H}_0$ of the electron system and its interaction with the radiation fields $\mathcal{H}'$ of (6.6). In addition to this, the electronic system interacts with other degrees of freedom such as the radiation field vacuum and phonon fields. These systems can be treated as reservoirs and these effects can be taken into account as relaxation constants of the electronic system. Then the density matrix of the relevant electronic system $\varrho(t)$ can be expanded in terms of the eigenstates of this electronic system as follows:

$$\varrho(t) = \sum_{nn'} \varrho_{nn'} |n\rangle\langle n'| \,. \tag{6.8}$$

The equation of motion for $\varrho(t)$, i.e.,

$$\frac{\partial \varrho}{\partial t} = \frac{1}{i\hbar}[\mathcal{H}_0 + \mathcal{H}', \varrho] + \left(\frac{\partial \varrho}{\partial t}\right)_{\mathrm{relax}} \tag{6.9}$$

is solved as a perturbational expansion in $\mathcal{H}'$. By the last term of (6.9), the interaction of the electronic system with reservoirs is described in the form of relaxation. For example, the diagonal ϱ_{nn} and the off-diagonal $\varrho_{nn'}$ components of the electronic density matrix are described in terms of the longitudinal T_1 and transverse T_2 relaxation times, respectively, as follows:

$$\left(\frac{\partial \varrho_{nn}}{\partial t}\right)_{\mathrm{relax}} = -\left(\frac{1}{T_1}\right)_{nn} \left(\varrho_{nn} - \varrho_{nn}^{(0)}\right) , \tag{6.10}$$

$$\left(\frac{\partial \varrho'_{nn}}{\partial t}\right)_{\mathrm{relax}} = -\left(\frac{1}{T_2}\right)_{nn'} \varrho_{nn'} \,. \tag{6.11}$$

Here $\varrho_{nn}^{(0)}$ means the distribution in thermal equilibrium.

The density matrix can be expanded according to the order of $\mathcal{H}'$, the electron–radiation field interaction:

$$\varrho(t) = \varrho^{(0)} + \varrho^{(1)}(t) + \varrho^{(2)}(t) + \cdots + \varrho^{(l)}(t) + \cdots \,. \tag{6.12}$$

Then the equation of motion (6.9) is rewritten for every order of $\mathcal{H}'$ as

$$\frac{\partial \varrho^{(1)}}{\partial t} = \frac{1}{i\hbar}\left\{\left[\mathcal{H}_0, \varrho^{(1)}\right] + \left[\mathcal{H}', \varrho^{(0)}\right]\right\} + \left(\frac{\partial \varrho^{(1)}}{\partial t}\right)_{\text{relax}}, \tag{6.13}$$

$$\frac{\partial \varrho^{(2)}}{\partial t} = \frac{1}{i\hbar}\left\{\left[\mathcal{H}_0, \varrho^{(2)}\right] + \left[\mathcal{H}', \varrho^{(1)}\right]\right\} + \left(\frac{\partial \varrho^{(2)}}{\partial t}\right)_{\text{relax}}. \tag{6.14}$$

For a stationary response, the external field $\boldsymbol{E}$, its interaction with the electronic system $\mathcal{H}'$, and the mth order density matrix $\varrho^{(m)}(t)$ are Fourier transformed into the following forms:

$$\boldsymbol{E} = \sum_j \boldsymbol{e}_j E(\omega_j) \exp\left[i\left(\boldsymbol{k}_j \cdot \boldsymbol{r} - \boldsymbol{\omega}_j t\right)\right], \tag{6.15}$$

$$\mathcal{H}' = \sum_j \mathcal{H}'(\omega_j) e^{-i\omega_j t}, \tag{6.16}$$

$$\varrho^{(m)}(t) = \sum_j \varrho^{(m)}(\omega_j) e^{-i\omega_j t}. \tag{6.17}$$

Then (6.13) is rewritten in the following form:

$$\begin{aligned} -i\omega_j \varrho_{nn'}^{(1)}(\omega_j) &= \frac{1}{i\hbar}\left(E_n \varrho_{nn'}^{(1)} - \varrho_{nn'}^{(1)} E_{n'}\right) \\ &\quad + \frac{1}{i\hbar}\mathcal{H}'_{nn'}(\omega_j)\left(\varrho_{n'n'}^{(0)} - \varrho_{nn}^{(0)}\right) - \Gamma_{nn'}\varrho_{nn'}^{(1)}. \end{aligned} \tag{6.18}$$

Denoting $E_n - E_{n'} \equiv \hbar\omega_{nn'}$, this equation is solved as

$$\hbar\left(\omega_j - \omega_{nn'} + i\Gamma_{nn'}\right)\varrho_{nn'}^{(1)}(\omega_j) = \mathcal{H}'_{nn'}(\omega_j)\left(\varrho_{n'n'}^{(0)} - \varrho_{nn}^{(0)}\right). \tag{6.19}$$

Note that $\varrho_{gg}^{(0)} \equiv \varrho_g^{(0)} = 1$ and other terms vanish when considering the electronic excitation at room temperature.

First let us derive the linear polarizability of an N-electron system. The electronic polarization of this system is

$$P = -\sum_{m=1}^{N} e\boldsymbol{r}_m, \tag{6.20}$$

and the expectation value of the j-component of the induced polarization under the external field $E_k(\omega)$ polarized in the k-direction is obtained by using the solution of (6.19) as

$$\begin{aligned} P_j^{(1)}(\omega) &\equiv \chi_{jk}^{(1)}(\omega) E_k(\omega) = \mathrm{Tr} P_j \varrho^{(1)}(\omega) \\ &= \sum_n \left[\langle g|P_j|n\rangle \varrho_{ng}^{(1)}(\omega) + \langle n|P_j|g\rangle \varrho_{gn}^{(1)}(\omega)\right] \\ &= \frac{-Ne^2}{\hbar}\sum_n \left[\frac{(r_j)_{gn}(r_k)_{ng}}{\omega - \omega_{ng} + i\Gamma_{ng}} - \frac{(r_k)_{gn}(r_j)_{ng}}{\omega + \omega_{ng} + i\Gamma_{ng}}\right]\varrho_g^{(0)} E_k(\omega). \end{aligned} \tag{6.21}$$

The second-order density matrix $\varrho^{(2)}(\omega_1+\omega_2)$ is derived from (6.14) as

$$
\begin{aligned}
-i\left(\omega_1+\omega_2\right)\varrho_{nn'}^{(2)}\left(\omega_1+\omega_2\right) &= \frac{1}{i\hbar}\left(E_n\varrho_{nn'}^{(2)}-\varrho_{nn'}^{(2)}E_{n'}\right)-\Gamma_{nn'}\varrho_{nn'}^{(2)} \\
&+\frac{1}{i\hbar}\sum_{n''}\Big[\mathcal{H}'_{nn''}\left(\omega_1\right)\varrho_{n''n'}^{(1)}\left(\omega_2\right)-\varrho_{nn''}^{(1)}\left(\omega_2\right)\mathcal{H}_{n''n'}\left(\omega_1\right) \\
&+\mathcal{H}'_{nn''}\left(\omega_2\right)\varrho_{n''n'}^{(1)}\left(\omega_1\right)-\varrho_{nn''}^{(1)}\left(\omega_1\right)\mathcal{H}_{n''n'}\left(\omega_2\right)\Big]\,.
\end{aligned}
\tag{6.22}
$$

Consequently, the expectation value of the second-order polarization $P_j^{(2)}(\omega=\omega_1+\omega_2)$ is derived from (6.22) as

$$
\begin{aligned}
&P_j^{(2)}\left(\omega=\omega_1+\omega_2\right) \\
&\equiv \chi_{jkl}^{(2)}\left(\omega;\omega_1,\omega_2\right)E_k\left(\omega_1\right)E_l\left(\omega_2\right) \\
&= -Ne\left[\sum_{n\neq g}\left\{\left(r_j\right)_{gn}\varrho_{ng}^{(2)}\left(\omega_1+\omega_2\right)+\left(r_j\right)_{ng}\varrho_{gn}^{(2)}\left(\omega_1+\omega_2\right)\right\}\right. \\
&\quad\left.+\sum_{n,n'\neq g}\left(r_j\right)_{n'n}\varrho_{nn'}^{(2)}\left(\omega_1+\omega_2\right)\right] \\
&= -N\frac{e^3}{\hbar^2}\sum_{n,n'\neq g}\left[\frac{\left(r_j\right)_{gn}\left(r_k\right)_{nn'}\left(r_l\right)_{n'g}}{\left(\omega-\omega_{ng}+i\Gamma_{ng}\right)\left(\omega_2-\omega_{n'g}+i\Gamma_{n'g}\right)}\right. \\
&\quad+\frac{\left(r_j\right)_{ng}\left(r_l\right)_{gn'}\left(r_k\right)_{n'n}}{\left(\omega+\omega_{ng}+i\Gamma_{ng}\right)\left(\omega_2+\omega_{n'g}+i\Gamma_{n'g}\right)} \\
&\quad\left.-\frac{\left(r_j\right)_{n'n}\left(r_k\right)_{ng}\left(r_l\right)_{gn'}}{\left(\omega-\omega_{nn'}+i\Gamma_{nn'}\right)}\left(\frac{1}{\omega_2+\omega_{n'g}+i\Gamma_{n'g}}+\frac{1}{\omega_1-\omega_{ng}+i\Gamma_{ng}}\right)\right] \\
&\quad\times E_k\left(\omega_1\right)E_l\left(\omega_2\right)+(k,1)\rightleftharpoons(l,2)\,.
\end{aligned}
\tag{6.23}
$$

Here $(k,1)\rightleftharpoons(l,2)$ means the contribution from the process in which the roles of the first $\hbar\omega_1$ and the second $\hbar\omega_2$ photons, are interchanged. As a consequence, $P_j^{(2)}(\omega=\omega_1+\omega_2)$ consists of eight terms and Fig. 6.2 describes these Feynman diagrams.

These diagrams present the time-development of the electronic states upward. The left (ket) and right (bra) states both start from the ground state $|g\rangle\langle g|$. The double Feynman diagram of Fig. 6.2(a) describes the process in which the ω_2 photon is absorbed first and the electronic state makes a transition into an excited state $|n'\rangle$ on the left-hand side, and then the ω_1 photon is successively absorbed also on the left-hand side, inducing the transition to another excited state $|n\rangle$, and finally the electronic polarization with angular frequency $\omega=\omega_1+\omega_2$ is generated. This process corresponds to the first term of (6.23). When we interchange the temporal order of the ω_1 and ω_2 photon absorptions in the diagram Fig. 6.2(a), we have the contribution of

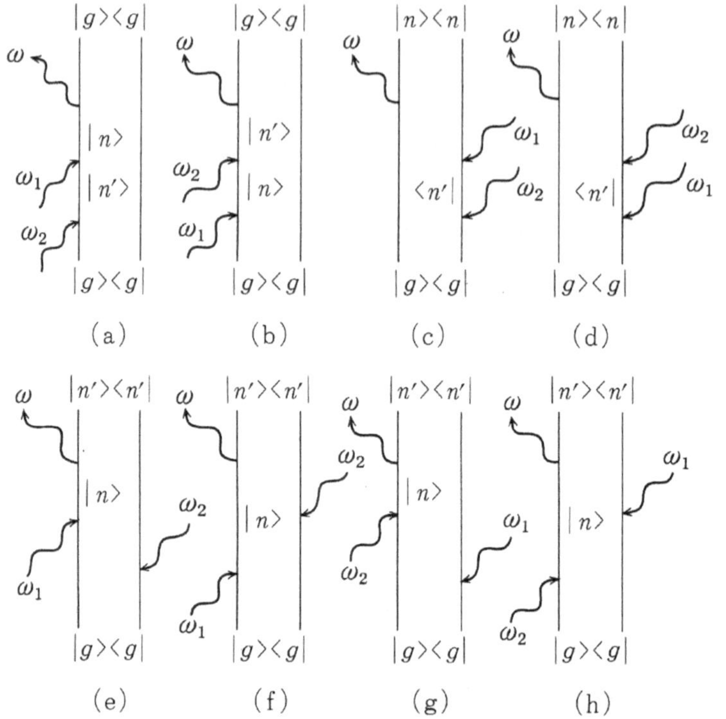

Fig. 6.2. Double Feynman diagrams which contribute to sum-frequency ($\omega = \omega_1 + \omega_2$) generation. The two vertical lines on the left- and right-hand sides describe the time development of the electronic states in the density operator, respectively, upward and downward. These eight diagrams mean that the nonlinear material absorbs ω_1 and ω_2 photons and subsequently induces the electronic polarization with angular frequency ω. Only diagrams (**a**) and (**b**) consist of resonant terms while the others contain antiresonant terms

Fig. 6.2(b). The second term of (6.23) is diagramatically drawn in Fig. 6.2(c). The vertical line on the right-hand side describes the time development of the right-hand (bra) state upward. In Fig. 6.2(c), the ω_2 photon is first absorbed, inducing the transition from the ground state $\langle g\,|$ into an excited state $\langle n'\,|$, then the ω_1 photon is absorbed, making the transition from $\langle n'\,|$ into another excited state $\langle n\,|$. The state on the right-hand vertical line develops upward obeying the time-reversed Schrödinger equation so that we may understand the state on the right-hand vertical line to propagate downward according to the conventional Schrödinger equation. Therefore, in Fig. 6.2(c), starting from the excited state $\langle n\,|$, the ω_1 photon is absorbed first, inducing the transition into $|\,n'\rangle$ and then the ω_2 photon is absorbed, making the transition into the ground state $|\,g\rangle$. The diagrams in Fig. 6.2, i.e., photon absorption and emission, were drawn according to the latter description. The diagram of

Fig. 6.2(d) describes the process in which the temporal order of the ω_1 and ω_2 photon processes was interchanged. The diagrams of Figs. 6.2(e) and (f) correspond to the third and fourth terms of (6.23), respectively, while those of Figs. 6.2(g) and (h) correspond to the processes in which the ω_1 and ω_2 photon are interchanged. The second-order polarizability $\chi^{(2)}(\omega;\omega_1,\omega_2)$ is defined as the coefficient of $E_k(\omega_1)E_l(\omega_2)$ as in the first line of (6.23). Note that only the first term of (6.23) and that in which the roles of the ω_1 and ω_2 photons are interchanged have the resonant enhancement effect and others have no such effect.

6.1.3 Conditions to Generate Second Harmonics

The first condition to observe SHG is the absence of inversion symmetry for the crystal. The expression for $\chi^{(2)}$ has the product $(P_j)_{gn}(P_k)_{nn'}(P_l)_{n'g}$ in the numerator. For example, $(P_j)_{gn}$ denotes the expectation value of the jth component of the transition dipole moment $\boldsymbol{P}$ between the ground state g and the excited state n. Under the operation of spatial inversion, the transition dipole moment $\boldsymbol{P}$ changes its sign so that the product of three transition moments changes only its sign. When the crystal has an inversion symmetry, $\chi^{(2)}$ itself should be constant under this operation. This means that $\chi^{(2)}$ should vanish for a crystal with inversion symmetry.

Both types of polydiacetylenes (a) and (b) shown in Fig. 6.3 have an inversion symmetry as long as the side-chains R and R$'$ on the left- and right-hand sides are the same as each other. Therefore these crystals do not show SHG. In order to obtain SHG, the microscopic unit of the crystals must have structures with broken symmetry. The polydiacetylene crystal with different kinds of side-chains R and R$'$ begin to generate strong second harmonics. The benzene molecule, which has an inversion symmetry, cannot show SHG, but this molecule can generate SH when one of the hydrogens is replaced by a donor substituent such as OH, NH_2, or $(CH_3)_2N$, or an acceptor substituent such as NO_2 or CN. According to the degree of symmetry breaking, the stronger donor or acceptor can produce the stronger SHG. But, we must arrange these asymmetric molecules so as to lose the inversion symmetry of the crystal to get SHG.

The second condition for obtaining an SHG signal is phase-matching. The condition of maximizing the third factor $\{2\sin(\Delta kl/2)/\Delta k\}^2$ in (6.5), i.e., $\Delta k \equiv k_1 + k_2 - k = 0$, is called a phase-matching condition. That is,

$$c\Delta k \equiv c\,(k_1 + k_2 - k) = \omega_1\,\{n\,(\omega_1) - n\,(\omega)\} + \omega_2\,\{n\,(\omega_2) - n\,(\omega)\} = 0\,. \tag{6.24}$$

Here $n(\omega) \equiv \sqrt{\epsilon(\omega)}$ is the refractive index at angular frequency ω. Let us consider, first, an isotropic material or cubic crystal. As $\omega = \omega_1 + \omega_2 > \omega_1, \omega_2$, the phase-matching condition (6.24) is not always satisfied both for normal dispersion as $n(\omega) > n(\omega_1), n(\omega_2)$ and for anomalous dispersion $n(\omega) <$

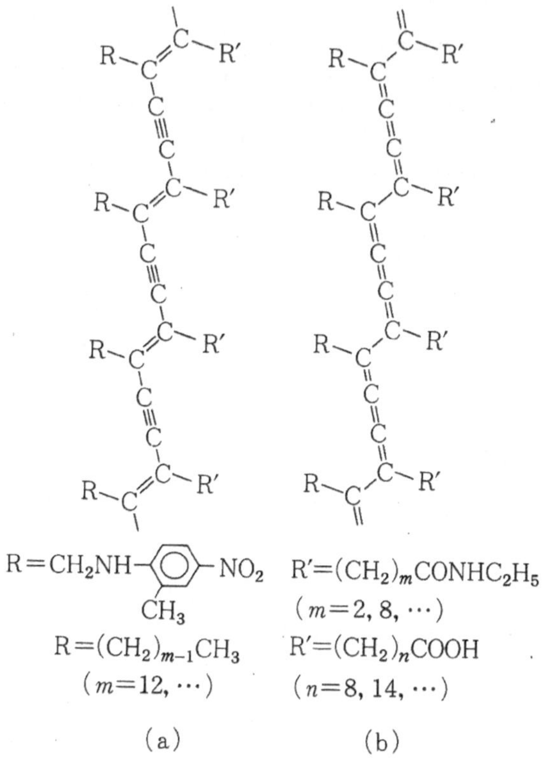

Fig. 6.3. Two chemical structures of polydiacetylene: (**a**) acetylene type and (**b**) butatolyene type. When side-chains R and R′ are different from each other, the polydiacetylene loses the inversion symmetry so that second-harmonic generation becomes possible

$n(\omega_1), n(\omega_2)$. On the other hand, the phase-matching condition can be satisfied by making use of the birefringence of crystals with lower symmetry. For example, let us consider a uniaxial crystal in which the refractive index for light polarized perpendicular to the optical axis is denoted by n_o and that parallel to the optical axis is n_e. When the wavevector of light makes an angle θ with the optical axis, the normal light, the polarization of which is perpendicular to the optical axis, has refractive index n_o, while the refractive index $n_e(\theta)$ of the extraordinary light perpendicular to the normal light is given by

$$\frac{1}{n_e(\theta)^2} = \frac{1}{n_o^2}\cos^2\theta + \frac{1}{n_e^2}\sin^2\theta\,. \tag{6.25}$$

For the case $n_e > n_o$, the ordinary second harmonic $2\omega_1$ can be induced by two fundamentals ω_1 with extraordinary polarization. That is, the phase-matching condition (6.24) is satisfied for the incident angle θ such that

$$n_o(2\omega_1) = n_e(\omega_1, \theta)\,. \tag{6.26}$$

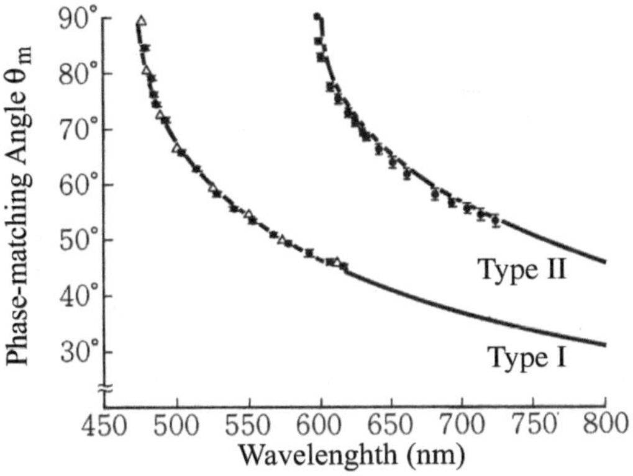

Fig. 6.4. Dependence of the phase-matching angle θ_m on the wavelength of an incident beam for second-harmonic generation from urea crystals. Solid lines describe the calculation and •, △ describe the observations [61]

This phase-matching angle θ_m, which satisfies (6.26), is obtained in terms of the expression (6.25) as

$$\sin^2\theta_m = \frac{n_o\left(2\omega_1\right)^{-2} - n_o\left(\omega_1\right)^{-2}}{n_e\left(\omega_1\right)^{-2} - n_o\left(\omega_1\right)^{-2}} \,. \tag{6.27}$$

This phase-matching condition of SHG is called type I. For type II phase-matching, we will be able to get the ordinary second harmonics $2\omega_1$ from one ordinary and one extraordinary fundamental. This phase-matching angle θ_m is obtained by solving $n_o(\omega_1) + n_e(\omega_1, \theta_m) = 2n_o(2\omega_1)$. The phase-matching angles θ_m for SHG and for the sum-frequency generation from a urea crystal $CO(NH_2)_2$ are drawn, respectively, in Figs. 6.4 and 6.5 [61].

The third condition for obtaining the effective SHG and the sum-frequency signals is that the crystal should be transparent to the fundamentals and the sum-frequency (the second harmonics). The amplitude of SHG decays as $\exp\{-(\alpha_\omega + \frac{1}{2}\alpha_{2\omega})l\}$ when we take into account the effects of the absorption coefficients α_ω and $\alpha_{2\omega}$ at the fundamental and the second harmonics, respectively. As the absorption edge of the urea crystal is at $\lambda = 210\,\text{nm}$, sum-frequency generation was successfully confirmed until the wavelength was as short as $\lambda = 228.8\,\text{nm}$, as shown in Fig. 6.5. Here two fundamentals consist of ordinary light at $\lambda = 1.06\,\mu\text{m}$ from a YAG laser and extraordinary light at $\lambda = 291.6\,\text{nm}$ from second harmonics of a rhodamine 6G dye laser, and the ordinary sum-frequency at $\lambda = 228.8\,\text{nm}$ was obtained under the normal incidence condition $\theta_m = 90°$. As high as 50% conversion efficiency was realized by using the urea crystal with thickness 15 mm.

A nonlinear optical crystal must be robust against strong incident laser power in order to obtain high-power second harmonics. This is the fourth condition of a crystal for SHG. The threshold power for a 10 ns incident pulse

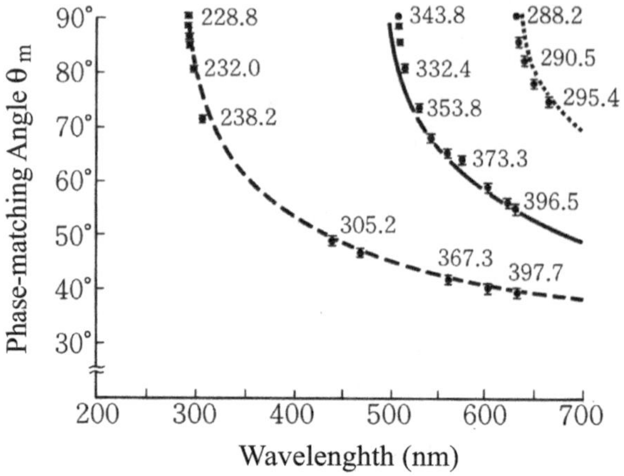

Fig. 6.5. Phase-matching angle θ_m of sum-frequency generation from a urea crystal ($n_e > n_o$), as a function of wavelength of one incident beam (extraordinary light). The broken line on the left-hand side and the solid line describe the cases in which the other incident beams are, respectively, 1.06 μm ordinary and extraordinary light. The dotted line on the right-hand side denotes the other incident beam, which is 532 nm extraordinary light. Numerical values in the figure mean the wavelength in nm of the sum-frequency [61]

at wavelength $\lambda = 1.064\,\mu\text{m}$ was observed $1.5\,\text{GW/cm}^2$ for the urea crystal, $0.2\,\text{GW/cm}^2$ for the KDP crystal and $0.03\,\text{GW/cm}^2$ for the $LiNbO_3$ crystal.

Finally a large, uniform, and good crystal is required to be grown cheaply and a mechanically and chemically stronger crystal is preferable. Recently inorganic crystals of high quality such as BBO (beta-barium-borate, β-BaB_2O_4), and KTP (potassium titanyl phosphate, $KTiOPO_4$) have been available and are attracting attention as nonlinear optical materials which are superior to the urea crystal.

Quasi-phase matching is becoming more important from an engineering point of view. It is a technique for phase-matching nonlinear optical interactions in which the relative phase is corrected at regular intervals using the structural periodicity built into the nonlinear medium [62]. Engineered nonlinear materials were introduced with the successful implementation of quasi-phase matching by periodic inversion of ferroelectric domains in lithium niobate. Recently lithographic processing techniques enabled the fabrication of quasi-phase matched nonlinear chips using electric field poling of lithium niobate on the wafer scale [63]. As a result, it has made it possible to have nonlinear optical devices having a conversion efficiently close to unity.

6.1.4 Optical Parametric Amplification and Oscillation

A parametric phenomenon is a reverse process to second-harmonic ($\omega_1 + \omega_1 \rightarrow 2\omega_1$) and sum-frequency generation ($\omega_1 + \omega_2 \rightarrow \omega_3$). Under pumping at $\omega_3 =$

$\omega_1+\omega_2$, the radiation field with angular frequency ω_1 is amplified. This process is called parametric amplification. Under parametric oscillation, the signal $(\omega_1, \boldsymbol{k}_1)$ and the idler $(\omega_2 = \omega_3 - \omega_1,\ \boldsymbol{k}_2 = \boldsymbol{k}_3 - \boldsymbol{k}_1)$ spontaneously oscillate under pumping at ω_3 and $\boldsymbol{k}_3$ without another supply of incident field. This parametric oscillation is used for optical squeezing, as discussed already in Chap. 1.

Let us describe three kinds of radiation fields which are involved in the parametric phenomena as $\boldsymbol{E}(\omega_j) = E_j(z)\boldsymbol{e}_j \exp[i(\boldsymbol{k}_j\cdot\boldsymbol{r}-\omega_1 t+\phi_j)]$ $(j = 1, 2, 3)$. The envelope functions $E_j(z)$ obey the following coupled equations under the slowly varying envelope approximation:

$$\begin{aligned}\frac{\partial}{\partial z}E_1 &= \frac{i\omega_1^2}{k_{1z}}K^* E_2^* E_3 e^{i\Delta kz+i\theta_0}\,,\\ \frac{\partial}{\partial z}E_2^* &= \frac{-i\omega_2^2}{k_{2z}}K E_1 E_3^* e^{-i\Delta kz-i\theta_0}\,,\\ \frac{\partial}{\partial z}E_3 &= \frac{i\omega_3^2}{k_{3z}}K E_1 E_2 e^{-i\Delta kz-i\theta_0}\,.\end{aligned} \tag{6.28}$$

Here the coefficient of parametric amplification K is expressed in terms of the second-order polarizability $\chi^{(2)}$ as

$$\begin{aligned}K &= \frac{1}{2\epsilon_0 c^2}\boldsymbol{e}_3\cdot\chi^{(2)}\left(\omega_3;\omega_1,\omega_2\right):\boldsymbol{e}_1\boldsymbol{e}_2\,,\\ \Delta k &= k_{3z}-k_{1z}-k_{2z}\,,\\ \theta_0 &= \phi_3-\phi_1-\phi_2\,.\end{aligned} \tag{6.29}$$

The solution of (6.28) can be obtained in a way similar to (6.4), and describes how the pump field ω_3 at $z = 0$ is divided into the signal ω_1 and idler ω_2 fields as it propagates in the z-direction. Taking into account the conservation laws of energy and wavevector:

$$\omega_3 = \omega_1+\omega_2\,, \quad \boldsymbol{k}_3 = \boldsymbol{k}_1+\boldsymbol{k}_2 \tag{6.30}$$

the phase-matching condition is written as

$$\omega_3\left[n_3\left(\omega_3\right)-n_2\left(\omega_3-\omega_1\right)\right] = \omega_1\left[n_1(\omega_1)-n_2\left(\omega_3-\omega_1\right)\right]\,. \tag{6.31}$$

The oscillating frequency ω_1 will be fixed by solving (6.31) once we know $n_j(\omega_j)(j = 1, 2, 3)$. For the case $n_e < n_o$ of a uniaxial crystal, we have the following two possibilities for the frequency region of normal dispersion:

$$\begin{aligned}\text{Type I}:\quad & \omega_3 n_3^e\left(\omega_3,\theta\right) = \omega_1 n_1^o\left(\omega_1\right)+\omega_2 n_2^o\left(\omega_2\right)\,,\\ \text{Type II}:\quad & \omega_3 n_3^e\left(\omega_3,\theta\right) = \omega_1 n_1^o\left(\omega_1\right)+\omega_2 n_2^e\left(\omega_2,\theta\right)\,,\\ & \left(\text{or} = \omega_1 n_1^e\left(\omega_1,\theta\right)+\omega_2 n_2^o\left(\omega_2\right)\right)\,.\end{aligned} \tag{6.32}$$

Here θ is the angle which the relevant wavevector makes against the optical axis in the uniaxial crystal and the θ-dependence of $n^e(\omega, \theta)$ is given by (6.25)

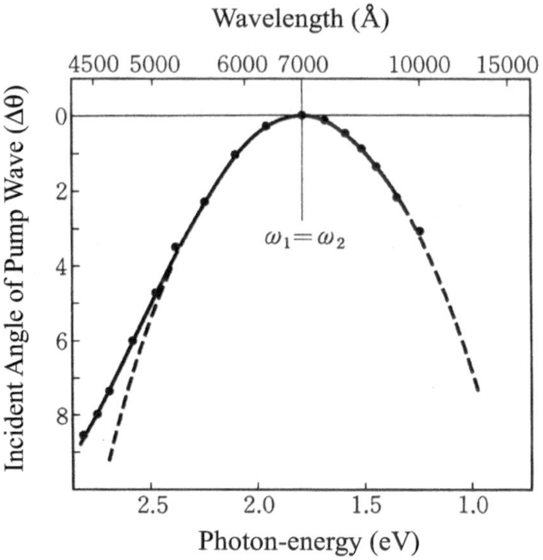

Fig. 6.6. Two wavelengths obtained in the parametric oscillation are demonstrated as a function of the incident angle ($\Delta\theta$) of the pump beam ($\lambda_p = 0.347\,\mu\mathrm{m}$) measured from the optical axis of an ADP crystal [64]

in terms of the refractive indices $n^e(\omega)$ and $n^o(\omega)$ of the extraordinary and ordinary light, respectively. The angular frequency ω_1 of the parametric oscillation or the maximum gain is determined by adjusting the incident angle θ from the optical axis or changing the crystal temperature. The oscillating frequency of the signal ω_1 or the idler ω_2 is drawn as a function of the incident angle $\Delta\theta$ in Fig. 6.6 [64]. The angle $\Delta\theta$, i.e., the coordinate in Fig. 6.6, is a deviation of incident angle from the optimum angle at which the signal frequency ω_1 becomes coincident with the idler frequency ω_2. It is shown in Fig. 6.6 that the wavelength of the signal can be changed from 4400 Å to 1 µm by changing the incident angle by $\Delta\theta = 0 \sim 8°$. We can also change the signal frequency as a function of crystal temperature for the case of $LiNbO_3$ as shown in Fig. 6.7 [65]. This figure also shows how much we can extend the region of signal frequency by varying the pump wavelength.

6.2 Third-Order Optical Response

A large number of nonlinear optical phenomena belong to the group of third-order optical processes. Three incident radiation fields with angular frequencies ω_1, ω_2, and ω_3 can produce the sum-frequency and difference-frequency as the fourth radiation field, which is called four-wave mixing. Third-harmonic generation with 3ω is also one of four-wave mixing and is possible even in

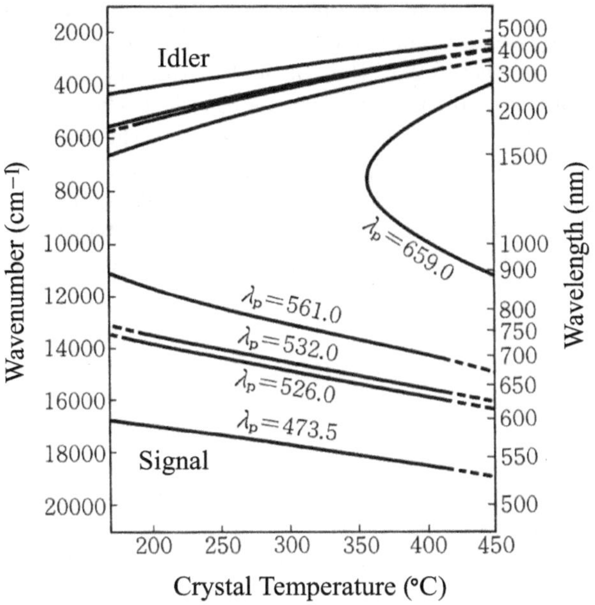

Fig. 6.7. The phase-matching condition of parametric oscillation is controlled by changing the crystal temperature of $LiNbO_3$. The oscillating wavelength or wavenumber is shown as a function of the crystal temperature with the pump wavelength λ_p as a parameter [65]

crystals with an inversion symmetry. This is in contrast to second-harmonic generation.

When frequency-variable lasers such as dye lasers are used as sources of incident light, one of the incident light beams and the sum- or difference-frequencies are chosen, possibly resonant to the elementary excitations. Then the nonlinear signal is not only resonantly enhanced but can also interfere with nonresonant terms. As a result, from this nonlinear spectroscopy, we will be able to study the elementary excitation and determine the magnitude and sign of the relevant transition dipole moments. As an example, we will introduce, in Sect. 6.2.1, CARS (coherent anti-Stokes Raman scattering) by which, for example, the frequency of the lattice vibration will be determined sensitively. Here we make the difference-frequency $\omega_1 - \omega_2$ of two incident beams ω_1 and ω_2 to be resonant to the lattice vibration. In Sect. 6.2.2, we introduce phase-conjugated wave generation, a degenerate four-wave mixing in which three incident frequencies are degenerate and resonant to an exciton. Here two colliding beams propagate in the nonlinear medium and the third beam is supplied to this medium. Then the time-reversal wave (the phase-conjugated wave) of the third beam is produced. The optical Kerr effect and absorption saturation come also from the process of four-wave mixing. In terms of this nonlinearity, optical bistability can be achieved in which the transmitted light intensity shows hysteresis against the change of incident light intensity. This will be discussed in Sect. 6.2.3. This optical bistability may be considered to be applied to optical information processing. Here the most desirable nonlinear materials must satisfy two requirements at the same time, i.e., large $\chi^{(3)}(\omega;\omega,-\omega,\omega)$ and rapid response time. We will describe the possibility of satisfying these requirements by using resonant pumping of the exciton in Sect. 6.3. In Sect. 6.4, we will discuss two-photon absorption spectroscopy, by which fruitful information about electronic structures supplementary to one-photon absorption is available.

6.2.1 Four-Wave Mixing – CARS

When we irradiate a crystal by two incident beams $(\omega_1, \boldsymbol{k}_1)$ and $(\omega_2, \boldsymbol{k}_2)$, the third-order polarization $P^{(3)}(2\omega_1 - \omega_2)$ can be induced. With this nonlinear polarization as a source, we can observe the signal with an angular frequency $2\omega_1 - \omega_2$ in the direction $2\boldsymbol{k}_1 - \boldsymbol{k}_2$. When we observe the signal at $2\omega_1 - \omega_2$ as a function of $\omega_1 - \omega_2$ by changing one of ω_1 or ω_2, or both of them, the signal reaches a maximum at the frequencies where $\omega_1 - \omega_2$ is resonant to the elementary excitation. This is called CARS and is schematically shown in Fig. 6.8. This CARS is a useful tool to observe rotations and vibrations in molecular gases and liquids, in addition to elementary excitations in solids. While the Raman signal is produced by spontaneous emission in conventional Raman scattering, the CARS signal is created coherently by induced emission at a difference-frequency $2\omega_1 - \omega_2$ of two coherent incident beams.

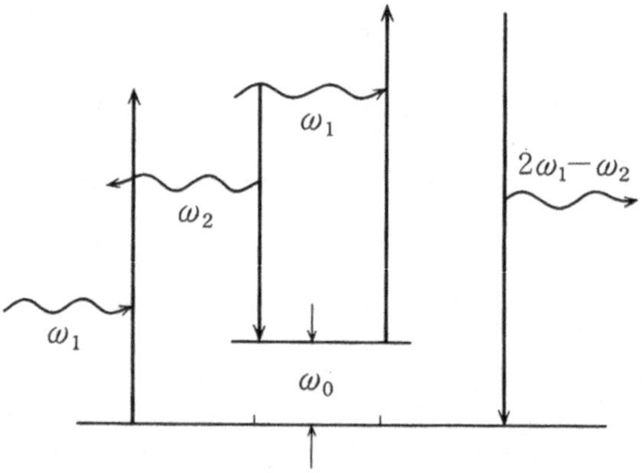

Fig. 6.8. Concept of CARS (Coherent Anti-Stokes Raman Scattering). When the difference-frequency $\omega_1 - \omega_2$ of two incident beams ω_1 and ω_2 becomes equal to the angular frequency of any elementary excitation ω_0 in solids, the CARS signal at $2\omega_1 - \omega_2$ shows strong enhancement

We decompose the third-order polarizability of the CARS process $\chi^{(3)}$ $(2\omega_1-\omega_2;\omega_1,-\omega_2,\omega_1)$ into the resonant part $\chi_R^{(3)}$ and the nonresonant part $\chi_{NR}^{(3)}$. The third-order electric polarization $P^{(3)}(\omega_s)$ with $\omega_s=2\omega_1-\omega_2$ and wavevector $2\boldsymbol{k}_1-\boldsymbol{k}_2$ under two incident beams $\boldsymbol{E}_1\exp[i(\boldsymbol{k}_1\cdot\boldsymbol{r}-\omega_1 t)]$ and $\boldsymbol{E}_2\exp[i(\boldsymbol{k}_2\cdot\boldsymbol{r}-\omega_2 t)]$, is given by

$$P^{(3)}(\omega_s)=\chi^{(3)}(\omega_s;\omega_1,-\omega_2,\omega_1):\boldsymbol{E}_1\boldsymbol{E}_1\boldsymbol{E}_2^*\,. \tag{6.33}$$

Inserting this nonlinear polarization into the right-hand side of (6.2), the signal $\boldsymbol{E}_s(\omega_s)\exp[i(\boldsymbol{k}_s\cdot\boldsymbol{r}-\omega_s t)]$ is obtained in the same way as in Sect. 6.1.1:

$$\begin{aligned}P^{(3)}(\omega_1,\omega_2)&=\frac{c\epsilon_0\sqrt{\epsilon(\omega_s)}}{2}\left|\boldsymbol{E}_s(\omega_s)\right|^2\\&=\frac{\omega_s^2}{8c\epsilon_0\sqrt{\epsilon(\omega_s)}}\left|\chi^{(3)}(\omega_s)\right|^2|E_1|^4|E_2|^2\frac{\sin^2(\Delta kl/2)}{(\Delta k/2)^2}\,.\end{aligned} \tag{6.34}$$

Here, $\Delta kl\equiv(2\boldsymbol{k}_1-\boldsymbol{k}_2-\boldsymbol{k}_s)\cdot\boldsymbol{l}$, and $\boldsymbol{l}$ is a path vector of the signal light within the medium. Equation (6.34) means that the CARS signal is proportional to $|\chi^{(3)}(\omega_s)|^2$ in the direction $\boldsymbol{k}_s=2\boldsymbol{k}_1-\boldsymbol{k}_2$. The nonresonant term $\chi_{NR}^{(3)}$ may in general be considered to be a constant and the resonant term $\chi_R^{(3)}$ is written as

$$\chi_R^{(3)}=\frac{a}{\omega_1-\omega_2-\omega_0+i\Gamma}\,. \tag{6.35}$$

Here ω_0 is the angular frequency of the relevant elementary excitation and Γ its relaxation constant. The spectrum of the CARS signal is proportional to

$$\begin{aligned}&\left|\chi^{(3)}(2\omega_1-\omega_2;\omega_1,-\omega_2,\omega_1)\right|^2\\&=\left\{\chi_{\mathrm{NR}}^{(3)}+\frac{a(\omega_1-\omega_2-\omega_0)}{(\omega_1-\omega_2-\omega_0)^2+\Gamma^2}\right\}^2+\frac{a^2\Gamma^2}{\{(\omega_1-\omega_2-\omega_0)^2+\Gamma^2\}^2}\,.\end{aligned} \tag{6.36}$$

When $a/\chi_{\mathrm{NR}}^{(3)}<0$ and $|a/\chi_{\mathrm{NR}}^{(3)}|\gg 2\Gamma$, $|\chi^{(3)}|^2$ has the peak value $(a/\Gamma)^2(\gg|\chi_{\mathrm{NR}}^{(3)}|^2)$ at $\omega_1-\omega_2=\omega_0$, and shows the dip value $(\chi_{\mathrm{NR}}^{(3)})^4(\Gamma/a)^2$ at $\omega_1-\omega_2=\omega_0-a/\chi_{\mathrm{NR}}^{(3)}$ as Fig. 6.9 shows. For the case of $a/\chi_{\mathrm{NR}}^{(3)}>0$, the relative positions of the signal peak and dip are reversed.

Levenson [66] irradiated a calcite crystal $CaCO_3$ by two dye laser beams ω_1 and ω_2 and observed the CARS signal with frequency $\omega_s=2\omega_1-\omega_2$ as a function of detuning $\omega_1-\omega_2$ as shown in Fig. 6.9. Comparing this figure with (6.36), we obtain $\omega_0=1088\,\mathrm{cm}^{-1}$, $a=-(8.5\pm1)\times10^{-2}\,\mathrm{cm}^3/\mathrm{erg}\cdot\mathrm{s}$, $\chi_{\mathrm{NR}}^{(3)}=(1.4\pm0.2)\times10^{-14}\,\mathrm{cm}^3/\mathrm{erg}$. When he used the calcite sample pasted with sapphire of thickness 0.25 mm , the dip frequency shift was observed as Fig. 6.9 shows. From this measurement $\chi_{\mathrm{NR}}^{(3)}=(1.14\pm0.15)\times10^{-14}\,\mathrm{cm}^3/\mathrm{erg}$ is obtained for the sapphire crystal.

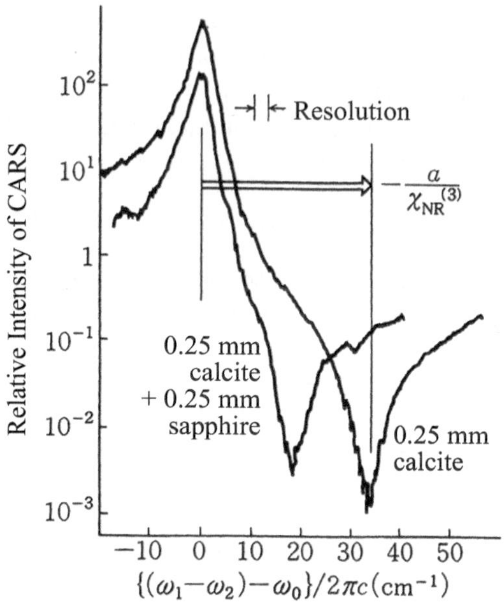

Fig. 6.9. CARS spectrum due to the 1088 cm^{-1} vibrational mode of a calcite crystal. (**a**) CARS signal from only 0.25 mm calcite crystal, and (**b**) that from a hybrid system of 0.25 mm calcite and 0.25 mm sapphire crystals [66]

6.2.2 Phase-Conjugated Waves

A great variety of third-order optical phenomena are available under three degenerate or nearly degenerate incident fields. First of all, we discuss the mechanism of phase-conjugated wave generation, i.e., the generation of a time-reversed wave.

As Fig. 6.10(a) shows, the nonlinear medium is irradiated by two colliding pump beams (ω_1, $\boldsymbol{k}_\mathrm{f} = \boldsymbol{k}_0$) and ($\omega_1$, $\boldsymbol{k}_\mathrm{b} = -\boldsymbol{k}_0$) and the third pump beam (ω_2, $\boldsymbol{k}_\mathrm{p}$) overlaps with the two colliding beams within the nonlinear medium. The density matrices of the electronic ground and excited states start to oscillate with wavevector $\boldsymbol{k}_\mathrm{f} - \boldsymbol{k}_\mathrm{p}$ and the angular frequency $\omega_1 - \omega_2$ as Fig. 6.10(b) shows.

This is called the population grating. The third pump beam (ω_1, $\boldsymbol{k}_\mathrm{b} = -\boldsymbol{k}_0$) is diffracted by the population grating and is scattered into the wavevector state $-\boldsymbol{k}_\mathrm{p}$ with angular frequency $2\omega_1 - \omega_2$. This wave is called the phase-conjugated wave and is described by

$$\chi^{(3)}\left(2\omega_1 - \omega_2; \omega_1, -\omega_2, \omega_1\right) \boldsymbol{E}_\mathrm{f} \boldsymbol{E}_\mathrm{b} \boldsymbol{E}_\mathrm{p}^* \exp\left[i\left\{-\boldsymbol{k}_\mathrm{p} \cdot \boldsymbol{r} - \left(2\omega_1 - \omega_2\right) t\right\}\right] . \tag{6.37}$$

This is just the phase-conjugated wave $[\boldsymbol{E}_\mathrm{p} \exp(i\boldsymbol{k}_\mathrm{p} \cdot \boldsymbol{r})]^* e^{-i\omega_1 t}$ when $\omega_1 = \omega_2$, i.e., this describes the time-reversal propagation of the probe beam

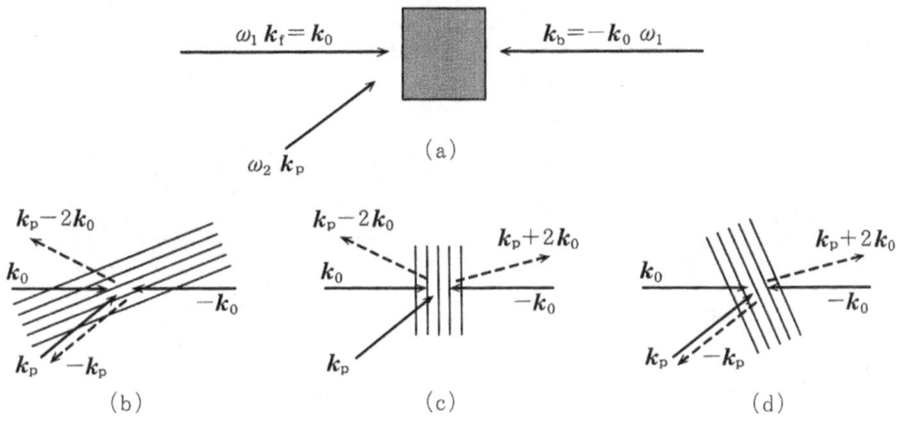

Fig. 6.10. (**a**) Three incident beams are irradiated in the nonlinear optical material to obtain the phase-conjugated wave. (**b**) The first example of phase-conjugated wave generation. The pump beam (ω_1, $\boldsymbol{k}_f = -\boldsymbol{k}_0$) and the probe beam ($\omega_2$, $\boldsymbol{k}_p$) make the population grating of the excitation, and the other pump beam (ω_1, $\boldsymbol{k}_b = -\boldsymbol{k}_0$) is diffracted by the population grating. The diffracted wave is the phase-conjugated wave ($2\omega_1 - \omega_2$, $-\boldsymbol{k}_p$). (**c**) Two pump beams (ω_1, $\boldsymbol{k}_f = \boldsymbol{k}_0$) and ($\omega_1$, $\boldsymbol{k}_b = -\boldsymbol{k}_0$) make the population grating, and the probe beam is diffracted into $\boldsymbol{k}_p \pm 2\boldsymbol{k}_0$. This is conventional four-wave mixing. (**d**) The pump beam (ω_1, $\boldsymbol{k}_b = -\boldsymbol{k}_0$) and the probe beam make the population grating and the other pump beam (ω_1, $\boldsymbol{k}_p$) is diffracted into $-\boldsymbol{k}_p$ and $\boldsymbol{k}_p + 2\boldsymbol{k}_0$. The former is the phase-conjugated wave and the latter is the conventional four-wave mixing

$\boldsymbol{E}_p \exp[(i\boldsymbol{k}_p \cdot \boldsymbol{r} - \omega_2 t)]$. In Fig. 6.10(c), excitations with $2\boldsymbol{k}_0$ and $-2\boldsymbol{k}_0$ are created by the two colliding beams and the probe beam is scattered into $\boldsymbol{k}_p \pm 2\boldsymbol{k}_0$. This is a conventional four-wave mixing process. Figure 6.10(d) shows that the incident beam (ω_1, $\boldsymbol{k}_0$) is scattered by the population grating made by (ω_1, $-\boldsymbol{k}_0$) and (ω_2, $\boldsymbol{k}_p$).

The fact that the phase-conjugated wave is a time-reversal propagation of the probe light was demonstrated by using semiconductor microcrystallites embedded in glass [67]. The laser light has good directionality, i.e., a well-defined wavevector so that the incident light is observed as a sharp spot as shown in Fig. 6.11(a). When this laser light passes through a frosted glass plate, the laser light suffers from aberration as shown in Fig. 6.11(b). If this transmitted light is reflected by a normal mirror and is sent back through the glass, the aberration grows further. However, in the case of a conjugated mirror instead of a conventional mirror, the aberration induced on the right-going path is completely eliminated on the backward propagation as demonstrated in Fig. 6.11(c). These processes are schematically summarized in Fig. 6.12. Here, these semiconductor microcrystallites embedded in glass under irradiation of two-colliding pump beams play the role of the phase-conjugated mirror, and the aberrated light plays the role of the probe light. This

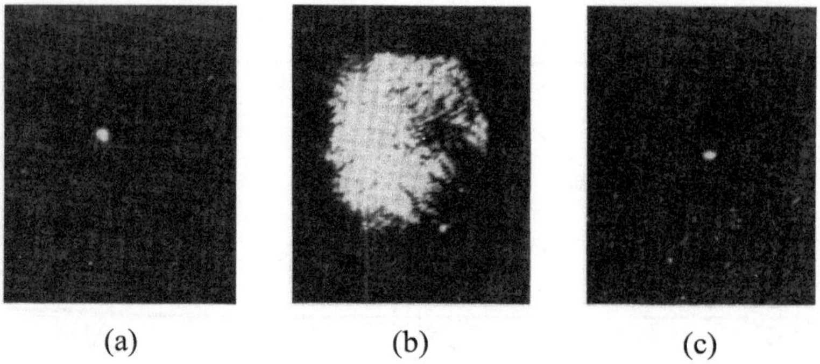

Fig. 6.11. (**a**) The laser spot of the incident beam, (**b**) the spot of the aberrated beam after passing through the frosted glass, and (**c**) the beam spot observed after the aberrated beam is reflected on the phase-conjugated mirror and this beam propagates backward through the same frosted glass [67]

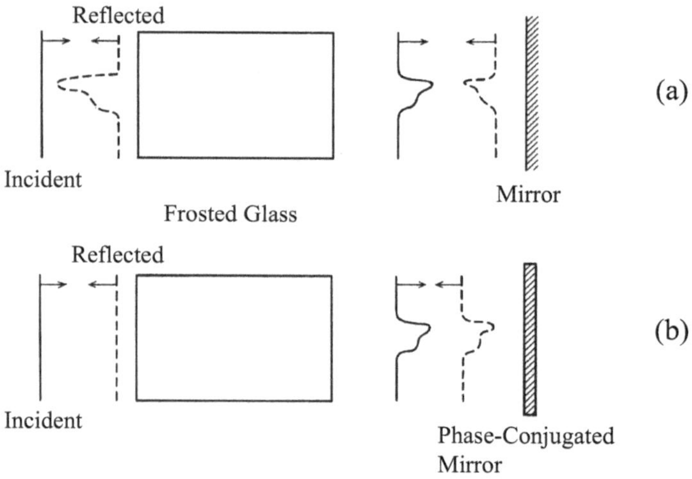

Fig. 6.12. Schematic demonstration of the difference between (**a**) a conventional mirror and (**b**) a phase-conjugated mirror. The aberration induced in the rightward propagation is completely eliminated by the backward propagation after the reflection on the phase-conjugated mirror (b), while the aberration is doubled in the case of the conventional mirror (a) [68]

phase-conjugation is a third-order optical process described by the nonlinear polarizability $\chi^{(3)}(2\omega_1 - \omega_2; \omega_1, -\omega_2, \omega_1)$.

6.2.3 Optical Bistability

We show an observed example of optical bistability in Fig. 6.13 [69]. Here the nonlinear material is an 11 μm thick CdS semiconductor plate containing

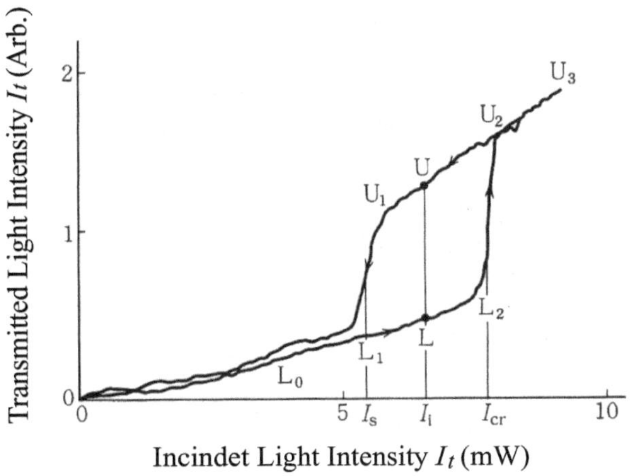

Fig. 6.13. Optical bistable response under resonant pumping of the bound exciton in CdS. The thickness of the CdS crystal is 11 μm and the neutral donor concentration is 10^{15} cm^{-3}. Note that the high transmitting state U and the low one L are possible when the incident light intensity I_i is between the sustain power I_s and the critical power I_{cr} [69]

neutral donors of $10^{15}/\text{cm}^3$ with dielectric mirror coated on both sides, which give a reflectivity of 0.9. The bound exciton at the neutral donor is resonantly pumped. When the incident light intensity I_i is varied, the transmitted light intensity I_t is observed to show hysteresis as Fig. 6.13 shows. Note here that the transmitted light I_t shows two stable states: high and low transmitting states U and L when the incident light intensity I_i is kept between I_s and I_{cr}. Because of these two stable states, this phenomenon is called optical bistability. Such a device has been used for optical information processing, making these U and L states correspond to the digital states 1 and 0, respectively.

This optical bistability originates in optical nonlinearity, such as the optical Kerr effect and feedback effect. The amplitude E_t of the transmitted light through the Fabry–Pérot resonator shown in Fig. 6.14 is described as a sum of a light field transmitting after no-, one-, ..., multiple extra round-trips within the sample:

$$E_t = e^{i\delta/2} tt' E_i (1 + r^2 e^{i\delta} + r^4 e^{2i\delta} + \cdots) . \tag{6.38}$$

Here t and t' are the amplitude transmitivities at the front and rear surfaces, r is the amplitude reflectivity with the relation $tt' = 1 - r^2$, and $\delta \equiv 4\pi n' l/\lambda$ is the phase-change after one round-trip over the sample thickness l. Here n' is a nonlinear refractive index at the wavelength λ, and $n' = n_0 + n_2|E|^2$ for the case of the optical Kerr effect with E the internal field. The coefficient n_2 is related to the third-order polarizability $\chi^{(3)}$ by

$$\chi^{(3)}(\omega; \omega, -\omega, \omega) = 2\epsilon_0 n_0 n_2 . \tag{6.39}$$

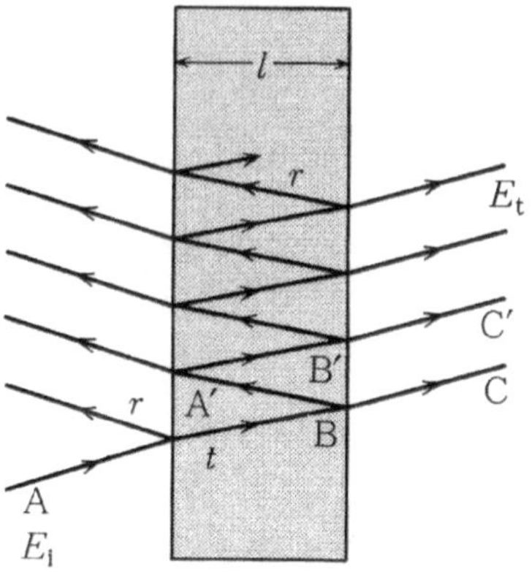

Fig. 6.14. Concept of the Fabry–Pérot resonator. t and r are, respectively, the amplitude transmitivity and reflectivity of the nonlinear optical material

The first term of (6.38) describes the contribution of direct transmission $A \to B \to C$ in Fig. 6.14, and the second term comes from the process $A \to B \to A' \to B' \to C'$ with one extra round-trip. We are considering normal incidence for optical bistability but the incident light E_i was drawn for visual clarity with a finite angle in Fig. 6.14.

From (6.38), the transmitted light intensity I_t is expressed in terms of the finnesse $F \equiv 4R/(1-R)^2$ of the Fabrt-Pérot resonator with $R \equiv |r^2|$ as follows:

$$I_t = \frac{1}{1 + F\sin^2(\delta/2)} I_i \,. \tag{6.40}$$

With the phase change $\delta \equiv 4\pi n'l/\lambda = 2\pi N$, with N an integer, the standing wave within the resonator constitutes the mode with nodes on both ends and the transmitivity is 100% as (6.40) shows. On the other hand, for the case of $\delta = 2\pi(N+1/2)$, the transmitivity is given by $I_t/I_i = 1/(1+F)$ and becomes very small for a system with a large value of finnesse F. The nonlinear refractive index is $n' = n_0 + n_2|E|^2$ for the dispersive type so that, even when $\delta = 2\pi(N+1/2)$ at the weak internal field $E \to 0$, the transmissivity increases through the change of δ in (6.40) when the incident light intensity I_i increases. As a consequence, the positive feedback effect works so as to increase the internal field and finally the transmitted light intensity I_t increases abruptly at $I_i = I_{\text{cr}}$ as Fig. 6.13 shows. On the other hand, as we decreases the incident light intensity I_i from the high transmitting state, the strong internal field which has already been standing within the resonator can persist below $I_i = I_{\text{cr}}$ and the high transmitting state is kept until $I_i = I_s$ ($< I_{\text{cr}}$) below which the transition is induced into the low transmitting state. The imaginary part of $\chi^{(3)}$ sometimes corresponds to absorption saturation and can also contribute to the optical bistability. The optical bistability observed in Fig. 6.13 originates in both the real and imaginary parts of $\chi^{(3)}$.

Optical bistability is being applied to parallel or planar information processing by putting a large number of these optically bistable devices on a plane. In this case, the weaker incident power is the more preferable both for the holding and switching incident powers. At the same time, the shorter switching time is also required to increase the efficiency of information processing. Therefore a nonlinear optical material with a larger $\chi^{(3)}$ value and a shorter switching time is being sought.

6.3 Excitonic Optical Nonlinearity

We introduce a strategy for obtaining the large third-order polarizability $\chi^{(3)}$ which can effectively induce four-wave mixing, generation of phase-conjugated waves, optical squeezing, and optical bistability. With the larger $\chi^{(3)}$ value and the shorter switching time τ, the nonlinear optical materials are more preferable. In general, however, we have an empirical law that these two values $|\chi^{(3)}|$ and $1/\tau$ must obey a trade-off relation as shown by the solid line in Fig. 6.15. This means that the figure of merit $|\chi^{(3)}|/\alpha\tau$ is almost constant. Here α is the absorption coefficient at the relevant frequency ω. Figure 6.15 demonstrates that the figure of merit is $|\chi^{(3)}|/\alpha\tau =$ constant, almost independently of the material and the excitation frequency ω. Certainly, this may be true as long as single-electron excitations are used for the origin of $\chi^{(3)}$. In this section we will demonstrate that we are free of this limitation, $|\chi^{(3)}|/\alpha\tau =$ constant, when the collective excitations such as excitons are resonantly pumped, that is, rapid switching and large $\chi^{(3)}$ can be satisfied simultaneously [70, 71].

A Frenkel exciton, i.e., a collective electronic excitation in a molecular crystal, has a macroscopic transition dipole moment

$$\boldsymbol{P}_{\boldsymbol{k}} = \sqrt{N}\boldsymbol{\mu}\delta_{\boldsymbol{k}\boldsymbol{K}}\,, \tag{6.41}$$

as mentioned already in Sect. 5.2.4. Here $\boldsymbol{\mu}$ is the transition dipole moment of a molecule, N the number of molecules in a crystal, and $\boldsymbol{K}$ the wavenumber vector of the incident light. On the other hand, a Wannier exciton in a semiconductor is made up of the superponsition of products of Bloch states around the bottom of the conduction band and those around the top of the valence band, and it has a transition dipole moment

$$\boldsymbol{P}_{\boldsymbol{k}} = \sqrt{N}\left(\frac{u^3}{\pi a_{\mathrm{B}}^3}\right)^{1/2}\mu_{\mathrm{cv}}\delta_{\boldsymbol{k}\boldsymbol{K}}\,. \tag{6.42}$$

Here u^3 is the volume of the unit cell, and a_{B} the exciton Bohr radius, i.e., the average distance between the electron and the hole composing a Wannier exciton. In general, the exciton Bohr radius a_{B} is much larger than the size u of the unit cell in semiconductors. Therefore, the transition dipole moment of the Wannier exciton has a reduction by $(u/a_{\mathrm{B}})^{3/2}$ but both Wannier and Frenkel

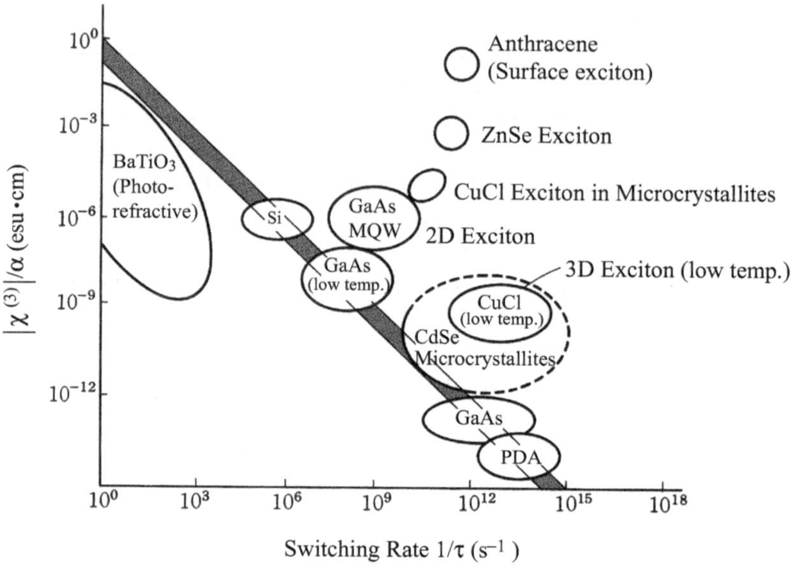

Fig. 6.15. The figure of merit of nonlinear optical materials is plotted by $|\chi^{(3)}|/\alpha$ vs. the switching rate $1/\tau$. The empirical law of constant figure of merit $|\chi^{(3)}|/\alpha\tau$ =const. is represented by the shaded straight line. Note that the figure of merit increases beyond the empirical law under nearly resonant pumping of the excitons

excitons commonly have a macroscopic enhancement $\sqrt{N}$. This is valid only in the limit that the coherent length of both excitons extends over the crystal volume $V = Nu^3$. In real crystals, however, the excitonic coherence length is restricted to finite size due to scattering by phonons and crystal defects. In the case of semiconductor microcrystallites such as CuCl and CdS, the coherent size of the exciton is determined by its size at low temperatures.

Let us derive $\chi^{(3)}(\omega;\omega,-\omega,\omega)$ under nearly resonant pumping of the exciton in a microcrystallite [68]. The interaction $\mathcal{H}'$ of this exciton with the radiation field is expressed in the electric dipole approximation as

$$\mathcal{H}' = -\boldsymbol{P}\cdot\boldsymbol{E}_\omega(t)\,, \tag{6.43}$$

where $\boldsymbol{E}_\omega(t) = \boldsymbol{E}\exp(-i\omega t)+$ c.c. The third-order polarization is evaluated as

$$\left\langle \boldsymbol{P}^{(3)}(\omega)\right\rangle = \mathrm{Tr}\left\{\boldsymbol{P}\rho^{(3)}(t)\right\}\,, \tag{6.44}$$

where the density matrix $\rho(t)$ is expanded to third order in the interaction Hamiltonian $\mathcal{H}'$, i.e., third order in the external radiation field E_ω. The lowest exciton level dominantly has the largest oscillator strength in most cases. When the radiation field ω is nearly resonant to this level ω_0, we may well accept the rotating wave approximation and safely neglect the process containing antiresonant electronic excitations. Although 48 terms contribute in

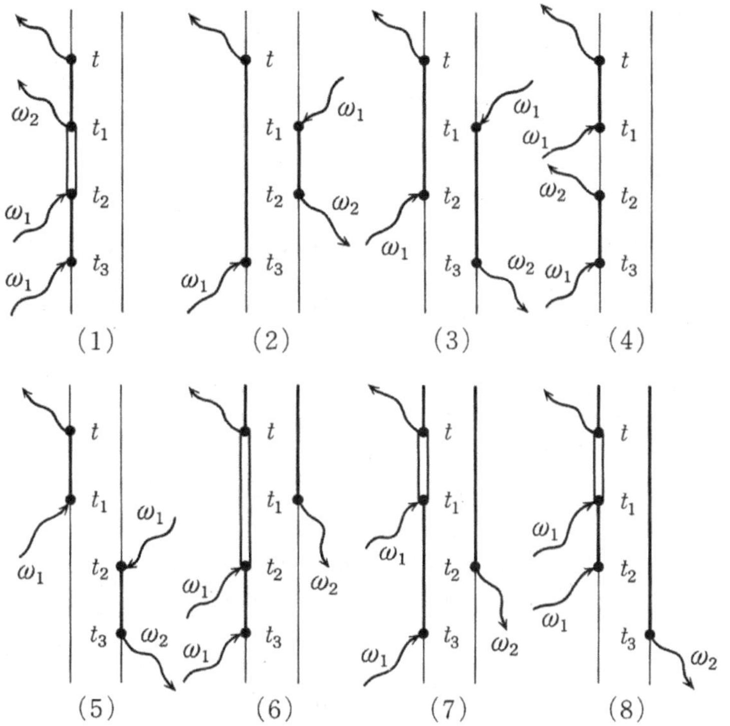

Fig. 6.16. Eight double Feynman diagrams contributing to $\chi^{(3)}(2\omega_1 - \omega_2; \omega_1, -\omega_2, \omega_1)$ of the excitonic system. Thin, thick, and double lines mean, respectively, the ground, a single exciton, and a double exciton state. See also Fig. 6.2

general to the third-order polarization, only the eight terms drawn in Figs. 6.16 and 6.17 are chosen as dominant terms contributing to the third-order polarizability $\chi^{(3)}(2\omega_1 - \omega_2; \omega_1, -\omega_2, \omega_1)$ under nearly resonant pumping of the lowest energy exciton ω_0. In this section, the ground state, one-photon, and two-photon excited states are described, respectively, as g, n, and m. In order to obtain the third-order polarization with angular frequency $2\omega_1 - \omega_2$, we must consider two first-order density matrices $\rho_{ng}^{(1)}(\omega_1)$ and $\rho_{gn}^{(1)}(-\omega_2)$ at first as shown in Fig. 6.17. From (6.13), $\rho_{ng}^{(1)}(\omega_1)$ obeys

$$\frac{\partial \rho_{ng}^{(1)}}{\partial t} = \frac{1}{i\hbar}\left\{\hbar\omega_{ng}\rho_{ng}^{(1)} + \mathcal{H}'_{ng}\rho_{gg}^{(0)}\right\} - \Gamma_{ng}\rho_{ng}^{(1)} . \tag{6.45}$$

Here $\mathcal{H}'_{ng}$ contains only the term with $\exp(-i\omega_1 t)$ under the rotating wave approximation so that the left-hand side of (6.45) can be replaced by $-i\omega_1\rho_{ng}^{(1)}$. As a result, the stationary solution is obtained as

$$\rho_{ng}^{(1)}(\omega_1) = \frac{\mathcal{H}'_{ng}(\omega_1)\rho_{gg}^{(0)}}{\hbar(\omega_1 - \omega_{ng} + i\Gamma_{ng})} . \tag{6.46}$$

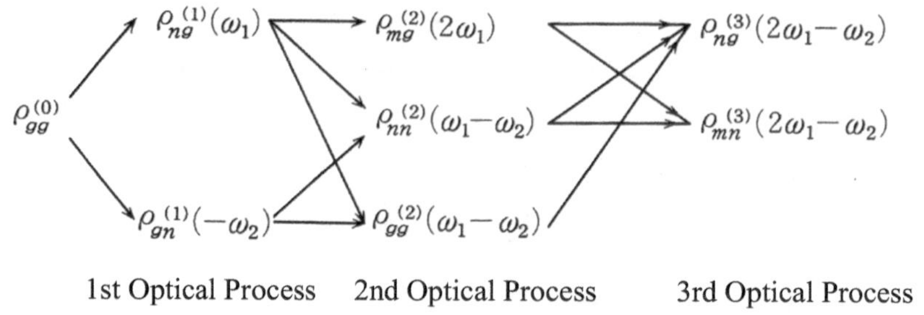

Fig. 6.17. Eight roots of the electronic density matrices contributing to $\chi^{(3)}(2\omega_1 - \omega_2; \omega_1, -\omega_2, \omega_1)$. These eight routes correspond to the eight diagrams of Fig. 6.16

The other first-order contribution is similarly calculated and

$$\rho_{gn}^{(1)}(-\omega_2) = \frac{\rho_{gg}^{(0)} \mathcal{H}'_{gn}(-\omega_2)}{\hbar(\omega_2 - \omega_{ng} - i\Gamma_{ng})} \,. \tag{6.47}$$

To second order in the external field, there are three density matrices for the present problem as shown in Fig. 6.17. The first one $\rho_{mg}^{(2)}$ obeys the following equation of motion:

$$\frac{\partial \rho_{mg}^{(2)}}{\partial t} = \frac{1}{i\hbar}\left[\hbar\omega_{mg}\rho_{mg}^{(2)} + \mathcal{H}'_{mn}(\omega_1)\rho_{ng}^{(1)}(\omega_1)\right] - \Gamma_{mg}\rho_{mg}^{(2)} \,, \tag{6.48}$$

where Γ_{mg} denotes the transverse relaxation rate of two-photon excited state m. Both $\rho_{ng}^{(1)}(\omega_1)$ and $\mathcal{H}'_{mn}(\omega_1)$ are accompanied by $\exp(-i\omega_1 t)$ so that $\rho_{mg}^{(2)}$ has a factor $\exp(-2i\omega_1 t)$. By replacing the left-hand side of (6.48) with $-2i\omega_1\rho_{mg}^{(2)}(2\omega_1)$, we obtain for the stationary response:

$$\rho_{mg}^{(2)}(2\omega_1) = \frac{\mathcal{H}'_{mn}(\omega_1)\mathcal{H}'_{ng}(\omega_1)\rho_{gg}^{(0)}}{\hbar^2(2\omega_1 - \omega_{mg} + i\Gamma_{mg})(\omega_1 - \omega_{ng} + i\Gamma_{ng})} \,. \tag{6.49}$$

The other two second-order density matrices $\rho_{nn}^{(2)}$ and $\rho_{gg}^{(2)}$ obey the following differential equations:

$$\frac{\partial \rho_{nn}^{(2)}}{\partial t} = \frac{1}{i\hbar}\left[\mathcal{H}'_{ng}(\omega_1)\rho_{gn}^{(1)}(-\omega_2) - \rho_{ng}^{(1)}(\omega_1)\mathcal{H}'_{gn}(-\omega_2)\right] - \Gamma_{n\to g}\rho_{nn}^{(2)} \,, \tag{6.50}$$

$$\frac{\partial \rho_{gg}^{(2)}}{\partial t} = \frac{1}{i\hbar}\left[\mathcal{H}'_{gn}(-\omega_2)\rho_{ng}^{(1)}(\omega_1) - \rho_{gn}^{(1)}(-\omega_2)\mathcal{H}'_{ng}(\omega_1)\right] + \Gamma_{n\to g}\rho_{nn}^{(2)} \,, \tag{6.51}$$

where $\Gamma_{n\to g}$ describes the longitudinal decay rate from the one-photon excited state n into the ground state g. Both $\mathcal{H}'_{ng}(\omega_1)$ and $\rho_{ng}^{(1)}(\omega_1)$ have the time dependence $\exp(-i\omega_1 t)$, while both $\mathcal{H}'_{gn}(-\omega_2)$ and $\rho'^{(1)}_{gn}(-\omega_2)$ contain the factor $\exp(i\omega_2 t)$. As a result, both $\rho^{(2)}_{nn}$ and $\rho^{(2)}_{gg}$ have the time dependence $\exp[-i(\omega_1-\omega_2)t]$. Therefore we may replace the left-hand sides of (6.50) and (6.51), respectively, with $-i(\omega_1-\omega_2)\rho^{(2)}_{nn}$ and $-i(\omega_1-\omega_2)\rho^{(2)}_{gg}$, and obtain for the stationary response:

$$\begin{aligned}\rho^{(2)}_{nn}(\omega_1-\omega_2) &= -\rho^{(2)}_{gg}(\omega_1-\omega_2)\\ &= \frac{\mathcal{H}'_{ng}(\omega_1)\rho^{(1)}_{gn}(-\omega_2)-\rho^{(1)}_{ng}(\omega_1)\mathcal{H}'_{gn}(-\omega_2)}{i\hbar\left[\Gamma_{n\to g}-i(\omega_1-\omega_2)\right]}\,. \qquad (6.52)\end{aligned}$$

There are two terms in the third-order density matrices for the external field, as Fig. 6.17 shows, and these obey

$$\begin{aligned}\frac{\partial\rho^{(3)}_{ng}}{\partial t} &= -i(\omega_{ng}-i\Gamma_{ng})\rho^{(3)}_{ng}+\frac{1}{i\hbar}\Big[\mathcal{H}'_{nm}(-\omega_2)\rho^{(2)}_{mg}(2\omega_1)\\ &\quad+\mathcal{H}'_{ng}(\omega_1)\rho^{(2)}_{gg}(\omega_1-\omega_2)-\rho^{(2)}_{nn}(\omega_1-\omega_2)\mathcal{H}'_{ng}(\omega_1)\Big]\,, \qquad (6.53)\\ \frac{\partial\rho^{(3)}_{mn}}{\partial t} &= -i(\omega_{mn}-i\Gamma_{mn})\rho^{(3)}_{mn}+\frac{1}{i\hbar}\Big[\mathcal{H}'_{mn}(\omega_1)\rho^{(2)}_{nn}(\omega_1-\omega_2)\\ &\quad-\rho^{(2)}_{mg}(2\omega_1)\mathcal{H}'_{gn}(-\omega_2)\Big]\,. \qquad (6.54)\end{aligned}$$

From the time dependence of the right-hand sides $\exp[-i(2\omega_1-\omega_2)t]$, the left-hand sides of (6.53) and (6.54) may be replaced by $-i(2\omega_1-\omega_2)\rho^{(3)}_{ng}$ and $-i(2\omega_1-\omega_2)\rho^{(3)}_{mn}$. As a result, the solutions of (6.53) and (6.54) are obtained as follows:

$$\rho^{(3)}_{ng}(2\omega_1-\omega_2)=\frac{\mathcal{H}'_{nm}(-\omega_2)\rho^{(2)}_{mg}(2\omega_1)-2\rho^{(2)}_{nn}(\omega_1-\omega_2)\mathcal{H}'_{ng}(\omega_1)}{\hbar(2\omega_1-\omega_2-\omega_{ng}+i\Gamma_{ng})}\,, \qquad (6.55)$$

$$\rho^{(3)}_{mn}(2\omega_1-\omega_2)=\frac{\mathcal{H}'_{nm}(\omega_1)\rho^{(2)}_{nn}(\omega_1-\omega_2)-\rho^{(2)}_{mg}(2\omega_1)\mathcal{H}'_{gn}(-\omega 2)}{\hbar(2\omega_1-\omega_2-\omega_{mn}+i\Gamma_{mn})}\,. \qquad (6.56)$$

Summing these results, the third-order electric polarization under stationary pumping is evaluated from (6.44) as

$$\left\langle P^{(3)}(2\omega_1-\omega_2)\right\rangle = P_{gn}\rho^{(3)}_{ng}(2\omega_1-\omega_2)+P_{nm}\rho^{(3)}_{mn}(2\omega_1-\omega_2)+\text{c.c.} \qquad (6.57)$$

Recently, microcrystallites of the semiconductor CuCl were crystallized in the insulator NaCl. The size of the CuCl microcrystallites can be well controlled from radius $R = 1.3\,\text{nm}$ to $10\,\text{nm}$ or even larger and the size-dependence of the excitonic superrradiance rate was observed as already mentioned in Sect. 5.2.4.

We can also expect a large third-order optical response under nearly resonant pumping of the exciton in microcrystallites. Let us apply our formula mentioned in the above paragraph to a system of CuCl microcrystallites embedded in the insulator. As the exciton Bohr radius in CuCl crystals is 0.67 nm, when the exciton is in the lowest excited state 1s, its center-of-mass motion is also well quantized in a microcrystallite with radius $R = 1.3 \sim 10\,\text{nm}$. The insulator NaCl outside the CuCl microcrystallite has a bandgap 7 eV so that the potential barrier for the CuCl exciton with 3.2 eV excitation energy may be approximated to be infinite outside the microcrystallite. Then the eigenenergies of the CuCl exciton are obtained as

$$E_n = E_g - Ry + \frac{\hbar^2}{2M}\left(\frac{\pi n}{R}\right)^2, \quad (n = 1, 2, \ldots). \tag{6.58}$$

Here E_g is the bandgap energy between the conduction and valence bands in the CuCl bulk crystal, Ry is the exciton binding energy 200 meV, and n is the principal quantum number of the exciton center-of-mass motion. The last term of (6.58) denotes the quantization energy of the center-of-mass motion with mass M and this quantization energy, i.e., the energy separation between the lowest $n = 1$ and the second lowest $n = 2$ states, is estimated to be of the order of 10 meV. The transition dipole moment to the lowest exciton state $n = 1$ in (6.58) is evaluated to be

$$P_n = \frac{2\sqrt{2}}{\pi}\left(\frac{R}{a_B}\right)^3 \frac{1}{n}\boldsymbol{\mu}_{cu}, \quad (n = 1, 2, \ldots). \tag{6.59}$$

Note that this value has a mesoscopic enhancement $(8R^3/\pi^2 a_B^2)^{1/2}$ and that the oscillator strength concentrates dominantly on the lowest excited state with quantum number $n = 1$. Here we will evaluate $\chi^{(3)}(\omega; \omega, -\omega, \omega)$ under nearly resonant pumping of the lowest exciton state $n = 1$ with the largest oscillator strength. Under this condition, we may take into account only the following three transition dipole moments:

$$P_{ng} \equiv P_1, \quad P_{mn} = \sqrt{2}P_1, \quad P_{nm} = \sqrt{2}P_1^*. \tag{6.60}$$

Here the factor $\sqrt{2}$ comes from the bosonic character of the excitons and we have neglected the spin degeneracy of the involving electrons. Denoting by $\hbar\omega_{\text{int}}$ the interaction energy between two excitons in a microcrystallite, the single-photon and two-photon excited states have the following excitation energies:

$$\begin{aligned} \omega_{ng} &= \omega_1, \quad \omega_{mg} = 2(\omega_1 + \omega_{\text{int}}), \\ \omega_{mn} &= \omega_{mg} - \omega_{ng} = \omega_1 + 2\omega_{\text{int}}. \end{aligned} \tag{6.61}$$

This interaction energy $\hbar\omega_{\text{int}}$ originates in exchange energies between two electrons and between two holes composing two excitons, which make the excitons deviate from ideal bosons. Here we neglect the bound state of two excitons, i.e., an excitonic molecule which is a topic in Sect. 6.4. This may be justified when $\omega \sim \omega_1$, i.e., nearly resonant pumping of the ω_1 exciton as 2ω is far enough off-resonant from an excitonic molecule in CuCl. However, we must take into account the relaxation effects of each excitation:

$$\begin{aligned} \Gamma_{n\to g} &= 2\gamma\,, \quad \Gamma_{ng} = \Gamma = \gamma + \gamma'\,, \\ \Gamma_{mg} &= 2(\gamma+\gamma')\,, \quad \Gamma_{mn} = \Gamma + 2\gamma = 3\gamma + \gamma'\,. \end{aligned} \tag{6.62}$$

Here the longitudinal relaxation $\Gamma_{n\to g} = 2\gamma$ consists of the superradiant decay described in Sect. 5.2.4 and the nonradiative decay, and γ' describes the pure dephasing rate. Γ_{mg} describes the transverse relaxation rate of two excitons while Γ_{mn} is the sum of a single exciton longitudinal decay rate and a single exciton transverse rate.

The third-order polarization of (6.57) is evaluated explicitly in terms of (6.60)–(6.62) and the third-order polarizability $\chi^{(3)}(\omega;\omega,-\omega,\omega)$ defined by

$$\left\langle P^{(3)}(\omega)\right\rangle = \chi^{(3)}(\omega;\omega,-\omega,\omega)E\,|E|^2\,e^{-i\omega t}\,, \tag{6.63}$$

is given by

$$\begin{aligned} \chi^{(3)}(\omega;\omega,-\omega,\omega) = {} & \frac{|P_1|^4}{\hbar^3}\frac{N_c}{(\omega-\omega_1+i\Gamma)(\omega-\omega_1-i\Gamma)} \\ & \times\left\{\frac{1}{\omega-\omega_1+i\Gamma} - \frac{1}{\omega-\omega_1-2\omega_{\text{int}}+i(\Gamma+2\gamma)}\right\} \\ & \times\left\{1+\frac{2\gamma'}{\gamma}+\frac{2i\Gamma-\omega_{\text{int}}}{\omega-\omega_1-\omega_{\text{int}}+i\Gamma}\right\}. \end{aligned} \tag{6.64}$$

Here N_c denotes the number density of microcrystallites, and $N_c \equiv 3r/(4\pi R^3)$ with r the ratio of the volume of the semiconductor microcrystallites to that of the insulating matrix. Let us consider the size dependence of $\chi^{(3)}$ mesoscopic enhancement under a constant volume ratio r. The result of (6.64) looks a little complicated so that we discuss $\chi^{(3)}$ for several limiting cases [70, 71].

(a) $\omega_{\text{int}} > |\omega-\omega_1| > \Gamma$:

$$\chi^{(3)} = \frac{2N_c\,|P_1|^4}{\hbar^3(\omega-\omega_1)^3}\left(1+\frac{\gamma'}{\gamma}\right) \propto \left(\frac{R}{a_B}\right)^3. \tag{6.65}$$

It is important to point out that $\chi^{(3)}$ increases in proportion to the volume of microcrystallite under a constant volume ratio r. This is because the fourth power of the transition dipole moment $|P_1|^4$ in the numerator in (6.65) has an R^6 dependence and overcomes the R^{-3} dependence of N_c [70]. In the linear response, the R-dependence of $|P_1|^2$ just cancels out that of N_c so that $\chi^{(1)}$ has no R-dependence as long as the volume ratio r is a constant. The interaction energy of two excitons with the same spin structure $\hbar\omega_{\text{int}}$ is given in the first Born approximation:

$$\hbar\omega_{\rm int} = \frac{13\pi}{3} Ry \frac{a_B^3}{v} = \frac{13}{4} Ry \left(\frac{a_B}{R}\right)^3 . \tag{6.66}$$

Note here that this interaction energy $\hbar\omega_{\rm int}$ is inversely proportional to the average volume $v \equiv 4\pi R^3/3$ of a microcrystallite. For a CuCl microcrystallite with $R = 3\,{\rm nm}$, $\hbar\omega_{\rm int} \sim 3\,{\rm meV}$ while $\hbar\Gamma = 0.02\,{\rm meV}$ at low temperatures so that a degree of off-resonance $|\omega - \omega_1|$ can be chosen over a wide frequency range for case (a).

(b) For the case of more nearly resonant pumping $|\omega - \omega_1| < \Gamma < \omega_{\rm int}$,

$$\chi^{(3)} = -i \frac{2N_c |P_1|^4}{\hbar^3 \Gamma^2 \gamma} \tag{6.67}$$

becomes pure imaginary and contributes to the absorption saturation. When the superradiant decay is not a dominant channel in the longitudinal 2γ and transverse Γ relaxation processes, both Γ and γ are independent of R so that $-{\rm Im}\chi^{(3)}$ is also proportional to R^3 and is accompanied by the mesoscopic enhancement. For example, for the case of the volume ratio 0.1% of CuCl microcrystallites with a radius 8 nm, and $\hbar\Gamma = 0.5\,{\rm meV}$, $2\hbar\gamma = 0.03\,{\rm meV}$, ${\rm Im}\chi^{(3)} = -10^{-3}\,{\rm esu}$. On the other hand, when 2γ is dominantly determined by the superradiant decay, $-{\rm Im}\chi^{(3)}$ becomes independent of the size of microcrystallites.

(c) Under off-resonant pumping $|\omega - \omega_1| > \Gamma > \omega_{\rm int}$,

$$\chi^{(3)} = \frac{2iN_c |P_1|^4 (2\gamma' + \gamma)}{\hbar^3 (\omega - \omega_1)^4} \propto \left(\frac{R}{a_B}\right)^3 . \tag{6.68}$$

In this case, $\chi^{(3)}$ is also pure imaginary and has the same size R dependence as in case (a). However, the absolute magnitude is $(2\gamma' + \gamma)/|\omega - \omega_1|$ (< 1) smaller than case (a).

Excitons are often treated as Bose particles in a bulk crystal. Although ideal Bose particles cannot show any nonlinearities, three factors make excitons deviate from ideal bosons so as to bring about a finite $\chi^{(3)}$ as (6.64) shows. The first factor is the exciton–exciton interaction $\hbar\omega_{\rm int}$, the second the longitudinal decay of the exciton 2γ, and the third the transverse relaxation $\Gamma = \gamma + \gamma'$ of the exciton. These three effects may be understood from the three cases (a), (b), and (c). In these cases, the mesoscopic enhancement of the exciton transition dipole moment works effectively, $|\chi^{(3)}|$ increases in proportion to R^3 and reaches the value $|\chi^{(3)}| \sim 10^{-3}\,{\rm esu}$ for CuCl microcrystallites with a radius $R = 8\,{\rm nm}$. This order of value was observed for these microcrystallites crystallized in the insulator NaCl and glasses [72]. In order to obtain the large $\chi^{(3)}$ value, we need not restrict our choice to microcrystallites, but we may expect such an enhancement from excitons in bulk crystals with a long coherent length at low temperature. For example, under resonant pumping of the lowest surface exciton level in an anthracene crystal [73] and the bulk exciton of a ZnSe crystal, the figures of merit of these crystals are

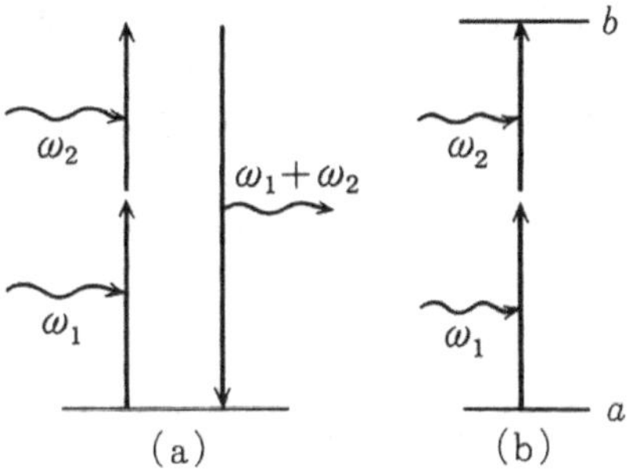

Fig. 6.18. Schematic diagrams of (**a**) sum-frequency generation and (**b**) two-photon absorption

found to be much larger beyond the conventional figure of merit line as shown in Fig. 6.15.

6.4 Two-Photon Absorption Spectrum

Two-photon absorption spectroscopy plays important roles in determining the electronic structure of crystals because it gives us information complementary to that obtained by one-photon absorption spectroscopy. Higher-harmonic generation and four-wave mixing are optical processes of dispersive type in which the electrons come back to the initial ground state after these processes. This is shown in Fig. 6.18(a). On the other hands, two-photon absorption is of dissipative type in which the electronic excitation remains after this process (see Fig. 6.18b).

Let us consider a crystal with inversion symmetry. One-photon absorption is electric dipole-allowed only between two electronic states with opposite parities. Conversely two-photon absorption is allowed only between two states with the same parity. From this fact we can understand that one- and two-photon absorption spectroscopies both with the common ground state as an initial state give complementary information on the electronic excited states. Furthermore, the more interesting point is that the two-photon absorption coefficient is very sensitive to the polarization directions of two incident beams $\boldsymbol{e}_1 = (l_1, m_1, n_1)$ and $\boldsymbol{e}_2 = (l_2, m_2, n_2)$. Here in terms of the angles of polarization relative to the three principal crystalline axes θ_{il}, θ_{im} and θ_{in} $(i = 1, 2)$, the directions of polarization are given as $l_i = \cos\theta_{il}$, $m_i = \cos\theta_{im}$, and $n_i = \cos\theta_{in}$. We can determine the symmetry of the electronic transition from the angle dependence of the two-photon absorption coefficient [74]. This spectrum is usually being observed by using a single laser light beam with a fixed frequency and conventional light, the frequency of which is changeable continuously. Here we measure the two-photon absorption coefficient from the degree of attenuation of the conventional light.

Let us irradiate the insulating crystal simultaneously by a laser beam with angular frequency ω_1 and polarization $\boldsymbol{e}_1$ and a conventional light beam with ω_2 and $\boldsymbol{e}_2$. We assume that both $\hbar\omega_1$ and $\hbar\omega_2$ are smaller than the bandgap so that single-photon absorption is negligible. The transition probability to the excited state $|e\rangle$ from the crystal ground state absorbing two photons $\hbar\omega_1$ and $\hbar\omega_2$ simultaneously is expressed as

$$\begin{aligned} W^{(2)} &= \frac{2\pi}{\hbar}\left(\frac{e}{m}\right)^4 \left(\frac{\hbar}{2\epsilon_0\epsilon_1 V\omega_1}\right)\left(\frac{\hbar}{2\epsilon_0\epsilon_2 V\omega_2}\right) \\ &\quad \times N_1 N_2 \left|A_{eg}^{(2)}\right|^2 \delta\left(E_{eg} - \hbar\omega_1 - \hbar\omega_2\right), \end{aligned} \tag{6.69}$$

$$A_{eg}^{(2)} = \sum_n \left[\frac{(\boldsymbol{P}_{en}\cdot\boldsymbol{e}_1)(\boldsymbol{P}_{ng}\cdot\boldsymbol{e}_2)}{\hbar(\omega_n-\omega_2)} + \frac{(\boldsymbol{P}_{en}\cdot\boldsymbol{e}_2)(\boldsymbol{P}_{ng}\cdot\boldsymbol{e}_1)}{\hbar(\omega_n-\omega_1)}\right]. \tag{6.70}$$

Here N_i $(i = 1,2)$ denote the photon numbers of the $\hbar\omega_i$ photon, and ϵ_i the crystal dielectric constant at angular frequency ω_i. The matrix elements $\boldsymbol{P}_{en}$ and $\boldsymbol{P}_{ng}$ are the expectation values of the crystal momentum $\boldsymbol{P} = \sum_j \boldsymbol{P}_j$, respectively, between the excited state $|e\rangle$ and the intermediate state $|n\rangle$ and between $|n\rangle$ and the ground state $|g\rangle$. The absorption coefficient $\alpha^{(2)}$ of conventional light ω_2 is obtained by dividing (6.69) by the ω_2 photon flux density $cN_2/\sqrt{\epsilon_2}V$ as

$$\alpha^{(2)} = \frac{2\pi}{\hbar}\left(\frac{e}{m}\right)^4 \left(\frac{\hbar}{2\epsilon_0\epsilon_1 V\omega_1}\right)\left(\frac{\hbar}{2\epsilon_0 c\sqrt{\epsilon_2}\omega_2}\right) N_1 \left|A_{eg}^{(2)}\right|^2 \rho\left(E_{eg}\right). \tag{6.71}$$

Here $\rho(E_{eg})$ is the density of states at the excitation energy $E_{eg} = E_e - E_g$. From group theoretical considerations of the matrix elements (6.70) for a two-photon transition, we will obtain the selection rules for 32 crystal point groups [74]. Instead of the sum over the intermediate states $|n\rangle$ in (6.70):

$$\Lambda(\omega_i) = \sum_n \frac{|n\rangle\langle n|}{\hbar(\omega_n - \omega_i)}, \tag{6.72}$$

we introduce the symmetric part $\Lambda^+ = \Lambda(\omega_1) + \Lambda(\omega_2)$ and the antisymmetric part $\Lambda^- = \Lambda(\omega_1) - \Lambda(\omega_2)$, and rewrite $A_{eg}^{(2)}$ of (6.70) as

$$A_{eg}^{(2)} = \boldsymbol{e}_1 \cdot \langle e\left|\left(\boldsymbol{P}\Lambda^+\boldsymbol{P}\right)_s + \left(\boldsymbol{P}\Lambda^-\boldsymbol{P}\right)_{as}\right| g\rangle \cdot \boldsymbol{e}_2 . \tag{6.73}$$

Here, $(\quad)_s$ and $(\quad)_{as}$ mean, respectively, the symmetric and antisymmetric parts of the tensor within $(\quad)$. The electronic elementary excitation in the crystal may be considered to conserve the total wavevector between the valence and conduction bands, because the photon wavevector is almost negligible compared with that of Bloch electrons in the Brillouin zone. Under this condition, the selection rule of (6.73), i.e., the two-photon absorption tensor, can be discussed in terms of the point group instead of the space group of the crystal. For example, two-photon absorption due to an exciton can be described in terms of the value $\varphi_{\mathrm{ex}}(0)$ of the wavefunction of the relative motion of the electron–hole pair within the exciton at the origin, as

$$A_{eg}^{(2)} = \varphi_{\mathrm{ex}}(0) \int u_{c0}^{*} \left\{ \boldsymbol{e}_1 \cdot \left(\boldsymbol{p}\Lambda^{+}\boldsymbol{p}\right)_{s} \cdot \boldsymbol{e}_2 + \boldsymbol{e}_1 \cdot \left(\boldsymbol{p}\Lambda^{-}\boldsymbol{p}\right)_{as} \cdot \boldsymbol{e}_2 \right\} u_{v0} d\boldsymbol{r} \,. \tag{6.74}$$

Here, $\boldsymbol{p}$ means the momentum operator of a single electron, and u_{c0} and u_{v0} the periodic parts of the conduction and valence band Bloch functions at the same extremum point in the Brillouin zone. For example, the two-photon excitation tensor at the Γ-point ($k = 0$) of the cubic crystal $\mathrm{O_h}$ is expressed by the following irreducible representation:

$$\begin{aligned}
&\boldsymbol{e}_1 \cdot (\boldsymbol{p}\Lambda^{+}\boldsymbol{p})_s \cdot \boldsymbol{e}_2 \\
&= \frac{1}{3}(l_1 l_2 + m_1 m_2 + n_1 n_2)(p_x \Lambda^{+} p_x + p_y \Lambda^{+} p_y + p_z \Lambda^{+} p_z) && \mathrm{A_{1g}} \\
&\quad + \frac{1}{2}(l_1 l_2 - m_1 m_2)(p_x \Lambda^{+} p_x - p_y \Lambda^{+} p_y) && \mathrm{E_g} \\
&\quad + \frac{1}{6}(l_1 l_2 + m_1 m_2 - 2 n_1 n_2)(p_x \Lambda^{+} p_x + p_y \Lambda^{+} p_y - 2 p_z \Lambda^{+} p_z) && \mathrm{E_g} \\
&\quad + \frac{1}{2}(m_1 n_2 + m_2 n_1)(p_y \Lambda^{+} p_z + p_z \Lambda^{+} p_y) && \mathrm{T_{2g}} \\
&\quad + \frac{1}{2}(n_1 l_2 + n_2 l_1)(p_z \Lambda^{+} p_x + p_x \Lambda^{+} p_z) && \mathrm{T_{2g}} \\
&\quad + \frac{1}{2}(l_1 m_2 + l_2 m_1)(p_x \Lambda^{+} p_y + p_y \Lambda^{+} p_x) \,, && \mathrm{T_{2g}}
\end{aligned} \tag{6.75}$$

$$\begin{aligned}
&\boldsymbol{e}_1 \cdot (\boldsymbol{p}\Lambda^{-}\boldsymbol{p})_{as} \cdot \boldsymbol{e}_2 \\
&\quad + \frac{1}{2}(m_1 n_2 - m_2 n_1)(p_y \Lambda^{-} p_z - p_z \Lambda^{-} p_y) && \mathrm{T_{1g}} \\
&\quad + \frac{1}{2}(n_1 l_2 - n_2 l_1)(p_z \Lambda^{-} p_x - p_x \Lambda^{-} p_z) && \mathrm{T_{1g}} \\
&\quad + \frac{1}{2}(l_1 m_2 - l_2 m_1)(p_x \Lambda^{-} p_y - p_y \Lambda^{-} p_x) \,. && \mathrm{T_{1g}}
\end{aligned} \tag{6.76}$$

Considering that the crystal ground state has the $\mathrm{A_{1g}}$ representation, it is understood that the states $\mathrm{A_{1g}}$, $\mathrm{E_g}$, $\mathrm{T_{1g}}$, and $\mathrm{T_{2g}}$ can be excited by two-photon transitions. At the same time, we can obtain from (6.75) and (6.76) the incident angle dependence of the two-photon absorption coefficients in the cubic crystals as

$$A_{1g} \to A_{1g} : (l_1 l_2 + m_1 m_2 + n_1 n_2)^2 = (\boldsymbol{e}_1 \cdot \boldsymbol{e}_2)^2 , \tag{6.77}$$

$$A_{1g} \to E_g : l_1^2 l_2^2 + m_1^2 m_2^2 + n_1^2 n_2^2 - (l_1 l_2 m_1 m_2 + m_1 m_2 n_1 n_2 + n_1 n_2 l_1 l_2) , \tag{6.78}$$

$$A_{1g} \to T_{1g} : 1 - (l_1 l_2 + m_1 m_2 + n_1 n_2)^2 = (\boldsymbol{e}_1 \times \boldsymbol{e}_2)^2 , \tag{6.79}$$

$$A_{1g} \to T_{2g} : 1 - (l_1^2 l_2^2 + m_1^2 m_2^2 + n_1^2 n_2^2) \\ +2 (l_1 l_2 m_1 m_2 + m_1 m_2 n_1 n_2 + n_1 n_2 l_1 l_2) . \tag{6.80}$$

One-photon absorption in a cubic crystal is allowed only for the transition $A_{1g} \to T_{1u}$, so that the absorption coefficient is isotropic against the incident angle. On the other hand, two-photon absorption coefficients to different symmetry states have a different angle dependence on two incident beams as shown in (6.77)–(6.80) for a O_h crystal. This means that four different excited states A_{1g}, E_g, T_{1g}, and T_{2g} can be identified by observing the $\boldsymbol{e}_1$ and $\boldsymbol{e}_2$ dependences of these two-photon absorption coefficients for a O_h crystal. As long as we use linearly polarized light as two incident beams, we cannot uniquely identify the four kinds of these excited states because the angle dependences of two-photon absorption (6.77)–(6.80) are written in terms of only three independent functions of the two incident angles. However, when we also use circularly polarized light as the incident beams, we will be able to identify these four levels.

We will demonstrate the experimental results in which the involving excited states could be identified from the angle dependences of the two incident beams. Two excitons are sometimes bound into an excitonic molecule. This consists of two electrons with different spin components in the conduction band and two holes also with different spin components in the valence band, and may be understood in analogy to a hydrogen molecule [75]. This excitonic molecule can be excited by two-photon absorption and its transition probability is written as

$$W^{(2)}(\omega) = \frac{2\pi}{\hbar} \left| \left\langle \mathrm{mol} \mid \mathcal{H}' \sum_n \frac{|n\rangle\langle n|}{\hbar(\omega_{ng} - \omega)} \mathcal{H}' \mid g \right\rangle \right|^2 \delta(2\hbar\omega - E_{\mathrm{mol}}) . \tag{6.81}$$

The sharp two-photon absorption line is observed more strongly by a factor 10^6–10^7 than the continuum background of band-to-band two-photon absorption. This abnormal enhancement of molecular two-photon absorption comes from (1) the giant oscillator strength and (2) resonance enhancement [76]. For the transition from the intermediate state $|n\rangle$ in which a single exciton exists to the final state of an excitonic molecule, we can choose any valence electron within the large molecular orbital around the first exciton in $|n\rangle$. This large freedom of the selection of the second valence electron (hole) results in the giant oscillator strength [76]. This is in contrast to the conventional band-to-band two-photon absorption, in which a single electron interacts with the first and second photon. The energy denominator of (6.81) is half the binding energy of the excitonic molecule, e.g., as small as 15 meV in a CuCl crystal. This small energy denominator enhances the two-photon absorption coefficient because this is in contrast to the two-photon band-to-band transition with the energy denominator of the order of eV. This is the second enhancement effect, i.e., the resonance enhancement which results also in the giant two-photon absorption due to the excitonic molecule. This absorption coefficient was estimated to become of the same order as the linear absorption coefficient by an exciton under 1 MW/cm^2 incident laser irradiation in a CuCl crystal.

This prediction of the giant two-photon absorption due to the excitonic molecule was proved experimentally by Gale and Mysyrowicz using a CuCl crystal [77]. Both the crystals CuCl and CuBr have a zinc-blende structure. However, the highest energy point on the valence band of CuCl is located at the Γ-point and has Γ_7 symmetry of the point group T_d, and the bottom of the conduction band has Γ_6 symmetry. As a result, a single level of the excitonic molecule with $\Gamma_1(A_1)$ was observed of the two-photon absorption spectrum. On the other hand, the top valence band of the CuBr crystal has a four-fold degenerate Γ_8 state. As a consequence, three levels $\Gamma_1(A_1)$, $\Gamma_3(E)$, and $\Gamma_5(T_2)$ of the excitonic molecule are observable by two-photon absorption [78]. Vu Duy Phach and R. Lévy [79] observed the two-photon absorption spectrum of these three kinds of excitonic molecule, and the incident angle dependence of these three peaks by changing the polarization angle $\boldsymbol{e}_2$ of the second incident beam with the first beam polarization $\boldsymbol{e}_1$ fixed as shown in Figs. 6.19(a)–(c). These three peaks were observed at $\hbar(\omega_1 + \omega_2) = 5.906\,\text{eV}$, $5.910\,\text{eV}$, and $5.913\,\text{eV}$ and the incident angle (θ) dependence of these peak intensities is shown in Fig. 6.20a–c. These three levels were assigned to $\Gamma_1(A_1)$, $\Gamma_5(T_2)$, and $\Gamma_3(E)$ from the low-energy side, from comparison with the theoretical curves calculated from theory of Inoue and Toyozawa [74]. This paper contains the dependence of the two-photon absorption intensity on the polarization angles of two incident beams for every irreducible representation of the elementary excitation for 32 point groups of the crystal.

6.5 Two-Photon Resonant Second-Harmonic Generation

Higher-harmonic generation under two- or three-photon resonant excitation of even or odd excited levels gives us fruitful information on the electronic structure. In this section, we will introduce some interesting features of constructive and destructive interference effects for SHG in the ferroelectric–antiferromagnetic crystals $R\text{MnO}_3$ (R = Y, Ho, Er) and the antiferromagnetic Cr_2O_3. These crystals have hexagonal and corundum structures, respectively. Therefore the 3d electrons on the Mn^{3+} and Cr^{3+} ions are relatively well localized so that ligand field theory gives a good starting point. This is in contrast to the perovskite structure where the itinerant nature of the excitations is inevitable because of the large overlap between the Cu ($3d_{x^2-y^2}$) and O ($2p_\sigma$)

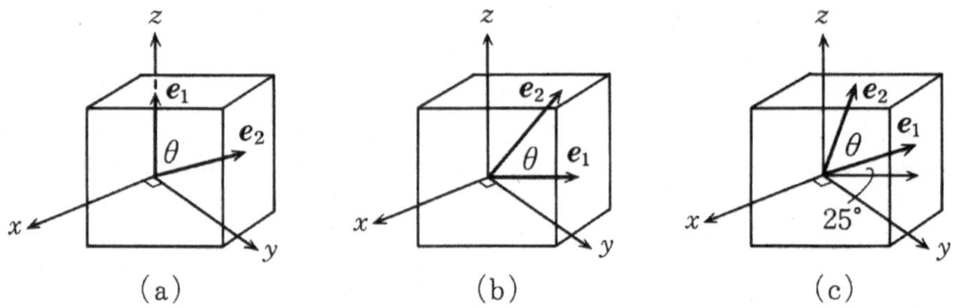

Fig. 6.19. Polarization directions e_1 and e_2 of two incident beams in the experiment of two-photon absorption due to excitonic molecules in CuBr. (a) $e_1 = (0, 0, 1)$, $e_2 = (-(\sin\theta)/\sqrt{2}, (\sin\theta)/\sqrt{2}, \cos\theta)$; (b) $e_1 = (-1/\sqrt{2}, 1/\sqrt{2}, 0)$, $e_2 = (-(\cos\theta)/\sqrt{2}, (\cos\theta)/\sqrt{2}, \sin\theta)$; (c) $e_1 = (-(\cos 25°)/\sqrt{2}, (\cos 25°)/\sqrt{2}, \sin 25°)$, $e_2 = (-\{\cos(\theta + 25°)\}/\sqrt{2}, \{\cos(\theta + 25°)\}/\sqrt{2}, \sin(\theta + 25°))$

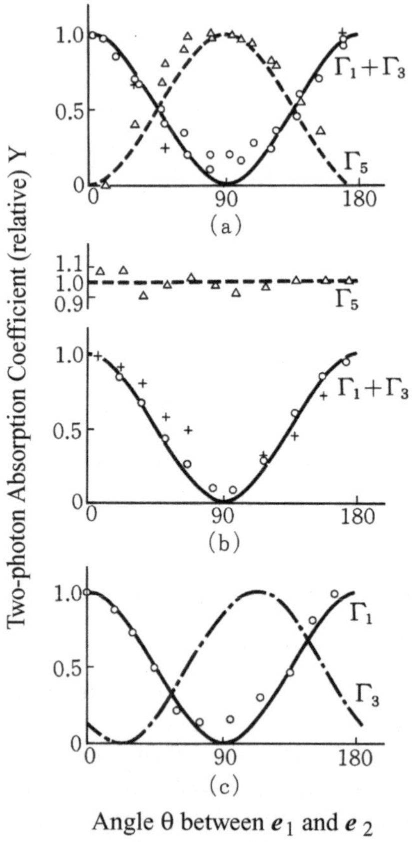

Fig. 6.20. Two-photon absorption intensity due to the excitonic molecule Γ_1, Γ_3, and Γ_5, as a function of the angle between e_1 and e_2 in Figs. 6.19(a)–(c) [79]

orbitals. This was treated by the excitonic cluster model [80–82]. However, even in the hexagonal and corundum structures, the propagation effects, i.e., the excitonic effects, should be taken into account on the excitations, even within the 3d-electron multiplets of a single ion.

In Sect. 6.5.1, we will introduce SHG under nearly resonant two-photon excitation of the $(3d)^4$ electronic levels $^5\Gamma_2$ and $^5\Gamma_1$ in $R\mathrm{MnO_3}$ with the magnetic crystal class $6\underline{mm}$, containing the magnetic space group $P\underline{6}_3c\underline{m}$ and $\underline{P}6_3\underline{c}m$. These levels are split into four levels $E_1^{\pm}$ and $E_2^{\pm}$ by the excitonic effect, i.e., Davydov splitting. SHG due to these two channels under nearly two-photon resonant excitation of the lower two levels(E_1^-) and (E_2^-) interfere constructively in $\mathrm{YMnO_3}$ and $\mathrm{HoMnO_3}$ ($T \leq 42$ K), but destructively in $\mathrm{ErMnO_3}$ and $\mathrm{HoMnO_3}$ ($42 < T < 70$ K). We will show, in Sect. 6.5.2, two kinds of second-order susceptibility: (1) which is invariant, and (2) changes sign under the time-reversal operation. For $\mathrm{YMnO_3}$, the former $\chi^{(i)}$ is linearly proportional to the electric polarization P_z while the latter $\chi^{(c)}$ is proportional to the product of P_z and $\langle S_x \rangle$ the sublattice magnetization [83]. Therefore we can determine the ferroelectric and magnetic domain structure by observing the external interference of SHG due to $\chi^{(i)}$, e.g., with SHG of a quartz crystal. The sign of the sublattice magnetization is determined by the internal interference of SHG due to $\chi^{(c)}$ and $\chi^{(i)}$ [84, 85].

The ferroelectric domain wall was found to be always accompanied by an antiferromagnetic Bloch wall [83, 86]. This can be explained in terms of the polarization-dependent spin anisotropy energy of the $\mathrm{Mn^{3+}}$ spins. In Sect. 6.5.3, we will discuss also the interference effects of SHG due to electric and magnetic dipole moments between the ground state $^4A_{2g}$ and the excited state $^4T_{2g}$ of $\mathrm{Cr^{3+}}$ ions in $\mathrm{Cr_2O_3}$. The SHG tensors χ^e and χ^m due to the electric and magnetic dipole moments are found to be of the same order of magnitude and the phases of χ^m and χ^e differ by $\pi/2$ under two-photon resonant excitation of $^4T_{2g}$. This comes from the propagation and relaxation effects of the magnetic dipolar excitation.

6.5.1 SHG Spectra in Hexagonal Manganites $R\mathrm{MnO_3}$

Optical second-harmonic spectroscopy has proved to be a powerful means for the determination of complex magnetic structures, for example, the non-collinear antiferromagnetic structure of the hexagonal manganites $R\mathrm{MnO_3}$ ($R =$ Sc, Y, Ho, Er, Tm, Yb, Lu) [84, 85]. These compounds are paraelectric above T_C (between 550 and 1000 K) with its space group $P6_3/mmc$, and ferroelectric below T_C with the group $P6_3cm$. They are antiferromagnetic below T_N around 80 K.

Take the case of $\mathrm{YMnO_3}$ as a first example. Below $T_N = 74$ K, second-harmonic generation (SHG) is observed in the region around 2.45 eV, which is described by the magnetic nonlinear susceptibility $\chi^{(c)}_{yyy}$. Here the first suffix describes the polarization direction of the second-harmonic signal while the second and third suffices describe those of the two incident fundamentals. This

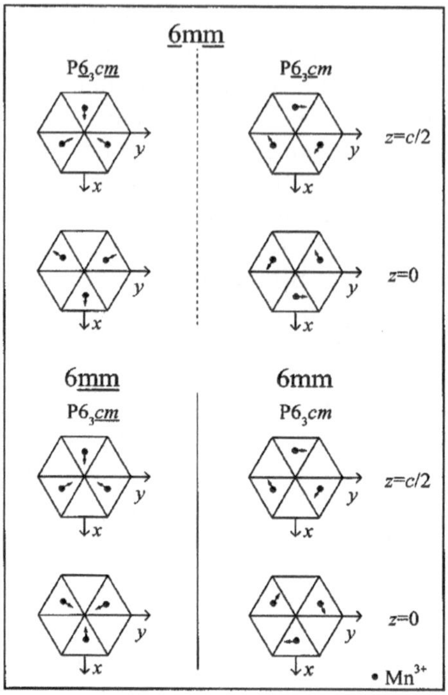

Fig. 6.21. Four different possibilities for the spin ordering of Mn^{3+} ions in the xy-plane. For the magnetic point group $\underline{6}m\underline{m}$ there are two possibilities with different space group symmetries

corresponds to the magnetic space group $P6'_3cm'$ ($P\underline{6}_3c\underline{m}$) and we know the spin ordering should be as shown on Fig. 6.21 [85]. Here and hereafter both a prime and an underbar mean the time-reversal operation. The peaks in the SHG spectra around 2.45 eV in Fig. 6.22(a) seem to indicate the existence of two excited levels 2.45 and 2.51 eV and constructive interference between the susceptibilities associated with each level [83].

On the other hand, in $ErMnO_3$, SHG below $T_N = 79\,K$ around 2.45 eV is observed only in the configuration corresponding to the nonvanishing susceptibility $\chi^{(c)}_{xxx}$. This means that the magnetic space group for this system is $P6'_3c'm$ ($P\underline{6}_3\underline{c}m$) and the spins of Mn ions are all rotated by an angle 90° compared to $YMnO_3$ as shown by Fig. 6.21. There are two peaks also in this case but with destructive interference with a dip between them as shown by Fig. 6.22(b). The spectra of $HoMnO_3$ are interesting in that they show both types of behavior described above, depending on the temperature. The peaks at 2.45 eV show constructive interference in $\chi^{(c)}_{yyy}$ below $T_R = 42\,K$ and become destructive in $\chi^{(c)}_{xxx}$ above T_R. See Figs. 6.22(a′) and (b′).

Observation of $\chi^{(i)}_{zyy}$ related to ferroelectricity gives the position of an excited level at 2.7 eV in all these systems, as shown in Fig. 6.23(a) [84]. The

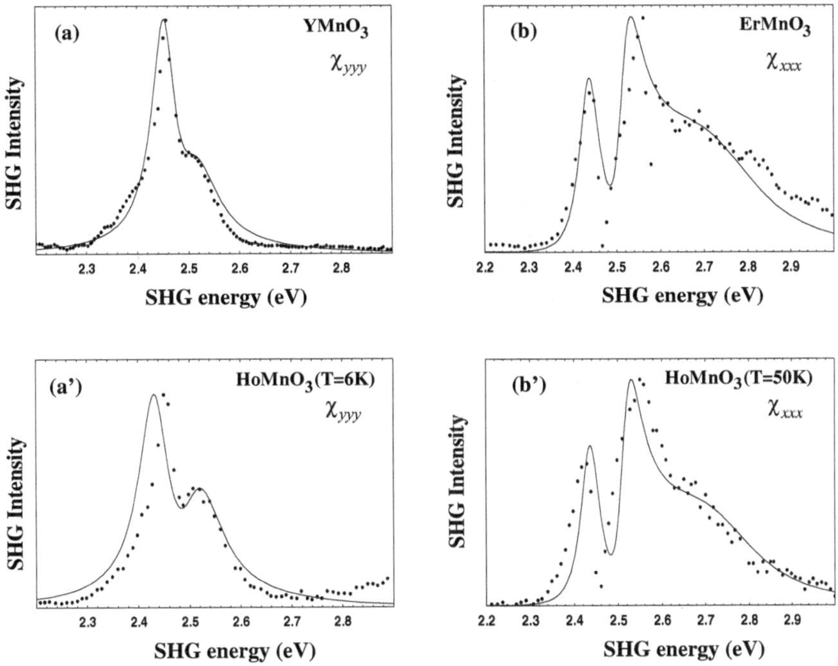

Fig. 6.22. The SHG spectra associated with the nonlinear magnetic susceptibilities of (**a**) $YMnO_3$, (**b**) $ErMnO_3$, (**a′**) $HoMnO_3$ ($T = 6\,K$), and (**b′**) $HoMnO_3$ ($T = 50\,K$). Dots show the experimental results and lines the numerical ones. The material constants have been fixed as shown in [83]

susceptibility $\chi^{(i)}_{zyy}$ is invariant against the time-reversal operation, while $\chi^{(c)}_{yyy}$ changes its sign under the time-reversal operation. Note that the SH signal due to $\chi^{(i)}_{zyy}$ has z-polarization, in contrast to the y-polarization of $\chi^{(c)}_{yyy}$ both under the y-polarization of two fundamentals.

The purpose of the present subsection is to try to understand these features of the SHG spectra as well as to clarify the relation between them and the magnetic structures of hexagonal manganites through the calculation of the susceptibilities $\chi^{(c)}_{yyy}$ and $\chi^{(c)}_{xxx}$, which are simply denoted as χ_{yyy} and χ_{xxx}, in the present subsection. Here, we describe the crystal magnetic structure of the present system. The environment of the Mn ions to be treated here is unusual in that the Mn^{3+} ions with total spin $S = 2$ are surrounded by the five coordinated (trigonal) bipyramid of O^{2-} ions. The electronic states will be discussed. There are six Mn sites in a unit cell of the antiferromagnetic phase. We describe how to correlate wavefunctions at different sites, and then find that the single-ion theory does not work well and develop the exciton theory for the excited states around 2.45 eV. The susceptibilities obtained in the exciton model turn out to be satisfactory. The exciton model predicts two excited levels near the single level expected in the single-ion theory and

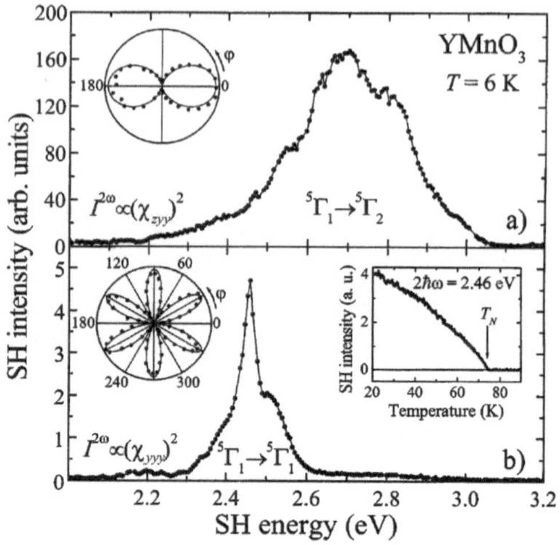

Fig. 6.23. SHG spectra of $YMnO_3$ due to the ferroelectric ordering (**a**) and due to the antiferromagnetic ordering (**b**) [84]

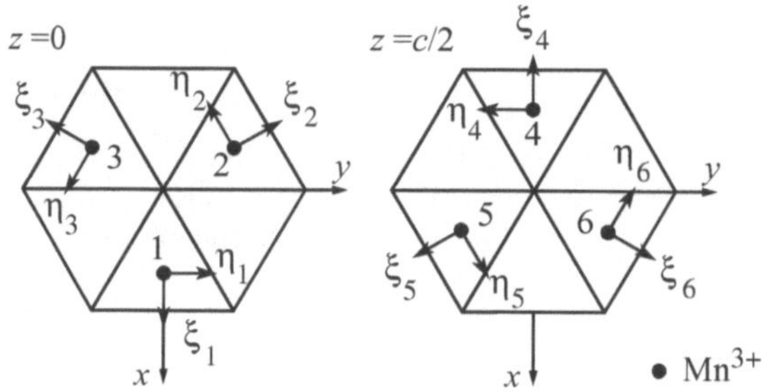

Fig. 6.24. Six Mn sites in the unit cell and the local coordinate axes. The filled circles denote manganese ions, and oxygen ions are located at every corner of the unit cell. The spins $S = 2$ of the Mn_i ion are in the directions ξ_1, ξ_2, ξ_3, $-\xi_4$, $-\xi_5$, and $-\xi_6$ in $YMnO_3$

quite different interference behavior for χ_{yyy} and χ_{xxx}. Finally we give a brief discussion of the possible cause of clamping of two order parameters, ferroelectric and antiferromagnetic. Comparison of the calculated spectra with the observed one is also made there.

The crystal structure of ferroelectric $R\mathrm{MnO_3}$ (R = Y, Ho, Er, Lu) is reported by Yakel et al. [87]. The x- and y-axes chosen in the present subsection coincide with theirs and those of Fröhlich et al. [84] as well. There are six Mn ions in a magnetic unit cell. Their sites in the unit cell and our choice of local axes are drawn in Fig. 6.24. As seen in Fig. 6.24, three Mn_i ions (i = 1, 2, 3) are supposed to lie in the $z = 0$ plane, while the other three with $i = 4, 5, 6$ are in the $z = c/2$ plane. Let us further assume that the coordinate of Mn_1 is given by $(d, 0, 0)$ with $d \sim 0.3a$ and that of Mn_4 as $(-d, 0, c/2)$ so that

$C_2(\tau)$ (with time reversal θ) carries Mn_1 into Mn_4 with its environment in the crystal.

The local coordinate $(\xi_i, \eta_i) = R(\theta_i)(x, y)$ indicates that the local ξ_i- and η_i-axes are obtained by rotating the global x- and y-axes through an angle θ_i, so that, for example, we have the following relations:

$$P_x = P_{\xi_i} \cos\theta_i - P_{\eta_i} \sin\theta_i, \tag{6.82}$$
$$P_y = P_{\xi_i} \sin\theta_i + P_{\eta_i} \cos\theta_i, \tag{6.83}$$

between the components of the electric dipole-moment operators at different sites.

We follow Fröhlich et al. [84] in the choice of symmetry operations of the two possible magnetic space groups $P6_3'cm'$ (spins of $Mn_1 \parallel \boldsymbol{x}$) and $P6_3'c'm$ (spins of $Mn_1 \parallel \boldsymbol{y}$). They are given by

$$\begin{aligned} (a) \quad P6_3'cm' : 6_3' &= \theta C_6(\boldsymbol{\tau}), \\ c &= \sigma_d(\boldsymbol{\tau}), \\ m' &= \theta\sigma_v \,, \end{aligned} \tag{6.84}$$
$$\begin{aligned} (b) \quad P6_3'c'm : 6_3' &= \theta C_6(\boldsymbol{\tau}) \,, \\ c' &= \theta\sigma_d(\boldsymbol{\tau}) \,, \\ m &= \sigma_v \,, \end{aligned} \tag{6.85}$$

where σ_d and σ_v are reflection in the yz- and xz-planes, respectively, and primes instead of underlines have been used here to denote anti-unitary operators. The vector τ is give by (0, 0, $c/2$) and θ is time reversal as usual.

The spin ordering corresponding to these magnetic space groups are drawn in Figs. 6.21 and 6.24 and the relevant electronic levels of the Mn^{3+} ion obtained from the polarization characteristics are shown in Fig. 6.25.

Then we find that (a) $\epsilon_0\chi_{yyy}$ is linearly proportional to the product of two order parameters, i.e., the sublattice magnetization $\langle S_x \rangle$ and the ferroelectric potential v_{zx} or $\bar{v}_z$ for $YMnO_3$ and $HoMnO_3$ ($T \leq 42$ K), while (b) $\epsilon_0\chi_{xxx}$ for $ErMnO_3$ and $HoMnO_3$ ($42 < T < 70$ K) is linearly proportional to the product of the sublattice magnetization $\langle S_y \rangle$ and the ferroelectric potential v_{zx} or $\bar{v}_z$. The matrix elements v_{zx} and $\bar{v}_z$ of V_{zx} and V_z are linearly proportional to the ferroelectric polarization P_z. In the ferroelectric phase, the Mn ion is surrounded by a distorted and tilted bipyramid of O^{2-} ions, the site symmetry being $C_s = m = \{E, \sigma_v\}$, where $\sigma_v = \sigma_y$ is reflection in the xz-plane. The effect of this ferroelectric phase is treated [82] as a perturbation on Mn ion due to fields V_m having symmetry lower than D_{3h}:$V_z = \sum_i Az_i$ and $V_{zx} = \sum_i Bz_ix_i$.

The ground state of the whole system is described as

$$\Psi_g = \prod_{n\beta} \psi_{n\beta}, \tag{6.86}$$

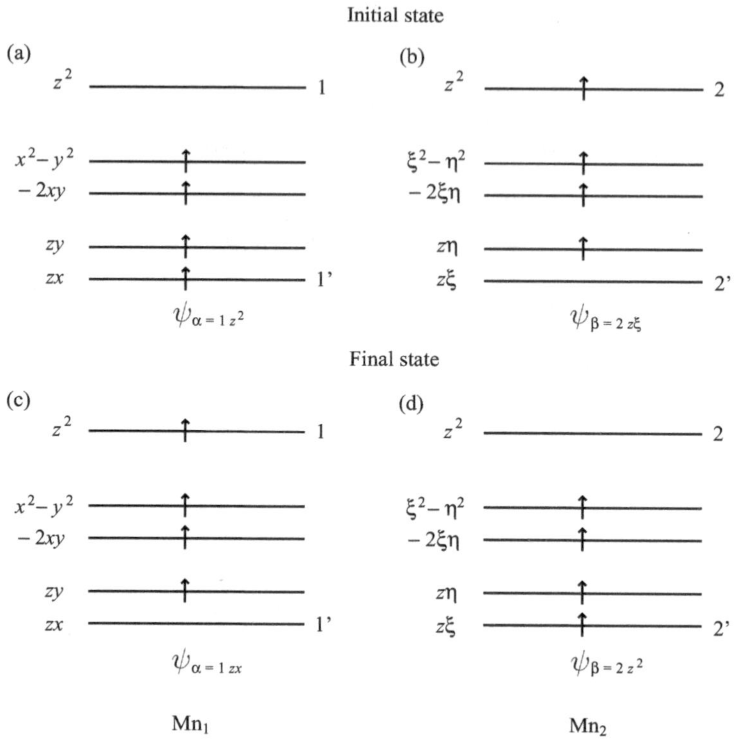

Fig. 6.25. Excitation transfer between ions Mn_1 and Mn_2

where $\psi_{n\beta}$ represents the ground state of the Mn ion at the β (=1,$\cdots$,6) site of the nth unit cell. For example, ψ_{n1} and ψ_{n2} are drawn in the upper left and the lower right, respectively, in Fig. 6.25. The coordinates of the β site are drawn in Fig. 6.24. When one of the Mn ions, i.e., that at (mα) is excited to the state $\psi_{m\alpha\lambda}$, we have the localized excited state

$$\Psi_{m\alpha\lambda} = \psi_{m\alpha\lambda} \prod_{n\beta}{}' \psi_{n\beta}. \tag{6.87}$$

Here $\lambda = 1$ and $\lambda = 2$ correspond to the excitation to the state zx (or $z\xi$) and zy (or $z\eta$) in Table 6.1, respectively.

First, in order to take into account the exciton effect, i.e., the Davydov splitting, we make the irreducible representation $\Psi_1(E_2x)$ of C_{6v} symmetry which is made up of a linear combination of $\Psi_1(\alpha)$ with $\Psi_{m\alpha 1} \equiv \Psi_{z\xi}(\alpha)$ ($\alpha = 1, \ldots, 6$), i.e.,

$$\Psi_1(E_2x) = \frac{1}{2}\{\Psi_1(2) - \Psi_1(3) + \Psi_1(5) - \Psi_1(6)\} \tag{6.88}$$

and $\Psi_2(E_2x)$ as a linear combination of $\Psi_2(\beta)$ with $\Psi_{m\alpha 2} \equiv \Psi_{z\eta}(\beta)$ ($\beta = 1, \ldots, 6$), i.e.,

Table 6.1. Wavefunctions of the ground and excited states of the Mn^{3+} ion in RMnO$_3$

states		wavefunction	excitation energy
ground	$^5\Gamma_1$ ($^5A_1$)	$\psi_{z^2} \equiv \lvert\varphi_{zx}, \varphi_{zy}, \varphi_{x^2-y^2}, \varphi_{-2xy}\rvert$	0
excited			
(2 photon)	$^5\Gamma_1$ (^{5}E$_1$a)	$\psi_{zx} \equiv \lvert\varphi_{zy}, \varphi_{x^2-y^2}, \varphi_{-2xy}, \varphi_{z^2}\rvert$	2.7 eV
	$^5\Gamma_2$ (^{5}E$_1$b)	$\psi_{zy} \equiv -\lvert\varphi_{zx}, \varphi_{x^2-y^2}, \varphi_{-2xy}, \varphi_{z^2}\rvert$	2.46 eV
(1 photon)	$^5\Gamma_1$ (^{5}E$_2$a)	$\psi_{x^2-y^2} \equiv \lvert\varphi_{zx}, \varphi_{zy}, \varphi_{-2xy}, \varphi_{z^2}\rvert$	} 1.6 eV
	$^5\Gamma_2$ (^{5}E$_2$b)	$\psi_{-2xy} \equiv -\lvert\varphi_{zx}, \varphi_{zy}, \varphi_{x^2-y^2}, \varphi_{z^2}\rvert$	

$$\begin{aligned}\Psi_2(E_2x) = \frac{1}{2\sqrt{3}}\{&2\Psi_2(1) - \Psi_2(2) - \Psi_2(3) \\ &+ 2\Psi_2(4) - \Psi_2(5) - \Psi_2(6)\}.\end{aligned} \tag{6.89}$$

These two states with the same symmetry are mixed by two kinds of excitation transfers among the sublattice ions, i.e., within the layer and between the neighboring layers:

$$\Psi(\nu E_2) = \nu_1 \Psi_1(E_2x) + \nu_2 \Psi_2(E_2x). \tag{6.90}$$

Hereafter, the lower and higher energy E_2 state will be distinguished by $\nu = -$ and $\nu = +$, respectively.

Similarly, two levels of the E_1 state are also mixed up as

$$\Psi(\mu E_1) = \mu_1 \Psi_1(E_1x) + \mu_2 \Psi_2(E_1x). \tag{6.91}$$

We also associate $\mu = -$ and $\mu = +$ with the lower and higher energy eigenvalues obtained here.

At this stage, we realize a possible interpretation of the structures of the observed χ_{yyy} for (a) YMnO$_3$ and χ_{xxx} for (b) ErMnO$_3$. The two lines on the lower energy side 2.46 eV are likely to correspond to ($\mu = -, E_1$) and ($\nu = -, E_2$), while the other two lines with higher energy 2.7 eV may be associated with ($\mu = +, E_1$) and ($\nu = +, E_2$).

All these four states are optically accessible, and the susceptibility for SHG are described for the cases (a) YMnO$_3$ and (b) ErMnO$_3$, respectively:

$$(a)\quad \varepsilon_0 N\chi_{yyy} = \sum_{\nu} \frac{\langle \Psi_g|P_y|\Psi(\nu E_2)\rangle\langle\Psi(\nu E_2)|P_yP_y|\Psi_g\rangle}{(E(\nu E_2)-2\hbar\omega)\Delta E} + \sum_{\mu} \frac{\langle \Psi_g|P_y|\Psi(\mu E_1)\rangle\langle\Psi(\mu E_1)|P_yP_y|\Psi_g\rangle}{(E(\mu E_1)-2\hbar\omega)\Delta E}, \tag{6.92}$$

$$(b)\quad \varepsilon_0 N\chi_{xxx} = \sum_{\nu} \frac{\langle \Psi_g|P_x|\Psi(\nu E_2)\rangle\langle\Psi(\nu E_2)|P_xP_x|\Psi_g\rangle}{(E(\nu E_2)-2\hbar\omega)\Delta E} + \sum_{\mu} \frac{\langle \Psi_g|P_x|\Psi(\mu E_1)\rangle\langle\Psi(\mu E_1)|P_xP_x|\Psi_g\rangle}{(E(\mu E_1)-2\hbar\omega)\Delta E}. \tag{6.93}$$

The matrix elements of (P_x, P_y) and (P_xP_x, P_yP_y) become finite only with the help of the spin–orbit interaction $\mathcal{H}_{so} = \lambda \boldsymbol{L}\cdot\boldsymbol{S}$ and the lower-symmetry crystalline field C_s, i.e., $V_m = V_z + V_{zx} + V_x$. Here $V = \sum_i v_i$, $v_{zi} = Az_i + \cdots$, $v_{zxi} = Bz_ix_i + \cdots$, $v_{xi} = Cx_i + D(x_i^2 - y_i^2) + \cdots$, and these fields are finite only in the ferroelectric phase. As a result, (a) χ_{yyy} is found to be proportional to the product of two order parameters $\langle S_x\rangle$ and v_{zx} or $\langle S_x\rangle$ and $\bar{v}_z$, and (b) χ_{xxx} is proportional to the product of $\langle S_y\rangle$ and v_{zx} or $\langle S_y\rangle$ and $\bar{v}_z$ [83]. Here $\langle S_x\rangle$ and $\langle S_y\rangle$ are the expectation value of the spin operators S_x and S_y in the electronic ground state at sublattice 1, and $v_{zx} \equiv \langle \Phi_{x^2-y^2}|V_{zx}|E_1a\rangle = -\langle \Phi_{-2xy}|V_{zx}|E_1b\rangle$, while $\bar{v}_z \equiv \langle A_1|V_z|\Phi_z\rangle$. It is noted here that both matrix elements v_{zx} and $\overline{v}_z$ are linearly proportional to the ferroelectric polarization P_z. We also introduce the relaxation rates $\Gamma(\nu E_2)$ and $\Gamma(\mu E_1)$ in such a way that the causality relation is satisfied and the best fitting is obtained between the calculated and observed SHG spectra. The results of SHG spectra are drawn in Fig. 6.22 and the chosen material constants and relaxation rates are listed in [83].

The first difference of (a) $YMnO_3$ (Fig. 6.22a) and (b) $ErMnO_3$ (Fig. 6.22b) is that the signals at $E(\nu = -,\ E_2) = 2.45$ eV and $E(\mu = -,\ E_1) = 2.51$ eV interfere constructively in (a), but destructively in (b) making the sharp dip of the SHG signal for $2\hbar\omega$ between 2.45 eV and 2.51 eV. The second difference is conversely that the SHG signals at $E(\mu = +,\ E_1)$ and $E(\nu = +,\ E_2)$ interfere constructively in (b) but that the SHG signal on the higher energy side of case (a) vanishes due to destructive interference as well as the larger relaxation rates. These differences originate from the different relative magnitudes of $\bar{v}_z/v_{zx}$.

We can determine both the amplitude and phase of χ_{yyy} or χ_{xxx} by measuring the interference effects of the SHG signal with that of a suitable reference material. In fact, Im χ_{yyy} of $YMnO_3$ was observed to change its sign when crossing the border of two antiferromagnetic domains where the sign of $\langle S_x\rangle$ changes, in agreement with the theoretical result. However, the sign of Im χ_{yyy} was found not to change when crossing the ferroelectric (FEL) domain boundary. To reconcile this experimental result with the bilinear form of two order parameters obtained theoretically, we should introduce the clamping model in which the direction of the sublattice magnetization is also reversed on the border of the FEL domains [83]. This is in accordance with the idea that the electronic states of the Mn_1 and Mn'_1 ions, each located near the boundary of the two FEL domains in contact, are connected by the operation of σ_h, including the spin as well as the orbital state. As a matter of fact, the reversal of the spin is energetically more favorable, as we have confirmed for the adopted model that the (FEL+, $\boldsymbol{S}//\boldsymbol{x}$) domain in contact with (FEL−, $\boldsymbol{S}//-\boldsymbol{x}$) can be lower in energy than (FEL+, $\boldsymbol{S}//\boldsymbol{x}$) with (FEL−, $\boldsymbol{S}//\boldsymbol{x}$). This point will be discussed in the next subsection [88, 89].

6.5.2 Ferroelectric and Magnetic Domains

The SH susceptibility $\chi^{(c)}_{yyy} \equiv \chi_{yyy}$, discussed for $YMnO_3$ in the last subsection, changes the sign, and $\chi^{(i)}_{zyy}$ is invariant under the time-reversal operation. These are derived explicitly [83] in terms of the order parameter of the ferroelectric polarization P_z and the sublattice magnetization $\langle S_x \rangle$ in the first sublattice in Fig. 6.24:

$$\epsilon_0 \chi^{(i)}_{zyy}(2\omega) \propto P_z \frac{1}{E_1 - 2\hbar\omega - i\Gamma_1}, \tag{6.94}$$

$$\epsilon_0 \chi^{(c)}_{yyy}(2\omega) \propto \langle S_x \rangle P_z \left(\frac{1}{E_2 - 2\hbar\omega - i\Gamma_2} + \frac{\gamma}{E_1 - 2\hbar\omega - i\Gamma_1} \right). \tag{6.95}$$

Here the first suffix z and y of χ means the polarization of the SHG signal, and the second and third suffices y are those of the fundamental. $E_1 \equiv E(^5E_1a) - E(^5A_1) = 2.7\,\mathrm{eV}$, $E_2 \equiv E(^5E_1b) - E(^5A_1) = 2.45\,\mathrm{eV}$, and γ is a constant much smaller than unity. In deriving (6.94) and (6.95), we have used as the basis functions the 3d orbitals in the paraelectric and paramagnetic phase, and taken into account the spin–orbit interaction and the lower-symmetry crystalline field V_{zx} and V_z due to the FEL displacement as the perturbation on the basis functions. The matrix elements of V_{zx} and V_z are linearly proportional to the order parameter of the FEL polarization P_z. The magnetic unit cell consists of six Mn^{3+} ions as shown in Fig. 6.24 and we have chosen the sublattice magnetization $\langle S_x \rangle$ of the Mn_1 ion as the AFM order parameter because those of other sublattices are transposed onto that of sublattice 1 by the symmetry operations in (6.84) and (6.85).

It is to be noted here that $\chi^{(i)}$ is linearly proportional to P_z and $\chi^{(c)}$ is linearly proportional to the product of $\langle S_x \rangle$ and P_z. As a result, we can determine the FEL domain structure by the interference effects of SHG due to $\chi^{(i)}$ and the external signal. When we observe the interference pattern of $\chi^{(c)}$ with the external one, the brightness of the interference pattern is determined by the sign of the product of $\langle S_x \rangle$ and P_z. In fact, these domain structures were observed by Fiebig *et al.* [86]. These examples are shown in Figs. 6.26(a) and (b). The bright (dark) domain region comes from the constructive (destructive) interference between the SHG signal and the external field. It looked strange in the beginning that $\chi^{(c)}$ does not change the sign

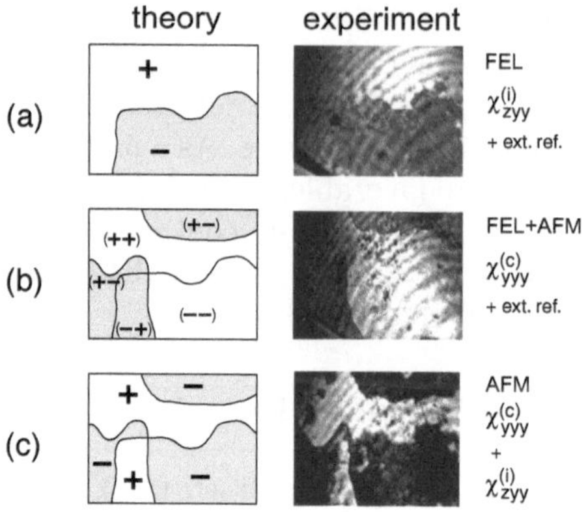

Fig. 6.26. Interference pattern of second-harmonic generation in ferroelectric and antiferromagnetic $YMnO_3$. (**a**) Interference of $\chi^{(i)}_{zyy}$ with external reference (abbreviated as ext. ref. in the figure) giving the FEL domain structure; (**b**) external interference of $\chi^{(c)}_{yyy}$ giving the product of FEL and AFM sign; and (**c**) the internal interference of $\chi^{(i)}_{zyy}$ and $\chi^{(c)}_{yyy}$ giving the AFM domain structure [83, 86]

at the FEL domain boundary (DB) in Fig. 6.26(b) in contradiction to the expression (6.95). We proposed that the clamping model of the AFM domain wall (DW) at the FEL DB can explain this phenomenon [83, 89]. In fact, when the internal interference of $\chi^{(i)}$ and $\chi^{(c)}$ SHG was observed, the sign of the sublattice magnetization $\langle S_x \rangle$ was found to change at the FEL DB as shown in Fig. 6.26(c). Here arises the question of why and how the coupling of two order parameters P_z and $\langle S_x \rangle$ is induced at the FEL DB. This also appeared at first to contradict the fact that the two critical temperatures 914 K and 74 K are so different from each other. This is because it is speculated from conventional Ginzburg–Landau (GL) theory that the coupling is very weak or negligible for such a case with large different values of the two critical temperatures as for $YMnO_3$ [90]. This mystery is resolved as follows. Starting from the microscopic Hamiltonian of the Mn^{3+} $(S = 2)$ spin system:

$$\mathcal{H} = -2 \sum_{\langle ij \rangle} J_{ij} \boldsymbol{S}_i \cdot \boldsymbol{S}_j - \sum_i \left(D_{\xi\xi} S_{i\xi}^2 + D_{\eta\eta} S_{i\eta}^2 \right) , \tag{6.96}$$

we obtain the Ginzburg–Landau (GL) free energy density [89]. In (6.96), J_{ij} represents the superexchange integral between the pair (i, j) of nearest neighbor Mn^{3+} ions, and $D_{\xi\xi} = 3\lambda^2/E_2$ and $D_{\eta\eta} = 3\lambda^2/E_1$ are obtained as the contribution of the second-order spin–orbit interaction $\sum_{(i)} \lambda \boldsymbol{S}_i \cdot \boldsymbol{L}_i$. Here, we adopt the continuum approximation and the classical spin model:

$$S_\xi(\boldsymbol{r}) = S \cos\phi(\boldsymbol{r}), \quad S_\eta(\boldsymbol{r}) = S \sin\phi(\boldsymbol{r}), \tag{6.97}$$

where the sublattice magnetization of the Mn_1 ion is chosen as the order parameter of the spin system, and then ξ and η are coincident with the global

x- and y-axis in Fig. 6.24, respectively. Solving the GL equation, we obtain solitons for both the electric polarization $P(y) \equiv P_z$ and the sublattice magnetization S_x:

$$P(y) = P_0 \tanh(y/\delta_p), \tag{6.98}$$
$$S_x(y) = S\cos\phi(y) = \pm S\tanh(y/\delta_B). \tag{6.99}$$

Here we chose the domain boundary along the x-axis, and the thickness of the FEL DB δ_p is of the order of the lattice constant, while that of the Bloch wall δ_B is estimated to be 20 times the lattice constant. We have two terms coupling two DBs, i.e., (1) the antisymmetric exchange interaction

$$\mathcal{H}'' = \sum_{\langle i,j\rangle} d_{ij}(\boldsymbol{S}_i \times \boldsymbol{S}_j)_z, \tag{6.100}$$

with

$$d_{ij} = 2\lambda\left[\frac{J_{ij}(20;00)}{E_1E_2}v_{z\xi}(i) - \frac{J_{ij}(00;20)}{E_1E_2}v_{z\xi}(j)\right], \tag{6.101}$$

and (2) the higher-order anisotropy energy

$$\mathcal{H}''_{P\phi} = -\sum_i D_{z\xi}\left(S_\xi S_z + S_z S_\xi\right)_i . \tag{6.102}$$

Here $v_{z\xi}(i) \equiv V^0 P_z$ is a matrix element of $V_{z\xi}$ at the i site and $D_{z\xi} = \sqrt{3}\lambda^2 v_{z\xi}/(E_1E_2)$. The first interaction is derived by associating the spin–orbit interaction with the Heisenberg Hamiltonian in (6.96) and the second is obtained by modifying the spin-anisotropy energy in (6.96) with the FEL crystalline field. These interactions are evaluated as perturbations working on the kink solitons (6.98) and (6.99). We have found that the polarization-dependent higher-order anisotropy energy $\mathcal{H}'$ is enough to stabilize the clamped FEL DB and AFM DW when the AFM domain size is larger than 10 μm [89].

We have shown in the last subsection that the FEL DB is always accompanied with the AFM DW. These results will be understood intuitively in this section by using the symmetry breaking of the crystal in the FEL and AFM phase transitions.

The hexagonal manganites $R\text{MnO}_3$ have space group $P6_3/mmc$ in the paraelectric and paramagnetic phase. When this crystal suffers an FEL phase transition around 1000 K, one of the mirror-reflection symmetries, σ_h (mirror-reflection in xy-plane), is lost below T_{C} and the crystal is deformed into the space group $P6_3cm$. The AFM phase transition depends on the species of R. In the case of YMnO_3, another mirror symmetry σ_v in the xz-plane is lost below $T_{\text{N}} = 74\,\text{K}$ and the magnetic space group is $P6'_3cm'$. Under this symmetry, the canting of the Mn_1 spin toward the z-axis is allowed so that P_z, $\langle S_x\rangle$ and $\langle S_z\rangle$ are finite but $\langle S_y\rangle$ vanishes. On the other hand, the third mirror symmetry σ_d in the yz-plane is lost below T_{N} in the case of ErMnO_3. Here the canting of the sublattice magnetization toward the z-axis is not allowed

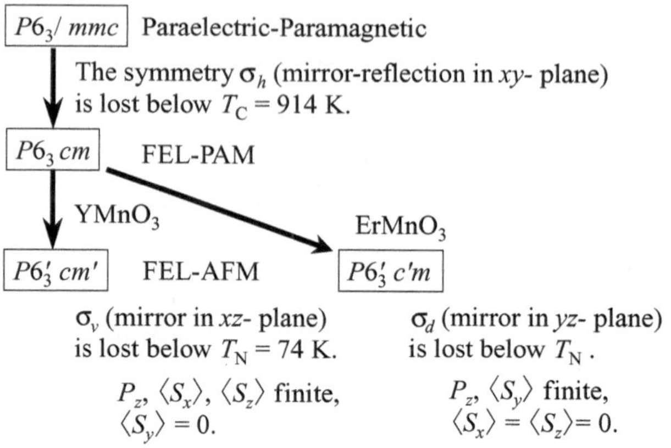

Fig. 6.27. Symmetry breaking and clamping of FEL and AFM order parameters at FEL DB in RMnO$_3$ [88]

and only P_z and $\langle S_y \rangle$ are finite, while $\langle S_x \rangle = \langle S_z \rangle = 0$. These processes of symmetry breaking are summarized in Fig. 6.27 and Table 6.2.

From these considerations, we will be able to point out two important facts. First, the FEL DB is always accompanied by the AFM DW in YMnO$_3$ while the AFM DW can exist independently of the FEL DB. Second, both DB and DW are stabilized by the anisotropy energy (6.102).

The FEL DB is made at $y = 0$ by operating with σ_h on the right-hand side ($y > 0$) of the crystal keeping the other side ($y < 0$) in the original state (P_z, S_x, S_z). Then the right-hand state is changed into ($-P_z$, $-S_x$, S_z) so that we have the FEL DB accompanied by the AFM DW around $y = 0$. On the other hand, when we operate with σ_v on the right-hand side ($y > 0$), keeping the

Table 6.2. Formation of the ferroelectric domain boundary (FEL DB) and the antiferromagnetic domain wall (AFM DW) in YMnO$_3$ [88]

FEL DB		AFM DW	
$y < 0$	$y > 0$	$y < 0$	$y > 0$
	operates σ_h		operates σ_v
P_z, S_x, S_z	$-P_z$, $-S_x$, S_z	P_z, S_x, S_z	P_z, $-S_x$, $-S_z$
At FEL DB, both (P_z, S_x) change sign simultaneously.		Only S_x and S_z changes sign. S_z is a hidden order-parameter.	
The clamping of (P_z, S_x) at FEL DB is stabilized by $\mathcal{H}''_{P\phi}$.		The AFM DW can exist independently of FEL DB.	

other side ($y < 0$) in the original state (P_z, S_x, S_z), the right-hand state is changed into (P_z, $-S_x$, $-S_z$). This describes a single AFM DW at $y = 0$. These processes are summarized in Table 6.2. As mentioned above, the observed clamping can be explained as due to the coupling terms between P_z and S_x by (6.102). The clamping corresponds to the simultaneous change of the spin- and polarization-direction across the FEL DB and will be realized under the operation σ_h as described here. If we accept the microscopic mechanism proposed in (6.102), we find that this is equivalent to assuming that σ_h always operates both on P_z and $\langle S_x \rangle$ in the same way as when FEL DB is crossed over. We are then tempted to say that, from symmetry considerations, the FEL DB is always accompanied by the AFM DW while the AFM DW can exist by itself. Both the FEL DB accompanied by the AFM DW and the single AFM DW are stabilized by the higher-order anisotropy energy (6.102). Here, however, nobody has yet observed the sign of the spin-canting S_z which plays the role of a hidden order parameter.

Note that this is a story for $YMnO_3$ and it is not simply applicable to other crystals, e.g., to $ErMnO_3$ where the sublattice magnetization $\langle S_y \rangle$ of the Mn_1 ion is parallel to the y-axis. Furthermore, $\langle S_x \rangle$ and $\langle S_z \rangle$ vanish here, so that the higher-order anisotropy energy in question will not be able to clamp two order parameters in $ErMnO_3$. Only the antisymmetric exchange interaction (6.100) will favor the clamping of the AFM DW to the FEL DB but that energy will not be large enough to compensate the formation energy of the AFM DW.

6.5.3 SHG from the Corundum Structure Cr_2O_3

Corundum, Al_2O_3, is a parent crystal of the ruby laser Cr:Al_2O_3 and Ti-sapphire laser Ti:Al_2O_3. Lasing was observed for the first time in the former, while the latter is the most popular crystal to achieve a short-pulse, high-power laser. The crystal Cr_2O_3 also has a corundum structure and shows absorption and emission spectra similar to Cr:Al_2O_3. This means that we will be able to understand the spectra of Cr_2O_3 also in terms of ligand field theory. However, coherent nonlinear spectroscopy is a very sensitive tool to detect the electronic structure. In fact, we will understand in this subsection that the excitonic effect, i.e., the effect of propagation of optical excitation, is inevitable in understanding the SHG spectrum also from Cr_2O_3 [91,92].

The ground state and excited states, in the visible region, of the Cr_2O_3 crystal are well described by the multiplet terms of the Cr^{3+} ion. This crystal loses its spatial inversion symmetry below the Néel temperature 307.5 K, so that the crystal has symmetry $R\bar{3}'c'$. As Fig. 6.28 shows, we choose the three-fold C_3 and two-fold C_2 axes on the z- and x-axes, respectively. We choose the basis functions in the cubic field to describe the wavefunctions of the $(3d)^3$ electrons on the Cr^{3+} ion. In this subsection, we will discuss SHG under nearly two-photon resonant excitation of ${}^4T_{2g}$ from the ground state ${}^4A_{2g}$.

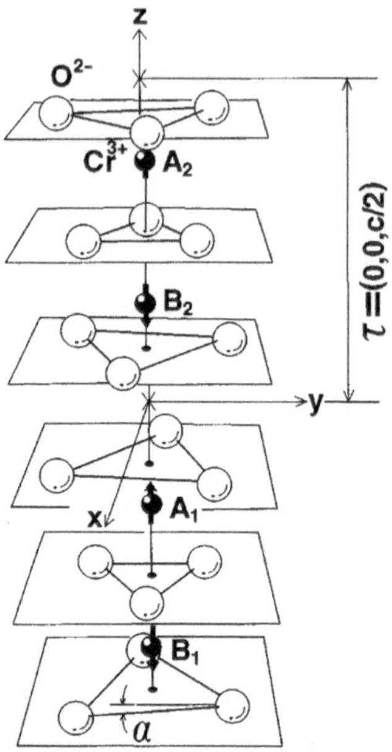

Fig. 6.28. The crystal and magnetic structure of the Cr_2O_3 crystal below the Néel temperature. The black and white circles represent the Cr^{3+} and O^{2-} ions, respectively. The C_3 axis of the crystal is chosen as the z-axis, and the C_2 axis is chosen as the x-axis. The yz-plane is the mirror plane. The origin of coordinates is the inversion center. The arrows indicate the directions of the spins

There are two channels of SHG through the magnetic-dipole and the electric-dipole, assisted by the crystalline field $V_{\mathrm{twist},u}[T_{2u}x_0]$. The SHG spectrum and the interference effect of these two channels are the main subjects of this subsection [91,92].

Four Cr^{3+} ions are contained within a unit cell in the AF phase, as shown in Fig. 6.28. Once the wavefunction on the A_1 site is known, the other three wavefunctions on the B_1, B_2, and A_2 sites are also obtained by the following three symmetry operations of $R\bar{3}'c'$ as follows [91]:

$$\Psi_i[B_1] = C_{2x}(\tau)\Psi_i[A_1], \tag{6.103}$$

$$\Psi_i[B_2] = \Theta I \Psi_i[A_1], \tag{6.104}$$

$$\Psi_i[A_2] = \Theta\sigma_d(\tau)\Psi_i[A_1]. \tag{6.105}$$

Here $C_{2x}(\tau)$ means a π rotation around the x-axis and subsequent translation in the z-direction by $\tau = (0,0,c/2)$. Θ and I are the temporal and spatial inversions, respectively, and $\sigma_d(\tau)$ is the mirror reflection in the yz-plane and subsequent translation by τ. Therefore once we evaluate the contribution to the magnetic dipole SHG tensor χ^m and the electric dipole χ^e from the A_1 ion, we will be able to obtain χ^m and χ^e for the whole crystal.

Second-Order Susceptibilities

The matrix elements of an operator $\hat{A}$ at different sites are correlated to each other through the equations

$$\langle R\Psi|\hat{A}|R\Psi'\rangle = \langle\Psi|R^{-1}\hat{A}R|\Psi'\rangle, \tag{6.106}$$
$$\langle\Theta R\Psi|\hat{A}|\Theta R\Psi'\rangle = \langle\Psi|\Theta^{-1}R^{-1}\hat{A}R\Theta|\Psi'\rangle^*, \tag{6.107}$$

where R stands for any of the symmetry operations $C_{2x}(\tau)$, I, and $\sigma_d(\tau)$. Since both orbital and spin states are involved in the present problem, operator R as well as Θ act upon both of them. The matrix elements of the x-components of the magnetic and electric dipole moments, M_x and P_x, at the B_1, A_2, and B_2 sites are related to those at A_1 by

$$M_x[B_1] = M_x[A_1], \qquad P_x[B_1] = P_x[A_1], \tag{6.108}$$
$$M_x[A_2] = -M_x[A_1]^*, \quad P_x[A_2] = -P_x[A_1]^*, \tag{6.109}$$
$$M_x[B_2] = -M_x[A_1]^*, \quad P_x[B_2] = -P_x[A_1]^*. \tag{6.110}$$

These equations enable us to correlate the values of SH susceptibilities at B_1, B_2, and A_2 to that of A_1. On summing up the contributions from the four Cr^{3+} ions in the unit cell by using relations (6.106)–(6.110), we have χ^m and χ^e below T_N, keeping only the term with the resonance enhancement, as

$$\chi^m = \frac{4Nn}{\epsilon_0 c\hbar^2}\sum_{i,m,k}\rho_i\frac{i\,\mathrm{Im}(MPP)_{imki}}{(\omega_{mi}-2\omega)(\omega_{ki}-\omega)}, \tag{6.111}$$
$$\chi^e = \frac{4N}{\epsilon_0\hbar^2}\sum_{i,m,k}\rho_i\frac{i\,\mathrm{Im}(\bar{P}PP)_{imki}}{(\omega_{mi}-2\omega)(\omega_{ki}-\omega)}. \tag{6.112}$$

Here N is the number density of the unit cells, n is the refractive index for the fundamentals, and $\rho_i = \rho(^4A_{2g}M_s)$ is the thermal distribution in the electronic ground state, $^4A_{2g}$. In the paramagnetic phase, we have the expressions

$$\chi^m = \frac{4Nn}{\epsilon_0 c\hbar^2}\sum_{i,m,k}\rho_i\frac{(MPP)_{imki}}{(\omega_{mi}-2\omega)(\omega_{ki}-\omega)}, \tag{6.113}$$
$$\chi^e = 0. \tag{6.114}$$

These are obtained by making use of the equations

$$M_x[B_1] = M_x[A_2] = M_x[B_2] = M_x[A_1], \tag{6.115}$$
$$P_x[B_1] = -P_x[A_2] = -P_x[B_2] = P_x[A_1], \tag{6.116}$$

which follow from the relations in the paramagnetic phase

$$\Psi_i[B_1] = C_{2x}(\tau)\Psi_i[A_1], \tag{6.117}$$
$$\Psi_i[B_2] = I\Psi_i[A_1], \tag{6.118}$$
$$\Psi_i[A_2] = \sigma_d(\tau)\Psi_i[A_1]. \tag{6.119}$$

In (6.111)–(6.113), we used the following abbreviations for the numerators:

$$(MPP)_{imki} = (M_x)_{im}(P_x)_{mk}(P_x)_{ki}, \tag{6.120}$$
$$(\bar{P}PP)_{imki} = (\overline{P}_x)_{im}(P_x)_{mk}(P_x)_{ki}. \tag{6.121}$$

We are interested in nearly two-photon resonant excitation of ${}^4T_{2g}\ x_m$, i.e.,

$$2\hbar\omega \sim \hbar\omega_{mg} = E({}^4T_{2g}) - E({}^4A_{2g}).$$

Then the matrix element $(M_x)_{gm}$ of the magnetic dipole moment is presented in terms of the Bohr magneton μ_B as

$$\langle {}^4A_{2g}M_s|M_x|{}^4T_{2g}M_s x_\mp\rangle = \mp\sqrt{2}i\mu_B. \tag{6.122}$$

In evaluating the matrix element describing two-photon excitation $(P_xP_x)_{mki}/(\omega_{ki}-\omega)$, we choose the 4p state of the Cr^{3+} ion as the intermediate state $|k\rangle$ with excitation energy $\Delta E_0 \equiv \Delta E(pd) \sim 10$ eV. In order to obtain the finite contribution to χ^m and χ^e, $(P_xP_x)_{mki}$ should not be pure imaginary. For this purpose, we should take into account the perturbations due to the lower symmetry crystalline field $V_{\text{twist},g}[T_{1g}a_0]$ and the spin–orbit interaction $\lambda\boldsymbol{L}\cdot\boldsymbol{S}$ working on $(P_xP_x)_{mki}$.

The electric dipole transition between ${}^4A_{2g}$ and ${}^4T_{2g}$ becomes possible with the help of $V_{\text{twist},u}[T_{2u}x_0]$, i.e.,

$$\langle {}^4A_{2g}M_s|\bar{P}_x|{}^4T_{2g}M_s x_m\rangle = -\frac{m}{6}\langle {}^4A_{2g}||\bar{P}_x[T_1]||{}^4T_{2g}\rangle. \tag{6.123}$$

The matrix element of two-photon excitation $(P_xP_x)_{mki}/(\omega_{ki}-\omega)$ becomes finite between ${}^4A_{2g}$ and ${}^4T_{2g}$ with the help of $\mathcal{H}^z_{so} = \lambda L_z \cdot S_z$. As a result, χ^m and χ^e are written in the following simple forms [92]:

$$\chi^m = \chi^m_1 + \chi^m_2, \tag{6.124}$$

$$\chi^m_1 = \frac{iA^m}{\omega_- - 2\omega - i\Gamma_-} + \frac{iA^m}{\omega_+ - 2\omega - i\Gamma_+}, \tag{6.125a}$$

$$\chi^m_2 = \frac{iB^m}{\omega_o - 2\omega - i\Gamma_0}. \tag{6.125b}$$

Here both A^m and B^m are real and A^m is proportional to $V_{\text{twist},g}[T_{1g}a_0]$ while B^m is proportional to $V_{\text{twist},g}$ and λ^2. Similarly for χ^e,

$$\chi^e = \chi^e_1 + \chi^e_2, \tag{6.126}$$

$$\chi^e_1 = \frac{iA^e_-}{\omega_- - 2\omega - i\Gamma_-} + \frac{iA^e_+}{\omega_+ - 2\omega - i\Gamma_+}, \tag{6.127a}$$

$$\chi^e_2 = \frac{iB^e}{\omega_0 - 2\omega - i\Gamma_0}. \tag{6.127b}$$

Here both $A^e_\pm$ and B^e are real, and $A^e_\pm$ are proportional to $V_{\text{twist},u}[T_{2u}x_o]$ and λ while B^e ia proportional to $V_{\text{twist},u}[T_{2u}x_0]$ and λ^2, and

$$\omega_+ = E\left({}^4T_{2g}\frac{3}{2}x_+\right) - E\left({}^4A_{2g}\frac{3}{2}\right) = 2.164\,\text{eV},$$

$$\omega_- = E\left({}^4T_{2g}\frac{3}{2}x_-\right) - E\left({}^4A_{2g}\frac{3}{2}\right) = 2.145\,\text{eV},$$

$$\omega_0 = E\left({}^4T_{2g}\frac{3}{2}x_0\right) - E\left({}^4A_{2g}\frac{1}{2}\right) = 2.08\,\text{eV}.$$

The magnitudes of χ^m and χ^e are estimated and compared with each other:

$$|\chi_1^m| \sim |\chi_0^m| \frac{\langle V_{\text{twist},g}\rangle}{|\Delta E_2|}, \tag{6.128a}$$

$$|\chi_1^e| \sim |\chi_0^e| \frac{\lambda}{|\Delta E_1|} \frac{\langle V_{\text{twist},u}\rangle}{|\Delta E_0|}, \tag{6.128b}$$

where χ_0^m and χ_0^e are ideal quantities in which all the matrix elements of M, P, and $\bar{P}$ in (6.111) and (6.112) take their nonvanishing values. The relative magnitude of $|\chi^m/\chi^e|$ is then estimated to be of the order of unity:

$$\left|\frac{\chi^m}{\chi^e}\right| \sim \frac{n\mu_B}{cea_0} \frac{\Delta E(pd)}{\langle V_{\text{twist},u}\rangle} \frac{\langle V_{\text{twist},g}\rangle}{\lambda} \frac{|\Delta E_1|}{|\Delta E_2|} \sim 2. \tag{6.129}$$

Here we used the following values of the material constants: $\Delta E_0 = \Delta E(pd) = 10\,\text{eV}$, the spin–orbit coupling constant $\lambda = 10\,\text{meV}$, $\langle V_{\text{twist},g}\rangle \sim \langle V_{\text{twist},u}\rangle \sim 0.1\,\text{eV}$, $|\Delta E_1| = E({}^4T_{1g}) - E({}^4T_{2g}) \sim 1\,\text{eV}$, $|\Delta E_2| = E({}^4T_{2g}) - E({}^4A_{2g}) \sim 2\,\text{eV}$, and refractive index $n \sim 1$. The absolute value of $|\chi^e|$ at $2\omega = [E({}^4T_{2g}x_\pm) - E({}^4A_{2g})]/\hbar$ is estimated to be $1 \times 10^{-12}\,\text{m/V}$. This is by one or two orders of magnitude smaller than the value $5 \times 10^{-11}\,\text{m/V}$ of $LiNbO_3$ at $\lambda = 1.064\,\mu\text{m}$. Here we used the values $4N = 3.3 \times 10^{28}\,\text{m}^{-3}$, $e = 1.6 \times 10^{-19}\,\text{C}$, relaxation constant $\Gamma_m \sim 0.1\,\text{eV}$ and $a_0 = 0.53 \times 10^{-10}\,\text{m}$.

The tensors χ^m and χ^e obey the following symmetry relations under the point group $\underline{\bar{3}m}$ of the Cr^{3+} ion in the crystal:

$$\chi^m \equiv \chi^{mee}_{xxx} = -\chi^{mee}_{xyy} = -\chi^{mee}_{yxy} = -\chi^{mee}_{yyx}, \tag{6.130}$$

$$\chi^e \equiv \chi^{eee}_{xxx} = -\chi^{eee}_{xyy} = -\chi^{eee}_{yxy} = -\chi^{eee}_{yyx}. \tag{6.131}$$

As a result, the source term S of the SH:

$$S = \mu_0 \left(\nabla \times \frac{\partial \boldsymbol{M}}{\partial t} + \frac{\partial^2 \boldsymbol{P}}{\partial t^2}\right) \tag{6.132}$$

is written in the following form:

$$\begin{pmatrix} S_x \\ S_y \\ S_z \end{pmatrix} = \frac{4\omega^2}{c^2} \begin{pmatrix} 2\chi^m E_x E_y - \chi^e (E_x^2 - E_y^2) \\ \chi^m (E_x^2 - E_y^2) + 2\chi^e E_x E_y \\ 0 \end{pmatrix}, \tag{6.133}$$

$$\begin{pmatrix} S_+ \\ S_- \\ S_0 \end{pmatrix} = \frac{4\sqrt{2}\omega^2}{c^2} \begin{pmatrix} (-i\chi^m - \chi^e)E_-^2 \\ (i\chi^m - \chi^e)E_+^2 \\ 0 \end{pmatrix}. \tag{6.134}$$

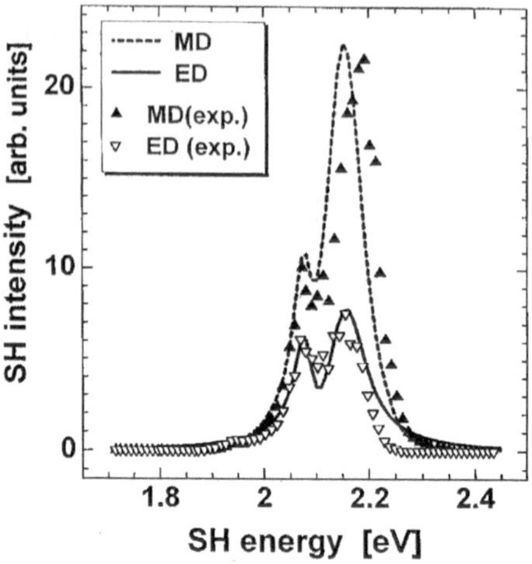

Fig. 6.29. Calculated spectra of SHG through the magnetic dipole (broken line) and the electric dipole (solid line) as a function of the signal frequency in eV near the transition from ${}^4A_{2g}$ to ${}^4T_{2g}$ at low temperatures [91, 93]

When we use the linearly polarized fundamental E_x, the SH signal with x-polarization is produced by χ^e and the SH signal with y-polarization by χ^m, as (6.133) shows. The observed [93] and calculated $\chi^m(2\omega)$ and $\chi^e(2\omega)$ spectra are drawn in Fig. 6.29. The weaker peak on the low-energy side originates in the two-photon resonance excitation of ${}^4T_{2g}(M_s = 1/2)x_0$ with the excitation energy $E({}^4T_{2g}1/2x_0) - E({}^4A_{2g}3/2) = 2.08\,\text{eV}$ while the stronger peak on the high-energy side comes from the superposition of two excitations of $E({}^4T_{2g}3/2x_-) - E({}^4A_{2g}3/2) = 2.145\,\text{eV}$ and $E({}^4T_{2g}3/2x_+) - E({}^4A_{2g}3/2) = 2.164\,\text{eV}$.

Interference Effects [92]

From (6.134), we can see that the interference effect between the χ^m and χ^e SHG channels will be observable by using circularly polarized light as the fundamentals. The signal intensity $I \propto |S|^2$ for the circularly polarized light is expressed as

$$\begin{aligned}|S|^2 \propto\ & (|\chi^m|^2 + |\chi^e|^2)(|E_+|^4 + |E_-|^4) \\ & -2(\chi'_m\chi''_e - \chi''_m\chi'_e)(|E_+|^4 - |E_-|^4),\end{aligned} \tag{6.135}$$

where $\chi^m \equiv \chi'_m + i\chi''_m$ and $\chi^e \equiv \chi'_e + i\chi''_e$. The interference of the second harmonics generated by the magnetic and electric dipole moments is described by $\Delta \equiv -2(\chi'_m\chi''_e - \chi''_m\chi'_e)$, which is proportional to the sublattice magnetization, e.g., in the A_1 sublattice. Therefore we can detect the magnetic domains of the crystal through this interference factor Δ by using circularly polarized light as the pump source. When the magnetic domain to pump is fixed, the total signals of the second harmonics show different spectra against the positively and negatively circularly-polarized fundamentals, due to the second term of (6.135).

The interference effect will be most pronounced when the value of Δ becomes of the same order of magnitude as that of $|\chi^m|^2 + |\chi^e|^2$. If we define the phase angles θ_m and θ_e by

$$\chi^m = |\chi^m| \exp(i\theta_m), \tag{6.136}$$

$$\chi^e = |\chi^e| \exp(i\theta_e), \tag{6.137}$$

we find

$$\Delta = 2|\chi^m||\chi^e| \sin(\theta_m - \theta_e), \tag{6.138}$$

so that perfect interference will be attained when (1) $|\chi^m| = |\chi^e|$ and (2) $\theta_m - \theta_e = \pm\pi/2$. From the observed spectra shown in Fig. 6.30, we find that these conditions are almost satisfied experimentally. It was pointed out by (6.129) that the first condition $|\chi^m| = |\chi^e|$ is almost satisfied. The second condition $\theta_m - \theta_e = \pm\pi/2$ is also nearly satisfied when the exciton propagation by the magnetic dipolar interaction

$$\mathcal{H}' = \sum_{i>j} K_{ij} \mu_B^{-2} M_i \cdot M_j \tag{6.139}$$

and the subsequent relaxation of the electronic excitation are taken into account. This effect of (6.139) is evaluated by the local-field corrections on the

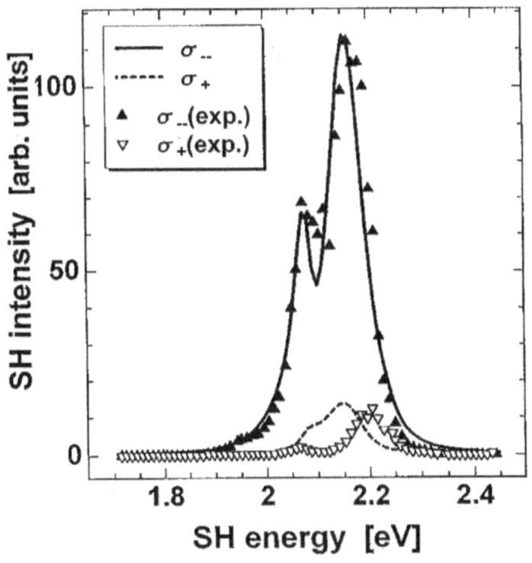

Fig. 6.30. Calculated second-harmonic signals under σ_- (solid line) and σ_+ (broken line) circularly polarized incoming laser light as a function of the signal frequency in eV. The difference shows the interference effect. The phase of χ^m has been changed by $\pi/2$ rather arbitrarily [91]

SH and the fundamental. The calculated spectra of $|\chi^m|^2$ and $|\chi^e|^2$, and the interference spectra are drawn in Figs. 6.29 and 6.30, respectively. We used the relevant constants listed in [91,92]. The agreement between the calculated and observed spectra is reasonable and the phase difference $\theta_m - \theta_e$ reaches 70° around the peaks of $|\chi^m|^2$ and $|\chi^e|^2$.

7

Nonlinear Optical Responses II

Some conventional nonlinear optical responses were summarized in Chap. 6. In the present chapter, we will discuss the recent development of nonlinear optical responses. First, phase-matched harmonic conversion of infrared and visible light into extreme ultraviolet (XUV) or soft X-rays is introduced. Here ultrashort laser pulses shown in Chap. 5 can generate even shorter bursts of coherent XUV or soft X-rays. We are, however, not allowed to describe these phenomena by the perturbational methods used in Chap. 6. These higher order harmonic generations will be discussed in Sect. 7.1. Second, these XUV or X-rays in the frequency region are possibly converted into a train of attosecond pulses or a few bursts of attosecond pulses. These will be discussed in Sect. 7.2. Atoms are used as a material system for high-harmonic generation (HHG) of XUV and soft X-rays and for generation of their attosecond laser pulses. Although the efficiency of HHG is rather small, i.e., of the order of 10^{-6}, the high-order stimulated Raman scattering (HSRS) is much stronger by several orders of magnitude than that of HHG. Usually the rotational and vibrational modes of molecules are used as a material system. Third, the second nonperturvative phenomena of nonlinear optics come from resonant optical pumping of a specified Raman-active phonon mode of molecular rotation and vibration by two incident beams of femtosecond laser pulses. This will be discussed in Sect. 7.3. Fourth, it is possible, in the excitation of crystals, to combine HHG and HSRS. This combined effect can be observed by resonant excitation of the Raman-active mode of lattice vibration in crystals by two intense near-IR femtosecond laser pulses generated from an optical parametric system prepared by Ti:sapphire lasers. Here the frequency difference of the signal and idler beams is chosen nearly equal to the phonon mode. The efficiency of the HSRS is further enlarged by this method. This will be discussed in Sect. 7.4. It is again seen that these nonlinear optical responses cannot be described by the perturbational treatment discussed in Chap. 6.

7.1 Enhanced Higher-Harmonic Generation

Laser technology – the ability to generate intense and coherent light with controllable properties – is one of the most significant achievements of 20th century science. In recent years, nonlinear optical techniques that convert one frequency of light to another have played an increasingly pivotal role in laser technology. Optical frequency doubling or parametric amplification, for instance, converts laser light into coherent radiation tunable over the near-IR, visible, and near-UV regions of the spectrum, as shown already in Chap. 6. Recent years have also seen the development of ultrashort pulse technologies, as seen in Sect. 5.1.5. The uncertainty principle $\Delta E \Delta t \geq \hbar/2$ dictates that a very wide spectral bandwidth is inevitable to obtain a short pulse. Generating such a short pulse also requires that the generated components in that broad spectrum have a well-defined phase relationship with each other; that is, the coherence must span the entire spectrum.

By using the techniques of nonlinear optics and ultrashort pulse generation to an extreme limit, one can generate coherent light at even shorter wavelengths using a process called high-harmonic generation (HHG). This process can convert femtosecond laser light from the near-IR (1–2 eV) to the extreme UV (XUV; tens to hundreds of eV) and soft X-ray (up a keV) regions of the spectrum [94]. The XUV is a difficult region of the spectrum for nonlinear optics because traditional frequency-conversion techniques generally rely on crystalline solids as the nonlinear medium, and solids are not transparent in the XUV. Nevertheless, good scientific and industrial requirements are driving the development of new light source in that spectral range as a structural and chemical spectroscopic probe and as a tool for nanoscale lithography. Synchrotron sources were originally developed to address such applications, but those machines are large and have limited access. High-harmonic generation produces tunable, laser-like light with high spatial and temporal coherence from an assembly of components small enough to sit on a tabletop.

In nonlinear optics, electrons in the material act like driven oscillators that respond to the laser's electric field. Ordinarily, these electrons remain bound but are driven strongly enough that the potential that binds an electron to its atomic core is no longer a purely parabolic, harmonic-oscillator potential. The motion of the electrons themselves becomes anharmonic, which gives rise to a time-dependent nonlinear polarization [95] that reradiates electromagnetic waves not only at the driving laser frequency, but also at higher harmonics of the driving laser field.

High-harmonic generation takes this concept to its extreme, i.e., the laser's intensity is increased to the point where the electric field becomes strong enough to ionize an atom. Once stripped from the atom, the electron still moves in response to the oscillating field, and the electron can recollide with its parent ion during a single optical cycle and recombine with the ionized atom. Harmonic orders well exceeding 100 have been observed by L'Huillier and Balcou [96] from two lighter noble He and Ne gases using 1 ps, 1053 nm

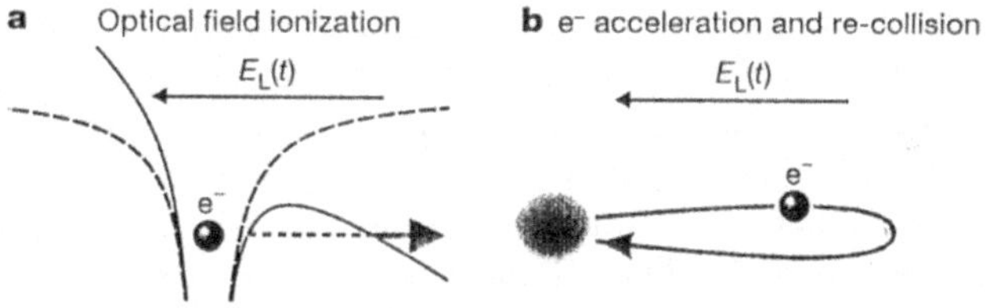

Fig. 7.1. Optical-field ionization and generation of coherent extreme ultraviolet and soft X-ray radiation from an atom exposed to a strong, linearly polarized pulse. (**a**) The effective Coulomb potential binding valance electrons to the atomic core (dashed curve) is temporarily suppressed around the oscillation peak of the laser electric field. A valance electron can tunnel through or escape above the potential barrier formed by the superposition of the atomic Coulomb field and the instantaneous laser field (solid line). (**b**) The freed electron moves away from the atomic core, is accelerated and then pulled back to it by a linearly polarized field. Recollision of the electron with its parent ion may trigger emission of an energetic (soft X-ray) photon

YAG laser pulses. In heavier noble gases, which have smaller ionization potentials, the number of harmonics which can be generated is less, although they have higher conversion efficiencies. Up to the 57th(in Ar) and 29th(in Xe) harmonics were observed from the same system of the YAG laser just mentioned above [96]. Macklin, Kmetec, and Gordon observed harmonics up to the 109th order in neon using 125 fs, 806 nm pulses of a Ti:sapphire laser, which gives the shortest wavelength harmonics reported to 1993 [97]. These observations are in agreement with theoretical predictions [98] that the photon energy of the highest harmonic emitted from a gas cannot exceed $I_p + 3.2U_p$, where I_p is the atomic ionization potential and $U_p \sim \lambda^2 I$ is the maximum ponderomotive potential that an electron may experience prior to detachment from the atom, and λ and I are the laser wavelength and intensity, respectively.

Krause, Schafer, and Kulander [98] used *ab initio* calculations of the Schrödinger equation in three dimensions to show that the breadth of the plateau in the harmonic spectrum obeys this cutoff rule. They also showed that this rule can be understood with the classical picture drawn in Fig. 7.1, where the electron detaches from the atom (Fig. 7.1a) and releases energy when it is recaptured by the atom after a laser cycle (Fig. 7.1b). The classical picture predicts that the maximum kinetic energy acquired by an electron from the field upon return to the nucleus is $3.2\,U_p$. A quantum-mechanical descriptions of high-harmonic generation give similar results under the adiabatic approximation [99], where the laser intensity varies slowly with respect to an optical period. This assumption may be allowed for laser pulses of duration greater than 100 fs.

Since the first HHG experiments [96, 97], two developments have dramatically changed the picture of the HHG. First, rapid advances in ultrashort-pulse lasers using Ti:sapphire led to instruments that produced high-power pulses with an unprecedented duration of 20 fs, or less than 10 optical cycles. This was made possible by making the best use of the broadest bandwidth of the laser material. Moreover, it is durable, with an energy capacity orders of magnitude greater than the laser dyes used in the 1980s. Ti:sapphire increased the average power of ultrafast laser systems from just 10 milliwatts to about 10 watts, while it dramatically reduced the pulse duration and shrank the overall size of the system [100]. The second advance came from the realization of matching the phase velocity of the laser fundamental to that of the harmonics. This has optimized the conversion efficiency.

First we will show high-harmonic generation from noble gases pumped by an 805 nm, 25 fs, Ti:sapphire laser [101]. The harmonic energies observed are unexpectedly high when compared with the results to date for longer excitation pulses. The efficiency of harmonic production is higher for shorter pulses. The wavelength of the harmonics can be tuned by adjusting the sign of the chirp of the excitation pulse, demonstrating a tunable, ultrashort (< 25 fs) pulse of soft X-ray source.

The group of Kapteyn investigated high-harmonics generated by a 25 fs, 10 Hz, 3 TW Ti:sapphire laser in various noble gases [101]. The bandwidth of the pulses is 32 nm, centered at a wavelength of 805 nm, and the laser system can provide up to 70 mJ of energy per pulse. The ultrashort nature of the 25 fs excitation pulses (10 optical cycles FWHM) implies that at the half maximum position of the temporal pulse envelope, the laser intensity changes by more than 25% during a single cycle. This denies the adiabatic assumption, which suggests that the atomic dipole moment undergoes quasiperiodic motion from cycle to cycle, with no dependence on the history of the pulse.

For the heavier noble gases, the group of Kapteyn observed harmonics with photon energies remarkably higher than previously seen. Figure 7.2(a) shows harmonics generated in argon, where orders up to the 61st are visible. This corresponds up to a photon energy of 93 eV. Figure 7.2(b) shows harmonics up to the 41st generated in krypton and harmonics up to the 29th generated in xenon. In neon, harmonics past the 105th could not be resolved, but light which may correspond to harmonic orders up to the 131st was observed as shown in Fig. 7.2(c). As the laser intensity was reduced, the short-wavelength edge gradually retreated to longer wavelengths, indicating that the shortest wavelength light is not an artifact.

The spectra seen in Figs. 7.2(a)–(c) were produced with a laser energy of 3.5 mJ per pulse, which for the focusing conditions described above corresponds to a peak intensity of approximately $(5 \pm 2) \times 10^{14}\,\mathrm{W/cm^2}$. This is significantly above ($\times 2$) the point where ionization should readily occur. The possibility therefore exists that the highest harmonics might arise from ions. It was observed that all of the harmonic peaks decreased in strength together as the pressure was gradually reduced from 5 to 1 torr. If the higher-order harmonic peaks were produced by ions while the lower were produced by neutral atoms, one would expect the harmonics to scale very differently with pressure because of a changing coherence length arising from free electrons. Thus these observations suggest that the harmonic peaks all arise from neutral atoms.

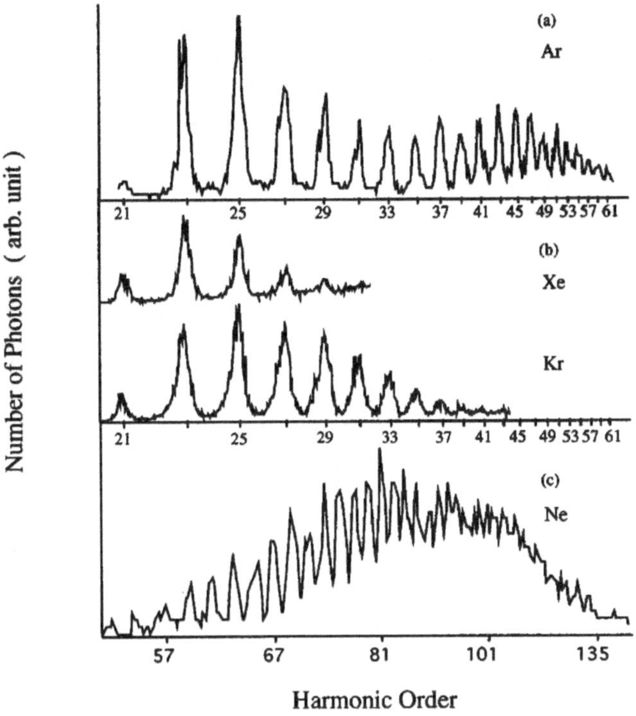

Fig. 7.2. High-harmonic generation (HHG) was observed in various noble gases pumped by a 25 fs, 10 Hz, 3 TW Ti:sapphire laser. (**a**) HHG from Ar (argon), (**b**) Xe (xenon), Kr (krypton), and (**c**) Ne (neon) gases [101]

As a conclusion, the ultrashort and ultrastrong laser pulses can produce a tunable, ultrashort pulse, 25 fs soft X-ray as short as 6 nm, from neon neutral gas.

Conventional nonlinear optics generally makes use of the birefringent properties of crystals to eliminate the phase mismatch between the laser and harmonic beams. However, HHG is implemented most often in a gas, to avoid absorption of the light by a solid. The gas system is isotropic so that the optical anisotropy such as the birefringence cannot be used for the phase-matching for the HHG.

Fortunately, guiding the light inside a gas-filled, hollow-core waveguide can achieve phase matching. In that scheme, as shown in Fig. 7.3, the laser beam propagates with a controlled intensity and phases as glancing reflections from the walls guide the light downstream [102]. Because the laser pulse slows down in a neutral gas but speeds up in a waveguide or plasma, the phase delay between the driving laser and the harmonic light can be manipulated.

When the level of ionization is small, phase matching of the laser and harmonic beams can be accomplished by adjusting the gas pressure so that

(a)

(b)

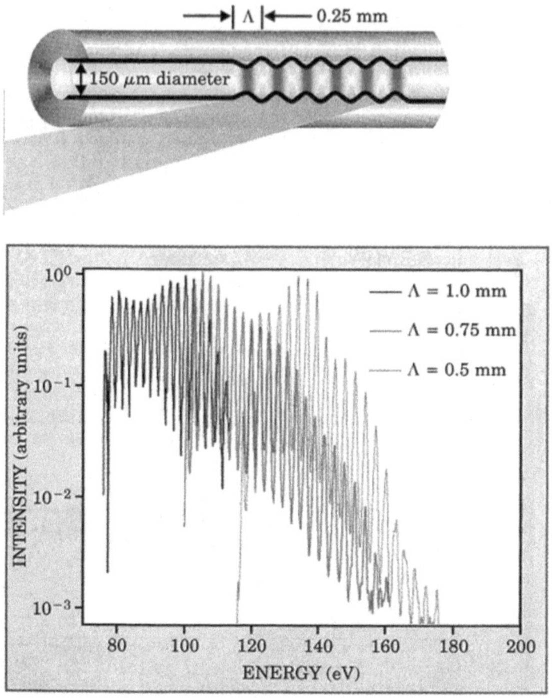

Fig. 7.3. (**a**) The high harmonics are generated in the hollow-core modulated waveguide with periodicity Λ, of inner diameter of 150 μm and modulation depth 10 μm. The waveguide, whose diameter changes periodically, can correct the mismatch of the phases of laser light and X-ray light by quasi-phase matching shown in (**b**) [94]

the waveguide dispersion balances the dispersion due to neutral atoms. That balance effectively adjusts the phase velocity of the fundamental laser beam to match that of the high-harmonic light. Because the harmonic light travels at a phase velocity, roughly, of the speed of light in vacuum due to its high frequency, the bandwidth of the pressure-tuned phase matching is very broad and encompasses many harmonic orders. Then the harmonic signal initially increases quadratically with interaction length. In the ionizing gas used for high-harmonic generation, plasma-induced dispersion causes the laser light to outrun. A waveguide whose diameter changes periodically can correct the mismatch using a technique called quasi-phase matching (QPM). The high harmonics are generated in the narrow regions of the waveguide where the laser intensity is highest. In between the narrow sections, the phases of laser light and X-ray light can realign, so that X-ray light always contributes in phase with the existing X-ray beam. A hollow-core modulated waveguide is characterized by an inner diameter of 150 μm, a modulation depth of 10 μm and several values of periodicity $\Lambda = 1.0\,\text{mm}$, 0.75 mm, 0.5 mm and 0.25 mm as shown in Fig. 7.3(a). The visible reflectance varies with the intensity of the light so that the bright spots illustrate how the modulation physically confines the intensity peaks to periodic regions in the guide as in Fig. 7.3(b).

Figure 7.3(b) shows that as Λ is decreased – to correct for the phase slip more often – the effect is to extend the X-rays to a higher energy. Thus we can generate higher harmonics more efficiently at higher laser intensities and ionization levels [94]. Recent experiments [103, 104] have extended QPM to the K-edge at 284 eV – a wavelength in the "water window" region of the soft X-ray spectrum, a region useful for ultra-high-resolution microscopy of biological samples.

7.2 Attosecond Pulse Generation

7.2.1 Attosecond Pulse Bunching

As introduced in the previous section, a frequency "comb" of extreme ultraviolet (XUV) odd harmonics can be generated by the interaction of subpicosecond laser pulses with rare gases. If the spectral components within this comb possess an appropriate phase relationship to one another, their Fourier synthesis results in an attosecond pulse train. Laser pulses spanning many optical cycles have been used for the production of such light bunching [105]. Some features of the generation process have remained inaccessible to direct experimental measurement. First, questions relating to the time profile of the harmonic emission are not easily resolved. Measurements of the harmonic pulse duration show that it is much shorter than that of the driving laser and lasts for only a few femtoseconds. Second, measurement difficulty involves beating between various harmonics; this occurs on a subfemtosecond time scale, and no streak camera or autocorrelator for this wavelength range can reach such a high resolution.

When one measures the field autocorrelation, which is equivalent to measuring the power spectrum of the harmonics, no information could be obtained about the relative phases of the harmonic components. It is exactly those phases that determine whether the harmonic field exhibits strong amplitude modulation, (i.e., forms an attosecond pulse train), rather than a frequency-modulated wave of approximately constant amplitude. Paul et al. [106] measured these phases through two-photon, two-color photoionization of atoms, i.e., the phase relation between the contributing harmonics by considering them in pairs. The periodic beat pattern of such a pair can be related to the phase of the infrared light from the driving laser according to how the combined fields ionize atoms.

According to Fermi's golden rule, the total transition probability from the initial ground state Ψ_i to the final state at the sideband energy $E_q = E_0 + q\hbar\omega$ is proportional to

$$S = \sum_f \left| M^{(+)}_{f,q-1} + M^{(-)}_{f,q+1} \right|^2 , \tag{7.1}$$

with

$$M^{(\pm)}_{f,q} = \left\langle \Psi_f \left| D^{\pm}_{\mathrm{IR}} \frac{1}{E_q - \mathcal{H}} D^{+}_{q} + D^{+}_{q} \frac{1}{E_{\pm 1} - \mathcal{H}} D^{\pm}_{\mathrm{IR}} \right| \Psi_i \right\rangle . \tag{7.2}$$

Here ω is the IR field frequency, $\mathcal{H}$ is the atomic Hamiltonian, and the dipole operators $D^+(D^-)$ correspond to the energy-increasing (decreasing) part of the electromagnetic perturbation:

$$\boldsymbol{E}(t) \cdot \boldsymbol{r} = D^{+}[\exp(-i\omega t)] + D^{-}[\exp(i\omega t)]. \tag{7.3}$$

The IR field is present as D^- in $M^{(-)}_{f,q+1}$, because of the emission of the IR photon, and it thus contributes a phase of the opposite sign with respect to $M^{(+)}_{f,q-1}$. Explicitly writing the phases $D^{\pm} = D_0 \exp(\pm i\psi)$, the interference terms in S of (7.1) becomes

$$A_f \cos(2\psi_{\mathrm{IR}} + \psi_{q-1} - \psi_{q+1} + \Delta\psi^{f}_{\mathrm{atomic}}), \tag{7.4}$$

where $A_f = 2\left|M^{(+)}_{f,q-1}\right|\left|M^{(-)}_{f,q+1}\right|$. Delaying the IR field by a time τ with respect to the harmonic fields sets $\psi_{\pm \mathrm{IR}} = \omega_{\mathrm{IR}}\tau$. By experimentally recording the magnitude of the sideband peak as a function of τ, and fitting a cosine to this, we determine $(\psi_{q-1} - \psi_{q+1})$. The phase $\Delta\psi^{f}_{\mathrm{atomic}}$, which can be obtained from established theory, is small. The experimental setup and the quantum paths are shown in Fig. 7.4 for the photoelectron generation from the second argon jet by mixed-color two-photon ionization.

A beam of a Ti:sapphire laser (800 nm, 40 fs, 1 kHz) is split by a mask into an outer, annular part (3 mJ) and a small central part (30 μJ). Both parts are focused into an Ar jet, where the smaller focus of the annular part generates harmonics (XUV). The annular part is then blocked by a pinhole, and only the central part of the IR pulse and its harmonics propagates. The light is refocused there by a spherical tungsten-coated mirror onto a second Ar jet, and the electrons resulting from photoionization in this jet are detected at the end of the time-of-flight (TOF) tube by microchannel plates (MCPs). Photoelectron spectra are shown for argon ionized by a superposition of odd harmonics from an IR laser in Fig. 7.5(A); sidebands are caused between the harmonic peaks by copropagating fundamental (IR) radiation in Fig. 7.5(B) and (C). Changing the time delay between IR and harmonics from −1.7 fs in (B) to −2.5 fs in (C), causes a strong amplitude change of the sidebands. The pairwise phase differences were also determined by measuring the first four sideband peaks as a function of the time delay between the IR pulse and the harmonics.

The temporal intensity profile of a sum of five harmonics is reconstructed from the measured phases and amplitudes as shown in Fig. 7.6. The period is 1.35 fs, half the cycle time of the driving laser and the full width at half maximum (FWHM) is ∼250 as.

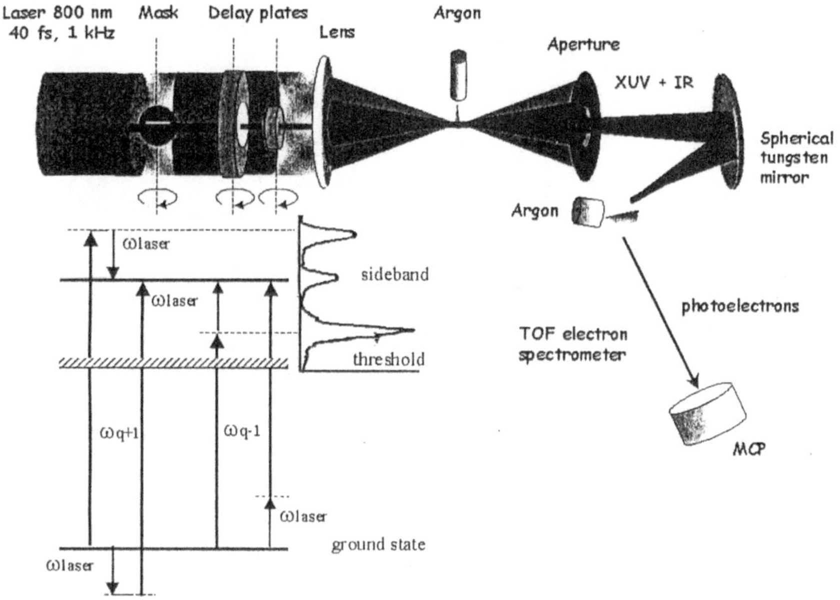

Fig. 7.4. The experimental setup (*top*). A beam of a Ti:sapphire laser (800 nm, 40 fs, 1 kHz) is split by a mask into an outer, annular part (3 mJ) and a small central part (30 μJ). Both parts are focused into an Ar jet, where the smaller focus of the annular part generates harmonics of XUV. The annular part is then blocked by a pinhole, and only the central part of the IR pulse and its harmonics propagates and is refocused by a spherical tungsten-coated mirror onto a second Ar jet. The electrons resulting from photoionization in this jet are detected at the end of the time-of-flight (TOF) tube by microchannel plates (MCP). The inset shows the quantum paths contributing to the photoelectrons generated in the second argon jet by mixed-color two-photon ionization; ω_{laser} is the IR field frequency and ω_q equals $q\,\omega_{\text{laser}}$ [106]

7.2.2 Direct Observation of Attosecond Light Bunching

The temporal characteristics of pulses was directly determined for the subfemtosecond regime, by measuring the second-order autocorrelation trace of a train of attosecond pulses [107]. In pico- and femtosecond laser laboratories, the pulse duration has for many years been routinely extracted to a satisfactory degree of accuracy from a measurement of the second-order autocorrelation trace. The extension of the approach to subfemtosecond XUV pulses poses several formidable problems, because attosecond pulses are spectrally much broader and in the nearly inaccessible UV–XUV spectral range, and are orders of magnitude weaker, thus requiring ultrasensitive nonlinear detectors with a flat broadband response.

In the experiment [107], harmonic generation takes place in a xenon gas jet using 130 fs laser pulses at wavelength $\lambda = 790\,\text{nm}$ of up to 10 mJ energy from a 10 Hz Ti:sapphire laser. For the second-order autocorrelation of this

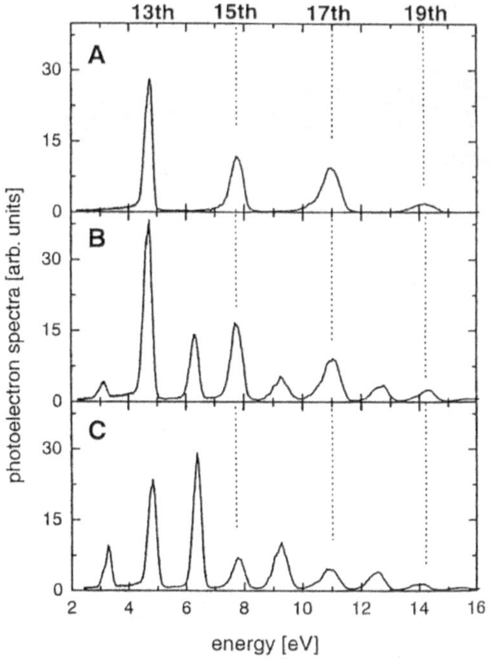

Fig. 7.5. (A) Photoelectron spectra of argon ionized by a superposition of odd harmonics from an IR laser. The fundamental radiation was added in (B) and (C), causing sidebands to appear between the harmonic peaks. Changing the time delay between IR and harmonics from −1.7 fs in (B) to −2.5 fs in (C) causes a strong amplitude change of the sidebands [106]

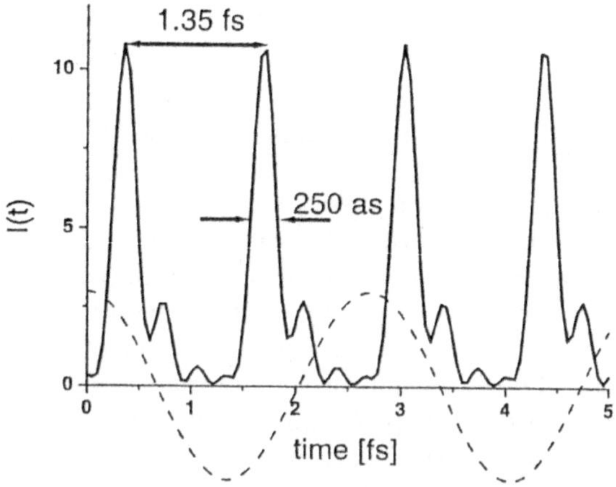

Fig. 7.6. Temporal intensity profile of a sum of five harmonics, as reconstructed from measured phases and amplitudes. The FWHM of each peak is ∼250 as [106]

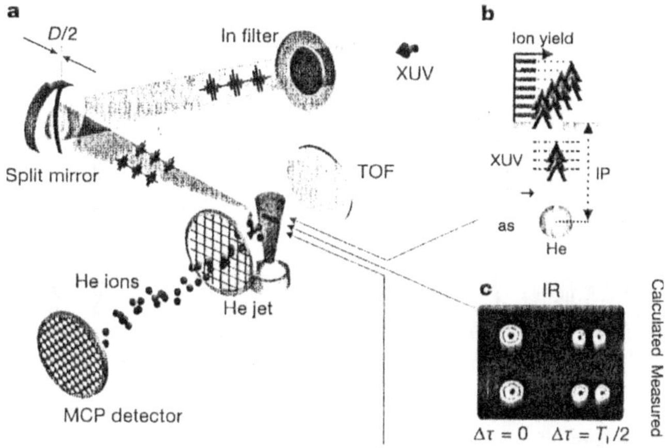

Fig. 7.7. (**a**) Schematic representation of the second-order XUV autocorrelator. Only the 7th to the 15th harmonics of the XUV radiation generated at the Xe jet are led through by the In filter and enter the volume autocorrelatror. The second-order XUV autocorrelator consists of a spherical mirror, split into two halves, serving as a focusing wavefront divider. As a nonlinear detector, a two-photon ionized He gas is used and the He ion yield, recorded by a time-of-flight (TOF) mass spectrometer, provides the autocorrelation signal. (**b**) The ionization occurs through two-XUV-photon nonresonant absorption from all possible combinations of the transmitted harmonics. (**c**) For zero and $\Delta\tau = T_L/2$ ($D = \lambda/2$) delay, the calculated and measured transverse intensity distribution for the IR laser frequency changes from a single spot to a double maximum distribution

harmonic superposition, they used a wavefront splitting arrangement consisting of a spherical mirror with 30 cm radius of curvature cut into two halves as shown in Fig. 7.7(a). The positioning of one of these halves is controlled by a piezocrystal translation unit with a resolution of ∼6 nm. The two parts of the bisected XUV pulse train are brought into a common focus in a helium gas jet. The ionization products are detected by a time-of-flight mass spectrometer as a function of the delay $\Delta\tau$ corresponding to a total displacement D between the two half-mirrors. For the laser intensities used, harmonics up to the 15th are generated. A 0.2 μm indium filter selects a group of harmonics from the 7th to the 15th, and blocks the residual fundamental. The recorded harmonic spectrum is shown before (Fig. 7.8a) and after (Fig. 7.8b) the In filter. The corrected relative intensity is 0.32: 1.0: 0.30: 0.11: 0.01 for the 7th to 15th harmonics as shown in Fig. 7.8(c). Ideal phase-locking of these harmonics (all phase differences equal to zero) would produce a train of attosecond pulses with full-width at half-maximum (FWHM) duration of 315 as. The exclusive contribution of a two-photon ionization was concluded from measurements of the ion yield vs. harmonic intensity for He^+ and for the rest gas H_2O^+ and Xe^+ ions observed in the recorded mass spectra as shown in Figs. 7.8(d)–(f).

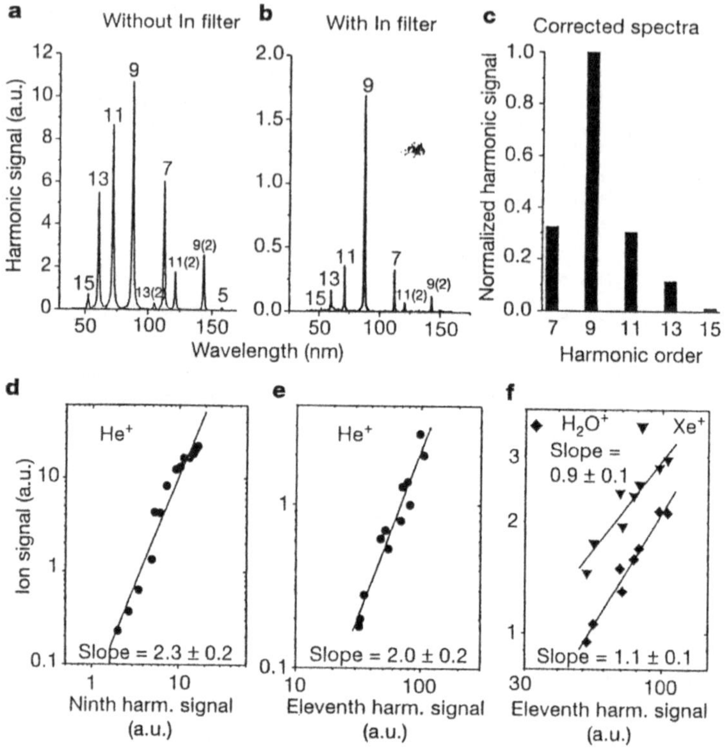

Fig. 7.8. Ion yield dependence on the XUV radiation. Higher-order harmonic generation spectra produced in the Xe jet, (**a**) measured without the In filter, (**b**) transmitted through the In filter, and (**c**) after correction. The slope of the He ion-yield as a function of the intensity of the 9th harmonic (**d**) and the 11th harmonic (**e**) is 2.3 ± 0.2 and 2.0 ± 0.2, respectively, in the log–log scale. In contrast, the slope of the ion-yield dependence for the 11th harmonic for H_2O and Xe (**f**) is 0.9 ± 0.1 and 1.1 ± 0.1, respectively. These results prove a two-XUV-photon ionization of He

First, ionization of He through the fundamental laser frequency only was observed, as Fig. 7.9(a) shows, for the calibration of the delay scale of the measured autocorrelation traces. The period of the observed oscillation is equal to the laser period, 2.63 fs. A second-order intensity autocorrelation trace of the superposition of the five harmonics (from 7th to 15th) is shown in Fig. 7.9(b) and (c). The average duration of the pulse train is estimated to be 780 ± 80 as while the pulse period is 1.3 fs.

7.2.3 Attosecond Control of Electronic Processes by Intense Light Fields

In pulses comprising just a few wave cycles, the amplitude envelope and carrier frequency are not sufficient to characterize and control laser radiation, because

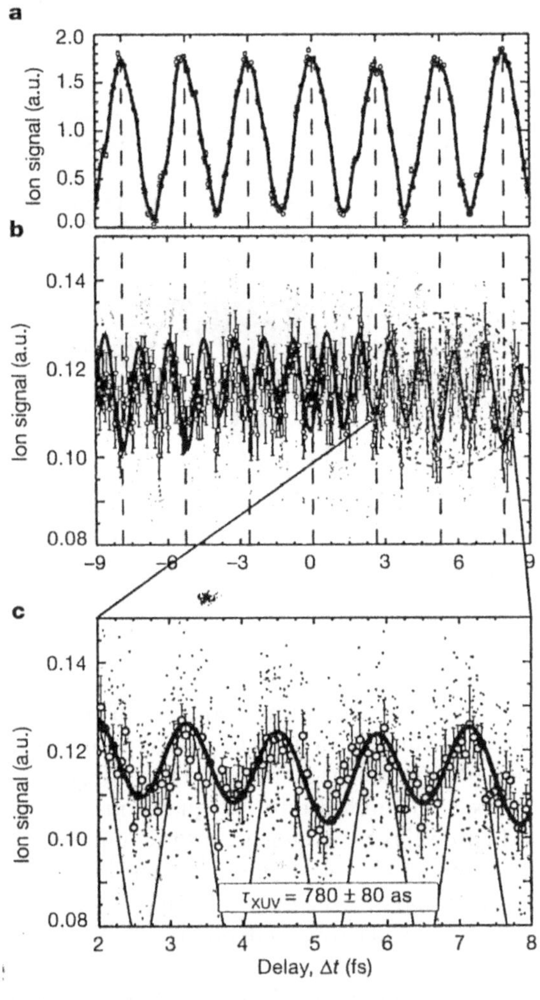

Fig. 7.9. (**a**) The measured higher-order autocorrelation trace of the fundamental laser field. The period of the observed oscillation is equal to the laser period, i.e., 2.63 fs. (**b**) A measured ~18 fs long second-order intensity volume autocorrelation trace of the superposition of the five harmonics, 7th to 15th. The error bars shown correspond to one standard deviation. A clear modulation with half the laser period is observable in the entire trace. (**c**) An expanded area of the trace in (b) gives an estimate of $\tau_{\mathrm{XUV}} = 780 \pm 80\,\mathrm{as}$ for average duration in the synthesis of the five harmonics

the evolution of the light field is also influenced by a shift of the carrier wave with respect to the pulse peak. Thus so-called carrier-envelope phase has been predicted and observed to affect strong-field phenomena. Random shot-to-shot shifts, however, have prevented the reproducible guiding of atomic process

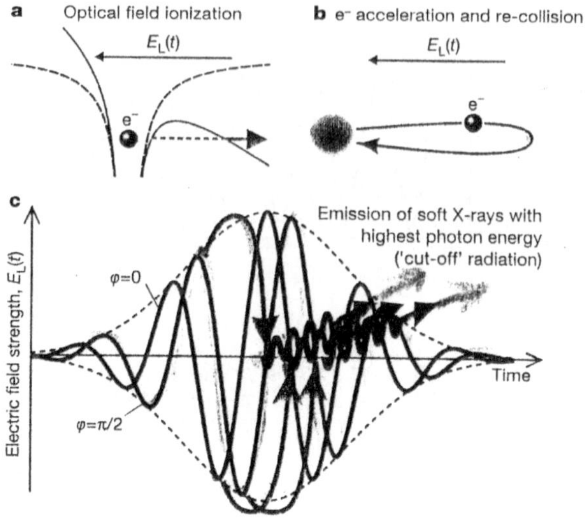

Fig. 7.10. Optical-field ionization of an atom (**a**), the freed electron is accelerated and then pulled back to the atomic core by a linearly polarized field (**b**). The highest-energy X-ray photons are emitted near the zero transitions of the laser electric field around the pulse peak, depending on the carrier-envelope phase φ (**c**) [108]

using the electric field of light. The combined team of Hänsch and Krausz [108] succeeded in generating intense, few-cycle laser pulses with a stable carrier envelope phase that permit the triggering and steering of microscopic motion with an ultimate precision limited only by quantum-mechanical uncertainty. Using these reproducible light waveforms, atomic currents in ionized matter are induced; the motion of the electronic wavepackets can be controlled on time scales shorter than 250 attoseconds. This has made it possible to control the attosecond temporal structure of coherent soft X-ray emission produced by the atomic currents.

An electronic wavepacket is set free around each oscillation peak of a laser electric field that is strong enough to overcome the effective binding potential (Fig. 7.10a). The ensuing motion of the wavepackets released by optical-field ionization (Fig. 7.10b) depends on the subsequent evolution of the driving laser field. A laser pulse consisting of many wave cycles launches a number of wavepackets at different instants. Each of these follows a differences in the initial conditions of their motion, preventing precise control of strong-field-induced electronic dynamics. Intense few-cycle light pulses with adjustable carrier-envelope (C-E) phase make the peak of the oscillating electric field coincide with the pulse peak (cosine wave) as shown by the solid line in Fig. 7.10(c), and the strength of the field is just sufficient to reach the ionization threshold at the pulse center. Then an isolated electronic wavepacket can be formed. The peak intensity of linearly polarized few-cycle laser pulses is enough strong as several electronic wavepackets in the vicinity of the pulse

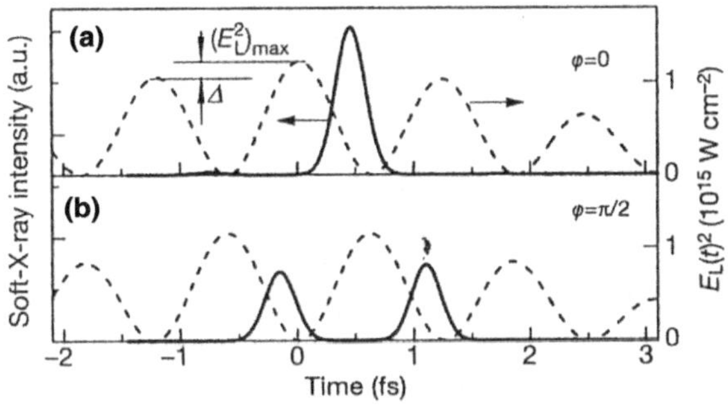

Fig. 7.11. The solid curves depict the temporal intensity profile of the cut-off harmonic radiation, whereas the dashed curves plot $E_L(t)^2$, for the carrier-envelope phase $\varphi = 0$ (**a**) and $\varphi = \pi/2$ (**b**) [108]

peak are set free. First they are removed from their parent ion, but within a laser period they are pulled back by the laser electric field (Fig. 7.10b). The highest-energy portion of the wavepacket recollides with the ion near the second zero transition of the laser electric field, and results in the emission of an energetic (soft-X-ray) photon as shown in Fig. 7.10(c). When the carrier-envelope phase ψ is chosen to be zero, the pump pulse has a cosine form and emits a single pulse of soft X-rays. In the case of $\psi = \pi/2$, the pump pulse has a sine form and emits two soft X-ray pulses. These are drawn in Fig. 7.11. The pulse width of these soft X-ray emissions is of the order of a subfemtosecond.

The distribution of high harmonics depends also on the carrier-envelope phase ψ as shown in Fig. 7.12.

7.3 Coherent Light Comb and Intense Few-Cycle Laser Fields

At present the shortest ultrafast light pulses are achieved using extreme ultraviolet (XUV) and soft X-ray radiation and enter the subfemtosecond or attosecond regime as described in the previous sections. High-order stimulated Raman scattering (HSRS) is accompanied by the potential to generate ultrabroad and, as a consequence, ultrashort light pulses with an energy conversion efficiency approaching unity [109, 110]. On the other hand, attosecond pulses generated using high-harmonic generation (HHG) have very low energy, as a consequence of the intrinsic low conversion efficiency of the process ($\sim 10^{-6}$). HSRS can be considered as an alternative to HHG, as it has the potential to produce pulses that are still in the sub-fs regime but much more energetic (by a factor of $\sim 10^4$) than those obtained through HHG [111]. In HSRS a key role is played by the temporal duration T of the driving pulses. The relevant characteristic time constant is the dephasing time T_2 of the molecular oscillations

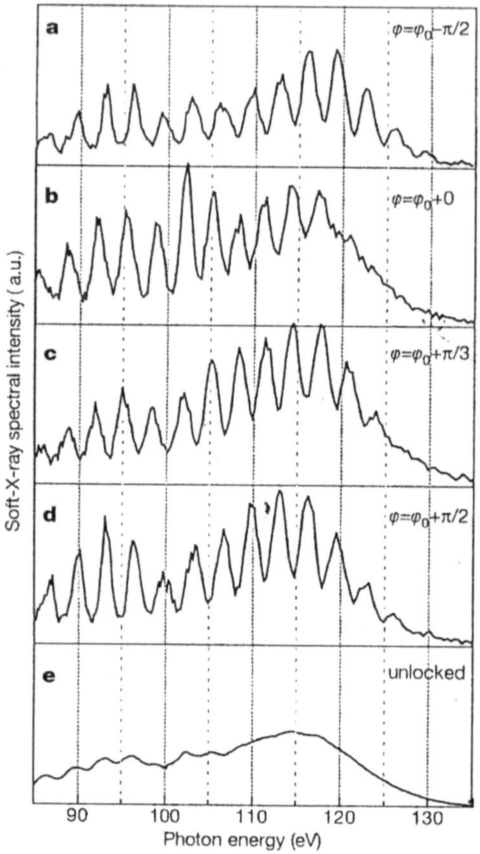

Fig. 7.12. Measured spectral intensity of few-cycle-driven soft X-ray emission from ionizing atoms. (**a**)–(**d**) Data obtained with phase-stabilized pulses for different carrier-envelope phase setting. (**e**) Spectrum measured without phase stabilization [108]

(rotation or vibrations). The relation between the pump pulse duration T and the dephasing time T_2 defines the different regimes of HSRS:

(a) In the quasistationary regime, the duration of the pump laser pulses is significantly longer than T_2.
(b) In the transient regime, the laser pulse duration is comparable to or shorter than T_2. Both terms "quasistationary" and "transient" refer only to the pulse duration and not to the molecular vibration.
(c) When the pulse duration is even shorter than the characteristic period T_ν of the molecular motion, the impulsive regime is entered. Femtosecond pulse sequences can also be used for optical manipulation of molecular

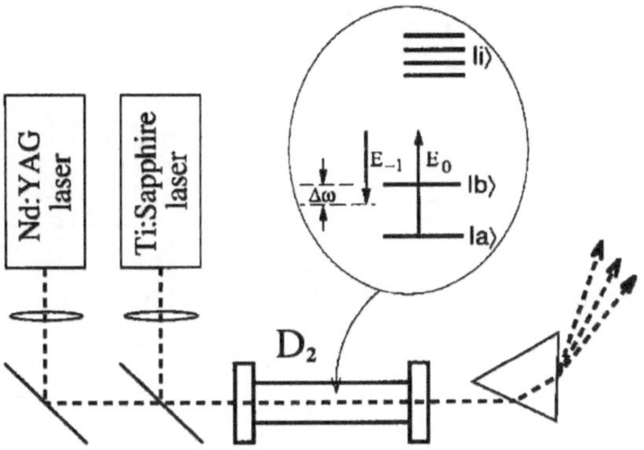

Fig. 7.13. Experimental setup and energy level diagram for coherent molecular excitation and collinear Raman generation. Raman detuning $\Delta\omega$ (positive as shown) is set by the driving laser frequencies [114]

motion [112,113]. This may be called multiple-pulse Impulsive Stimulated Raman Scattering (ISRS) excitation.

In this section, we will discuss these three cases separately.

7.3.1 Quasistationary Regime

Two driving lasers are necessary in this regime to prepare a single, highly coherent molecular state. Here we will introduce collinear generation of mutually coherent equidistant sidebands, covering 5000 cm^{-1} of spectral bandwidth and ranging from 2.94 μm to 195 nm in wavelength, which have been induced from molecular deuterium D_2 driven by two lasers [112]. The deuterium gas is pumped by two transform-limited laser pulses at wavelengths of 1.0645 μm and 807.22 nm, such that the (tunable) laser frequency difference is chosen to be approximately equal to the fundamental vibrational frequency in D_2. The first laser is a Q-switched injection-seeded Nd:YAG laser. Its output is attenuated to produce 100 mJ, 12 ns transform-limited pulses at a 10 Hz repetition rate. The second laser is a Ti:sapphire laser system, injection seeded from an external-cavity laser diode and pumped by the second harmonic of a separate Q-switched Nd:YAG laser. This laser produces 75 mJ, 16 ns transform-limited pulses at the seeding laser wavelength. The two driving laser pulses are synchronized by adjusting the delay between the two Nd:YAG laser Q-switched trigger pulses. The laser beams are combined on a dichroic beamsplitter and are loosely focused to a nearly diffraction-limited spot in a D_2 cell as shown in Fig. 7.13. The 1.06 μm laser spot size is 460 μm and the 807 nm laser spot size is 395 μm.

When the driving infrared lasers are tuned to within 1 GHz of the Raman resonance, a bright beam of white light is observed at the output of the

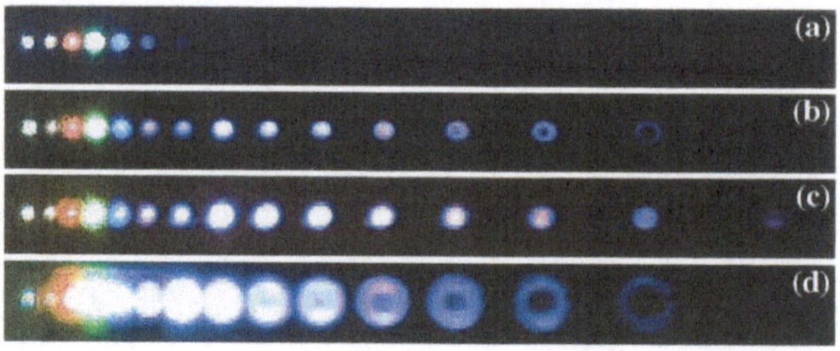

Fig. 7.14. Spectrum generated in the setup of Fig. 7.13 at (**a**) $P = 71\,\text{torr}$ and $\Delta\omega = -400\,\text{MHz}$, (**b**) $P = 71\,\text{torr}$ and $\Delta = 100\,\text{cm}^{-1}$, and (**c**) $\Delta\omega = 700\,\text{MHz}$, and (**d**) $P = 350\,\text{torr}$ and $\Delta\omega = 700\,\text{MHz}$ [114]

D_2 cell. When this light beam is diverted with a prism, 13 anti-Stokes and two Stokes sidebands are observed as shown in Fig. 7.14, in addition to the two driving frequencies. Starting from the left, the first two sidebands are the driving frequencies, and the next four are anti-Stokes sidebands in red, green, blue, and violet; beginning at the fifth anti-Stokes, the sidebands are in the ultraviolet and only fluorescence is visible. Figures 7.14(a)–(c) show the spectrum generated at a D_2 pressure of $P = 71$ torr and a Raman detuning of $\Delta\omega = -400 \pm 25\,\text{MHz}$ in part (a), $\Delta\omega = 100 \pm 25\,\text{cm}^{-1}$ in part (b), and $\Delta\omega = 700 \pm 25\,\text{cm}^{-1}$ in part (c). The smooth near-Gaussian beam profiles for nearly all sidebands, as shown in Fig. 7.14(a)–(c), demonstrate collinear anti-Stokes generation in a regime of high molecular coherence. At higher pressures the generation is no longer collinear and the anti-Stokes sidebands emerge in circles of increasing diameter. An example at a pressure of 350 torr and $\Delta\omega = 700\,\text{MHz}$ is shown in Fig. 7.14(d). These results have been analyzed by coupling a set of equidistant Raman sidebands with the Raman coherence of the first vibrational mode [109, 114, 115]. Katsuragawa et al. [116] observed a series of coherent anti-Stokes Raman scattering (CARS) signals due to the rotational mode of the hydrogen molecule (H_2) under pumping by a dual-wavelength injected-locked (6 ns pulsed) Ti:sapphire laser with $\lambda_0 = 763.180\,\text{nm}$ and $\lambda_{-1} = 784.393\,\text{nm}$. The frequency difference of these two modes, 10.631 THz, is slightly detuned by 600 MHz below the first rotational excitation. A train of ultrashort pulses was obtained by synthesizing these phase-coherent rotational-Raman sidebands in parahydrogen. It should be pointed out that self-induced phase matching was observed in parametric anti-Strokes stimulated Raman scattering from solid hydrogen [117]. This phenomenon is related to electromagnetically induced transparency [118] and was observed also using ns lasers so that these belong to the "quasistationary regime."

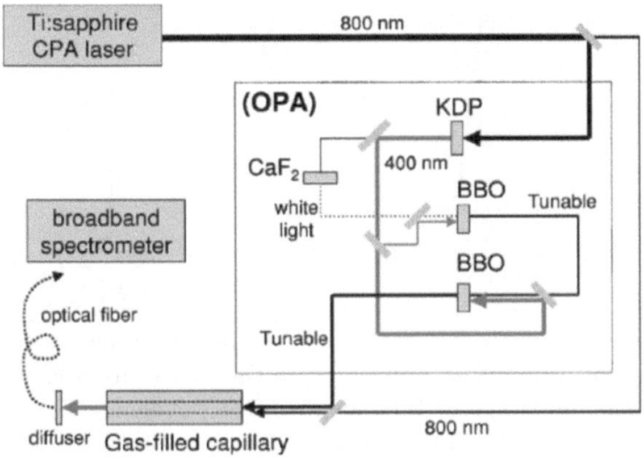

Fig. 7.15. Layout of the two-color stimulated scattering experiment [111]

7.3.2 Transient Regime

In this regime, the laser pulse duration is comparable to or shorter than T_2. The experiment of high-order stimulated Raman scattering (HSRS) by Sali et al. [111] belongs to a highly transient regime as two-color pumping with pulses of duration $T = 100$–400 fs was tuned to a vibrational Raman transition. That is, the pulse width $T = 100$–400 fs is much shorter than $T_2 = 2.6$ ns for H_2 at pressure $P = 10^5$ Pa, but longer than the vibrational period $T_\nu = 12$ fs for H_2. The experimental setup is shown in Fig. 7.15. The output of the Ti:sapphire chirped-pulse amplified system (CPA) (central wavelength $\lambda_1 = 800$ nm, pulse duration $T = 70$ fs, pulse energy $E = 20$ mJ, and repetition rate of 10 Hz) was divided by a beamsplitter. A small fraction of the pulse energy ($\sim$10%) was used to provide the first pump laser pulse at $\lambda_1 = 800$ nm. The largest part ($\sim$90%) of the CPA pulse was frequency-doubled to provide the pump pulse for an optical parametric amplifier (OPA), which in turn provided the second pump pulse for the high-order stimulated Raman scattering (HSRS) experiment.

The OPA was a double-stage collinear optical parametric amplifier seeded by white light. The white-light seed was produced in a calcium fluoride plate onto which was focused a small fraction of the infrared radiation which passes in the KDP. The seed was subsequently amplified in two 2 mm thick type-I BBO crystals. The output wavelength λ_2 was tunable from 450 to 800 nm and the pulse duration was estimated to be 250 fs.

The Raman-active molecular medium was confined in a 170 μm inner-diameter fused silica capillary. The capillary acts as a waveguide for the light and allows for a longer interaction length compared to the confocal parameter. Several measurements of HSRS were carried out in methane (CH_4) and hydrogen (H_2) at different pressures ranging from 100 mbar to 3 bar. In every experiment the wavelength λ_2 of the OPA pulse is tuned in such a way as

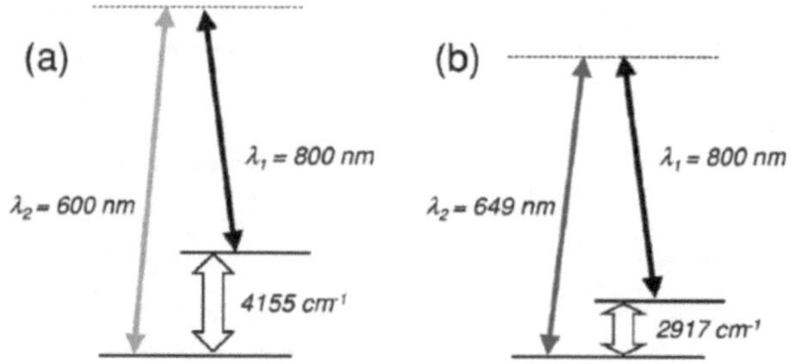

Fig. 7.16. The wavelength λ_2 of the OPA pulse is tuned in such a way as to obtain a condition of Raman resonance. (**a**) In the case of hydrogen the fundamental vibrational transition is $4155\,\mathrm{cm}^{-1}$, hence the OPA wavelength is set to $\lambda_2 = 600\,\mathrm{nm}$. (**b**) In the case of methane the fundamental vibrational transition is $2917\,\mathrm{cm}^{-1}$, and hence the OPA wavelength is set to $\lambda_2 = 649\,\mathrm{nm}$ [111]

to obtain a condition of Raman resonance as shown in Fig. 7.16. The relevant transition is the lowest-order vibrational transition, between the ground state ($\nu = 0$) and the first excited vibrational state ($\nu = 1$), whose width is $4155\,\mathrm{cm}^{-1}$ for H_2.

Effects of Gas Pressure

Figure 7.17 shows the generated spectra after propagating the two pump pulses through the capillary filled with H_2 gas at pressures from 100 mbar up to 3 bar (grey curves). A generated anti-Stokes sideband at $\lambda = 480\,\mathrm{nm}$ is already clearly present for a pressure as low as 100 mbar, and its intensity is of the order of a few percent of the transmitted pulses at the pump frequencies (Fig. 7.17a). The second anti-Stokes sideband at $\lambda = 400\,\mathrm{nm}$ is generated with a pressure of 500 mbar (Fig. 7.17b), and the third and fourth sidebands (at wavelengths of 343 nm and 300 nm, respectively) are visible in the spectrum measured with 800 mbar (Fig. 7.17c). The numbers of generated sidebands increased up to the fifth anti-Stokes ($\lambda = 267\,\mathrm{nm}$) at a pressure of 0.9 bar. At this pressure strong pump depletion was observed and up to five anti-Stokes sidebands were observed to have energies exceeding 10% of the transmitted pump pulse energies. For pressures of 1.7 bar and higher, the number of generated sidebands does not increase. What happens instead is that the sidebands get broadened and the generated spectrum progressively evolves from a comb of a few discrete sidebands to a broad supercontinuum (Figs. 7.17e and f).

Effects of Pump-Pulse Length

A similar set of measurements compared to those reported for H_2 was made using methane (CH_4) as the Raman-active medium as shown in Fig. 7.18. The

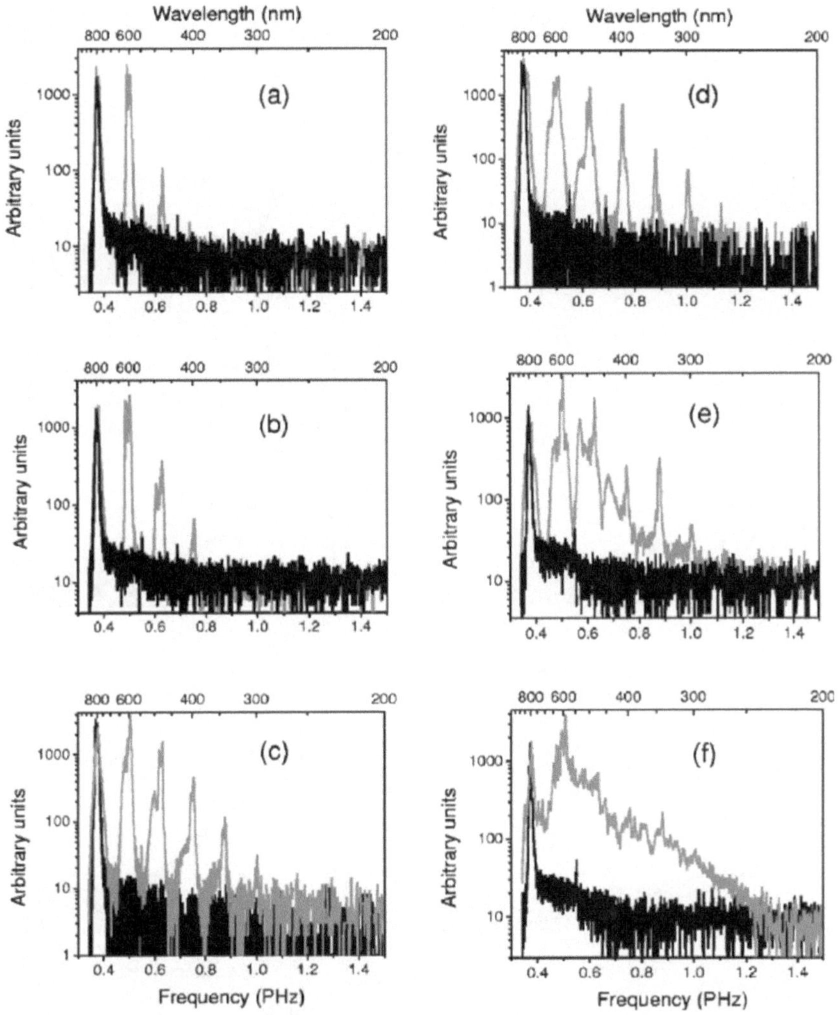

Fig. 7.17. HSRS (high-order stimulated Raman scattering) spectra generated in hydrogen at different pressures: (**a**) 100 mbar, (**b**) 500 mbar, (**c**) 800 mbar, (**d**) 900 mbar, (**e**) 1.7 bar, and (**f**) 3 bar. In each graph, the grey curve represents the output spectrum using two input pulses of λ_1 and λ_2, while the black ones represent the output spectra obtained using only the infrared λ_1 pump [111]

relevant Raman transition was the same lowest-order vibrational transition as in the case of H_2 and the frequency of the fundamental vibrational transition is 2917 cm^{-1}. Then the condition of Raman resonance is obtained by setting the OPA output wavelength to $\lambda_2 = 649$ nm as shown in Fig. 7.16(b).

Figure 7.18 shows the generated spectra after propagating the two pump pulses through the capillary filled with methane gas at pressures ranging from

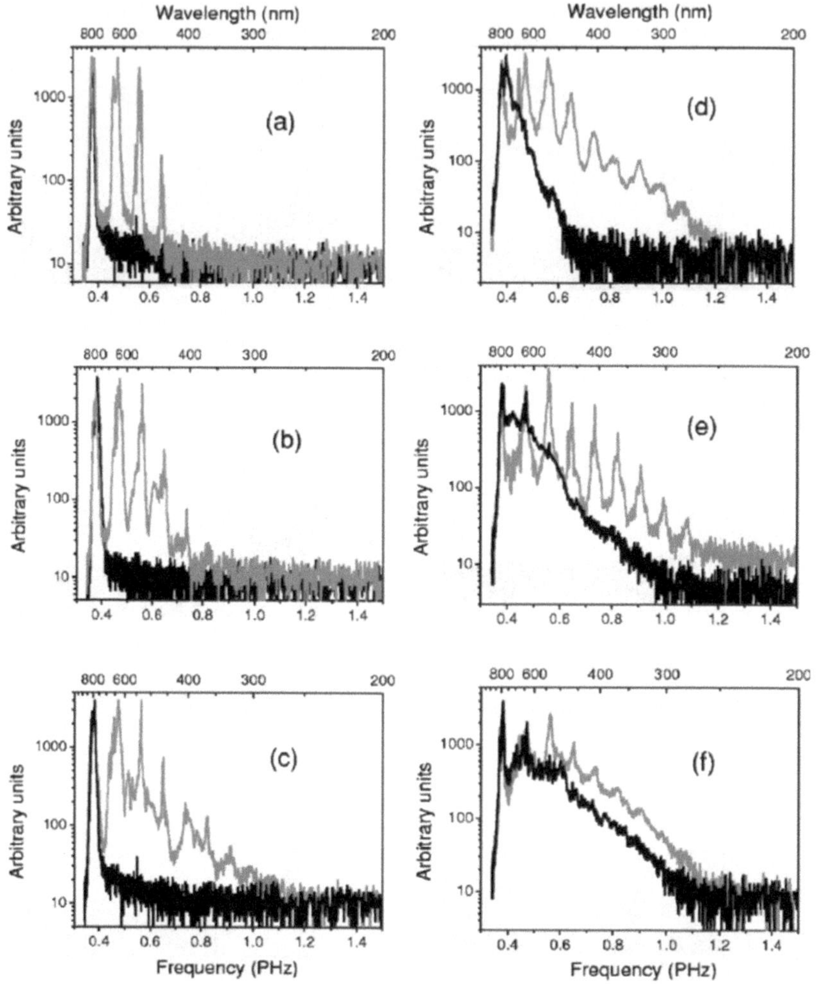

Fig. 7.18. HSRS spectra generated in methane at different pressures: (**a**) 300 mbar, (**b**) 500 mbar, (**c**) 700 mbar, (**d**)900 mbar, (**e**) 1.6 bar, and (**f**) 3 bar. The grey and black curves represent the same curves as in Fig. 7.17 [111]

300 mbar to 3 bar (grey curves). Due to the smaller Raman spacing (i.e., longer vibrational period) of methane compared to hydrogen, although more sidebands are generated with methane, in fact in both cases the generated frequency bandwidth is approximately the same. It should be noted that the bandwidth is as broad as the pump frequency value, i.e., $\Delta\nu \sim \nu_1$.

As with hydrogen, no Raman sidebands are produced with the single-pulse excitation (Fig. 7.18, black curves). This confirms the importance of the presence of the λ_2 field at the input for efficient generation of high-order sidebands, as expected. In these single-pump spectra, for pressures of 900 mbar

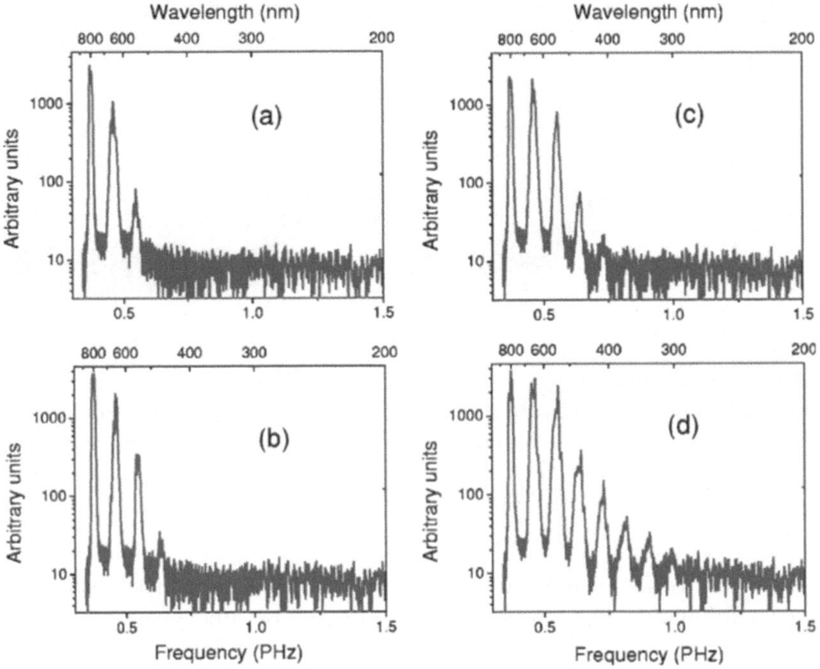

Fig. 7.19. HSRS spectra generated using two input pulses in methane at different pressures: (**a**) 100 mbar, (**b**) 300 mbar, (**c**) 600 mbar, and (**d**) 900 mbar. Differently from Fig. 7.18, the λ_1 pump pulse was stretched from $\tau \simeq 100$ fs to $\tau \simeq 400$ fs by propagating it through a 20 cm long fused silica block [111]

and higher, a blueshifted shoulder appears on the infrared pump pulse spectrum. This effect is attributed to self-phase-modulation (SPM). Its nonlinear coefficient n_2 of methane is larger by more than a factor 4 than that of hydrogen so that the effect due to SPM is more visible in Figs. 7.18(e) and (f).

This broadening effect is more and more important for higher pressures. As can be seen in Figs. 7.18(e) and (f), the broadening spans the spectral region of the first three or four anti-Stokes lines. When this happens, it is clearly possible to see weak Raman sidebands superimposed on the quasicontinuous structure. This could be due to the fact that the radiation generated by means of SPM starts stimulating the Raman process, giving rise to the generation of Raman sidebands. The resulting spectrum is therefore a combination of both SPM and SRS.

After propagation through a 20 cm long silica rod, the pulse duration was stretched from $T \simeq 100$ fs to $T \simeq 400$ fs. All the other parameters (pulse energies and λ_2 pulse duration) were left unchanged compared to the results mentioned above. The anti-Stokes spectra generated in this configuration are shown in Fig. 7.19 for different pressures up to 900 mbar. Many anti-Stokes sidebands are efficiently generated, with an efficiency comparable to the case

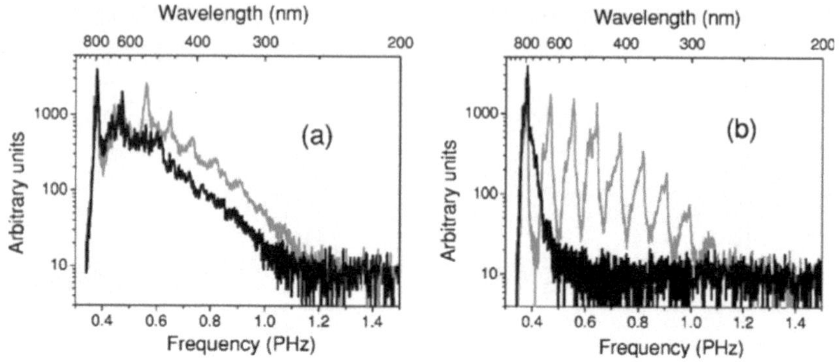

Fig. 7.20. HRSR spectra generated in methane at a pressure of 3 bar: (**a**) Output spectra using a short (100 fs) pulse and (**b**) a longer (400 fs) pulse. In each graph, the grey curve represents the output spectrum using two input pulses, i.e., the first two lines form the left, and the other components are Raman sidebands, while the black curves represent those obtained using only the infrared λ_1 pump pulse [111]

of the shorter λ_1 pump pulse of $T \simeq 100$ fs. In the present case, however, the generated sidebands do not appear to be affected by any spectral broadening, as was the case before. This is even more evident in Fig. 7.20(b), where the spectrum generated with a gas pressure of 3 bar is shown. The result obtained with the same pressure and shorter λ_1 pulse ($T \simeq 100$ fs) is also shown in Fig. 7.20(a) for comparison. The output spectrum obtained with the λ_1 pulse only is also shown in Fig. 7.20 without the λ_2 pulse for both values of T. The spectral broadening due to the SPM is almost negligible in the case of longer pulse duration compared to the case of shorter duration.

Efficient subfemtosecond pulse generation looks promising by superposing the vibrational sidebands of molecular hydrogen and methane.

7.3.3 Impulsive Regime

Generation of multiple phase-locked Stokes and anti-Stokes components is made possible in an impulsively excited Raman medium [113]. In this regime, all the nonlinear effects involved in the medium preparation (molecule excitation, nonlinear self-phase-modulation, etc.) are confined only within the pumping process, and the resulting spectral broadening of a relatively weak delayed (injection) pulse is linear in the field and therefore completely controllable. This process is schematically drawn in Fig. 7.21. An intense pump pulse $\varepsilon_p(\tau)$ with duration $T_p < T_\nu \equiv 2\pi/\Omega_\nu$ (Ω_ν is the vibrational or rotational frequency) performs impulsive preparation of the molecules, and a delayed (injection) pulse $\varepsilon_i(\tau)$, with intensity well below the threshold intensity for nonlinear self-action in the medium, experiences linear scattering on the temporal modulations of the susceptibility $\chi_{\mathrm{Ram}}(\tau)$. Temporally, the phase of the injection field is modulated with the sinusoidal modulation law,

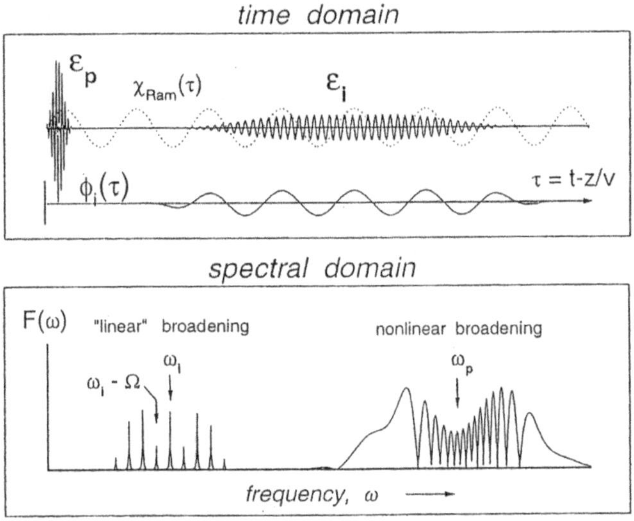

Fig. 7.21. A schematic of the linear regime of high-order SRS (stimulated Raman scattering). An intense pump pulse with duration $T_p < T_\nu = 2\pi/\Omega_\nu$ (Ω_ν is the vibrational or rotational frequency) performs impulsive preparation of the molecules, and a delayed (injection) pulse, experiences linear scattering on the temporal modulations of its susceptibility $\chi_{\rm Ram}(t)$. In the spectral domain, this leads to the generation of a comb of sideband components $\omega_n = \omega_0 \pm n\Omega_\nu$. In the regime of "linear" broadening, the intensity of a delayed incident pulse is well below the threshold intensity for nonlinear broadening [112]

$\partial\phi_i/\partial t = \alpha \sin\,(\Omega_\nu t)$, which leads, in the spectral domain, to the generation of a comb of sideband components $\omega_n = \omega_i \pm n\Omega_\nu$ ($n = 1, 2, \cdots$) with an intensity proportional to the squared nth order Bessel function [113]. In the experiments by Nazarkin et al. [113], they used a Ti:sapphire chirped pulse amplification laser system at 1 kHz repetition rate. The output pulses at 800 nm (1.5 eV) had an energy up to 500 μJ and a pulse width of 30 fs and could impulsively excite the symmetric vibrational mode A_{1g} of SF_6 with $\Omega_\nu = 755\,\text{cm}^{-1}$ (vibrational period $T_\nu = 44$ fs). A part of the fundamental was frequency doubled with a 1 mm thick type-I BBO (beta-barium-borate) nonlinear crystal. The used output energy at 400 nm was measured to be 3 μJ with a pulse width of 200 fs at full width half maximum (FWHM). This output was used as the delayed injection pulse with $\lambda_i = 400$ nm. The typical value of the pump pulse intensity in the gas (SF_6) filled waveguide (diameter 250 μ m, length 1 m) was about $10\,\text{TW/cm}^2$. The highly symmetric, spherical top SF_6 molecule exhibits no Raman rotational spectrum. At the same time, the totally symmetric vibrational mode A_{1g} of SF_6 gives rise to a very strong (compared to other two Raman modes) Raman line at $775\,\text{cm}^{-1}$. One can expect therefore that only one Raman-active mode of SF_6 ($775\,\text{cm}^{-1}$, vibrational period $T_\nu = 44$ fs) will be effectively excited by the impulsive pumping.

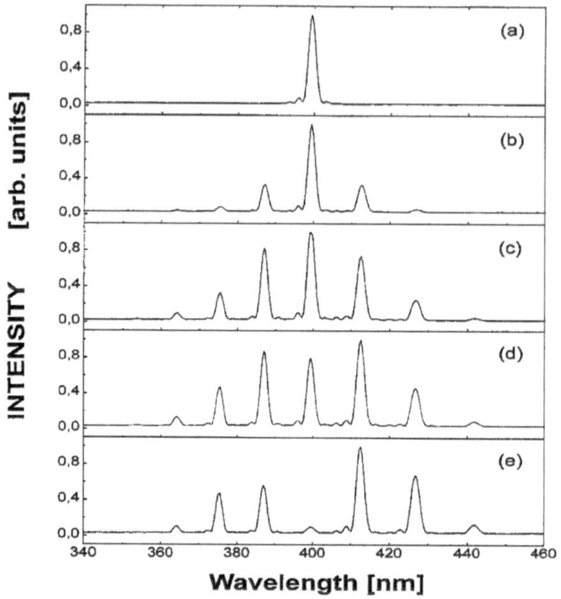

Fig. 7.22. The measured output Raman spectra of the 400 nm injection pulse for increasing values of SF_6 pressure: (**a**) $p = 0$, (**b**) $p = 346$ mbar, (**c**) $p = 395$ mbar, (**d**) $p = 410$ mbar, and (**e**) $p = 470$ mbar. The energy of the 800 nm 30 fs-pump pulse is 130 μJ [113]

The intense 800 nm, 30 fs-pump pulse (130 μJ) was focused on the tip of the waveguide filled with SF_6. The second (injection) pulse at 400 nm with a duration ∼200 fs and an energy much lower (∼30 μJ) than that of the pump, propagated with a temporal delay of several 100 fs. Figure 7.22 shows the output injection pulse spectra for increasing values of pressure $P = 0$–470 mbar. The well-pronounced equidistant character of the lines in Fig. 7.22 proves that the scattering of the injection field has been induced by the symmetric vibrational mode of 775 cm^{-1}.

In the present regime of SRS, the Stokes and anti-Stokes components are generated with nearly equal efficiency as shown in Figs. 7.22(a)–(d). Initially, with an increase of gas pressure, the number of components increases, and the intensity of the components falls off monotonically with the component order (see Figs. 7.22a–c). Further increase of gas pressure, however, breaks this monotonic distribution. The energy of the central component at the injection frequency is seen to be further converted to the sideband frequencies, even though its intensity is getting smaller than the intensity of the sideband components. As a result, a nearly 100% conversion efficiency into the Stokes and anti-Stokes components is achieved. These signals are converted into a train of short pulses.

Impulsive stimulated Raman scattering (ISRS) belongs also to the impulsive regime [119]. Here optical control over elementary molecular motion is enhanced with timed sequences of femtosecond pulses produced by pulse-shaping techniques. Appropriately timed pulse sequences are used to repetitively drive selected vibrations of a crystal lattice, e.g., an α-perylene crystal, to build up a large oscillation amplitude.

The experiment is arranged in a transient grating geometry. A colliding pulse mode-locked (CPM) ring dye laser and copper vapor laser-pumped dye amplifier system provide 75 fs, 620 nm, 5 μJ pulses at an 8.6kHz repetition rate. A small portion of the amplified output is split off to serve as the probe beam; the remaining portion is converted into a suitable terahertz-rate pulse sequence by the pulse-shaping apparatus [119] and split to yield the two excitation beams that overlap temporally and spatially inside the sample. The vibrational response is monitored by measuring the time-dependent diffraction intensity of the probe pulse, whose arrival time at the sample is varied with a stepping motor-controlled delay line. A sequence of femtosecond pulses with a repetition rate of 2.39 THz (419 fs between pulses) was produced though pulse-shaping techniques. Cross-correlation measurement of this sequence is shown Fig. 7.23(A). Figure 7.23(B) shows the ISRS data from the α-perylene organic molecular crystal driven by a sequence of pulses spaced at 419 fs to match the vibrational period of the 80 cm^{-1} mode. The diffracted signal from the mode grows stronger with each successive pulse. Selective amplification of the 80 cm^{-1} mode is demonstrated as a signal oscillating with twice the vibrational frequency. When a sequence of pulses spaced at 429 fs is slightly off resonance for the 80 cm^{-1} mode, the signal intensity is reduced, as shown in Fig. 7.23(C), relative to that of the resonant case (Fig. 7.23B). Figure 7.23(D) shows simulation of data in the case of Fig. 7.23(B). The zero of time is defined as the center of the input pulse train.

7.4 Multistep Coherent Anti-Stokes Raman Scattering in Crystals

Two intense femtosecond laser pulses having near-infrared frequencies are supplied as signal and idler from an OPO system pumped by 150 fs Ti:sapphire laser pulses. When the difference-frequency of the two pulses is set almost equal to the frequency of one of the Raman active phonon modes or two-phonon modes, coherent nonlinear optical responses are observed as novel phenomena characteristic of the crystals. These phenomena cannot also be described by conventional perturbational treatment.

The two incident pulses $(\omega_1, \boldsymbol{k}_1)$ and $(\omega_2, \boldsymbol{k}_2)$ excite resonantly the Raman active phonon mode $\omega_{\text{ph}} \sim \omega_1 - \omega_2$ with the wavevector $\Delta\boldsymbol{k} = \boldsymbol{k}_1 - \boldsymbol{k}_2$. Increasing the pump power of these pulses beyond a critical value, resonantly created phonons in a mode $(\omega_{\text{ph}}, \Delta\boldsymbol{k})$ persists as a coherently propagating

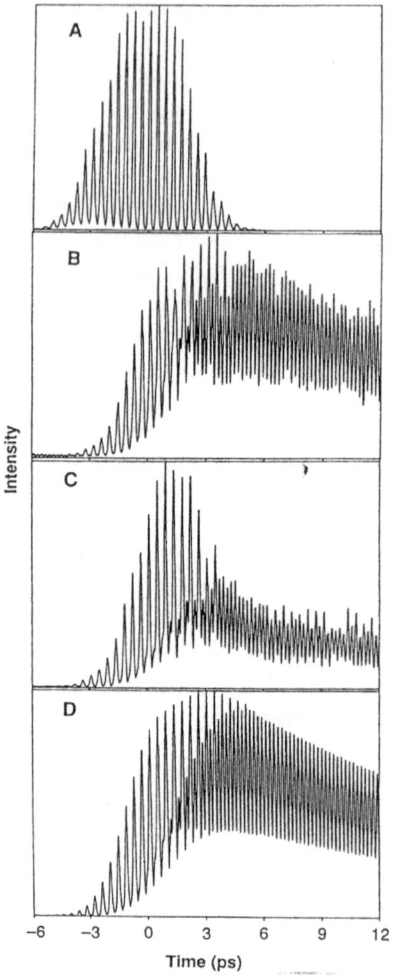

Fig. 7.23. (A) Cross-correlation measurement of a sequence of femtosecond pulses with a repetition rate of 2.39 THz (419 fs between pulses) produced through pulse-shaping techniques. (B) ISRS (impulsive stimulated Raman scattering) data from the α-perylene organic molecular crystal driven by a sequence of (b-polarized) pulses spaced at 419 fs to match the vibrational period of the 80 cm^{-1} mode. (C) ISRS data driven by a sequence of pulses spaced at 429 fs, slightly off resonance for the 80 cm^{-1} mode. (D) Simulation of data in (B), assuming a Gaussian temporal profile for the excitation pulse train [119]

wave or standing wave. The incident laser pulse is scattered repeatedly by the coherent phonon wave so that the multistep coherent anti-Stokes Raman scattering (CARS) signals are observable at $(\omega_1 + n\omega_{\mathrm{ph}},\ \boldsymbol{k} + n\Delta\boldsymbol{k})$. The case

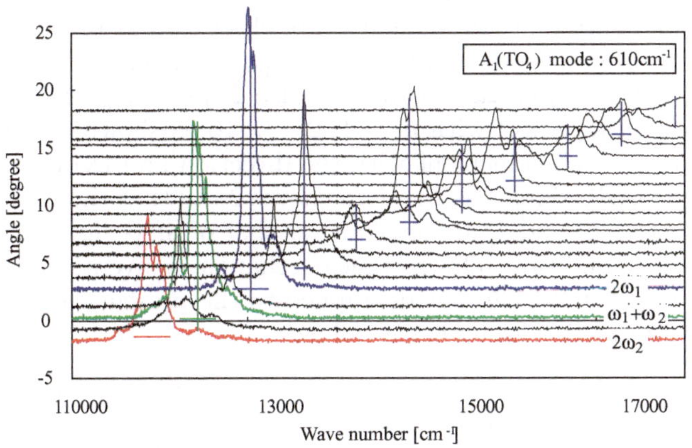

Fig. 7.24. Multistep CARS signals in addition to the second harmonics and sum-frequency generation signals are observed depending upon the observing angle in $KNbO_3$. The direction of the $\boldsymbol{k}_1$ incident beam is chosen as zero angle. Two incident beams: $\omega_1 = 6565\,\mathrm{cm}^{-1}$ ($3.0\,\mu$J/pulse) and $\omega_2 = 5952\,\mathrm{cm}^{-1}$ ($3.0\,\mu$J/pulse) have the polarization parallel to x-axis. A_{1g} phonon of $610\,\mathrm{cm}^{-1}$ is resonantly pumped [120]

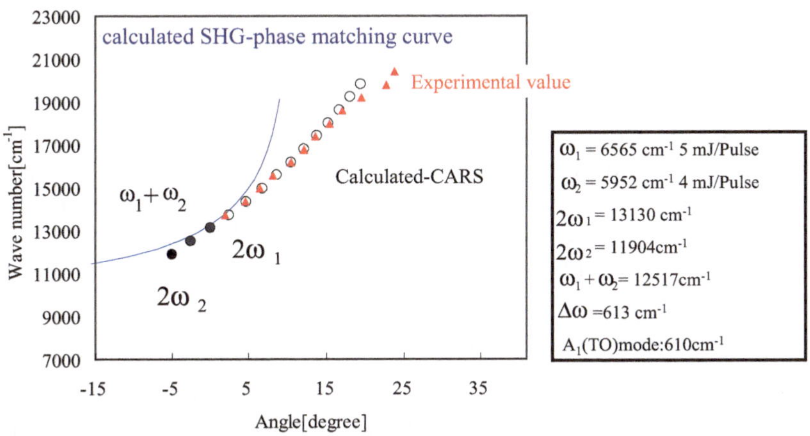

Fig. 7.25. The peak frequency of multistep CARS signals of $KNbO_3$ are plotted as a function of the observing angle. Ten CARS signals are observed in addition to SHG at $2\omega_1$ and $2\omega_2$ and the sum-frequency $\omega_1 + \omega_2$ [120]

of $n = 1$ was already discussed in Sect. 6.2.1. These multistep CARS signals were observed in $SrTiO_3$, $KTaO_3$, $YFeO_3$, $LiNbO_3$, and $KNbO_3$ crystals.

Ferroelectric crystals $LiNbO_3$ and $KNbO_3$ without inversion symmetry can show strong second-harmonic signals under the phase-matching condition. The signal pulse ($\omega_1 = 6561\,\mathrm{cm}^{-1}$) and the idler pulse ($\omega_2 = 5945\,\mathrm{cm}^{-1}$) are so focused as they overlap temporally and spatially inside the crystal and both wavevectors $\boldsymbol{k}_1$ and $\boldsymbol{k}_2$ satisfy the phase-matching condition for SHG. The frequencies of the parametric oscillator are chosen so as to resonantly excite the strongest Raman mode A_1(TO) with $\omega_{ph} = 610\,cm^{-1}$ in $KNbO_3$. Then not only $2\omega_2$, $\omega_1 + \omega_2$ and $2\omega_1$ signals but also $2\omega_1 + m\omega_{ph}$ are observed in the direction $k_1 + m(\boldsymbol{k}_1 - \boldsymbol{k}_2)$ with $m = 1, 2, \ldots, 10$ as shown in Fig. 7.24 [120]. These signals are well coincident with the direction and frequency calculated for multistep CARS signals as Fig. 7.25 shows. Note that the multistep CARS signals ($m = 1$, 2, 3, and 4) are as strong as that of SHG signals satisfying the phase-matching condition. This fact means that the conventional perturbation treatment in Sect. 6.2.1 can not be applied to the present case. Therefore we should understand this argument as follows:

(1) Two incident beams create the phonon grating of the A_1(TO) mode with the wavevector $\Delta\boldsymbol{k} = \boldsymbol{k}_1 - \boldsymbol{k}_2$.
(2) The incident beam (ω_1, $\boldsymbol{k}_1$) is dynamically scattered, e.g., into a ($\omega_1 + m\omega_{ph}$, $\boldsymbol{k}_1 + m\Delta\boldsymbol{k}$) photon by mth order scattering.
(3) This signal and that of the incident beam can produce the sum-frequency signal ($2\omega_1 + m\omega_{ph}$, $2\boldsymbol{k}_1 + m\Delta\boldsymbol{k}$).
(4) On the other hand, the SHG signal ($2\omega_1$, $2\boldsymbol{k}_1$) is also scattered dynamically m-times by the phonon grating, producing the same multistep CARS signals at ($2\omega_1 + m\omega_{ph}$, $2\boldsymbol{k}_1 + m\Delta\boldsymbol{k}$).

Both processes (3) and (4) give the same signals. Note, however, that the process (4) gives signals over a wide spectrum only when the phase-matching condition of the fundamental ω_1 and the SGH $2\omega_1$ is satisfied, as in Fig. 7.24, while the process (3) gives signals only over the spectrum range in which the phase-matching condition is satisfied for ω_1 and $\omega_1 + m\omega_{ph}$. This result suggests the possibility of generating subfemtosecond short pulses in the visible region when these signal beams are manipulated to be coaxial and the spider method is used to compensate the effect of dispersion.

Dynamical Symmetry Breaking

Quantum paraelectric crystals $KTaO_3$ and $SrTiO_3$ can show novel nonlinear optical phenomena far beyond the perturbational treatment. Although these crystals are almost ferroelectric, quantum fluctuation of ions prevents the phase-transition to the ferroelectric state. This means that the phonon modes have large anharmonicity. Both $KTaO_3$ and $SrTiO_3$ of cubic perovskite oxide have the highest symmetry, O_h^1, as a crystal. Therefore, all (seven) Γ-point optical phonons are of odd mode, so that they cannot be observed by one-phonon Raman scattering (RS). Although single phonons are all Raman-inactive, two-phonon Raman scattering is allowed. The process originates for the most part from the Brillouin zone (BZ) edge because the density of states of phonons diverges there. In both crystals, strong RS signals due to simultaneous excitation of two phonons at the opposite edges of the BZ have been strongly observed, reflecting also strong anharmonicity of phonons [121, 122]. The RS peaks have been well assigned, and single-phonon frequencies at the

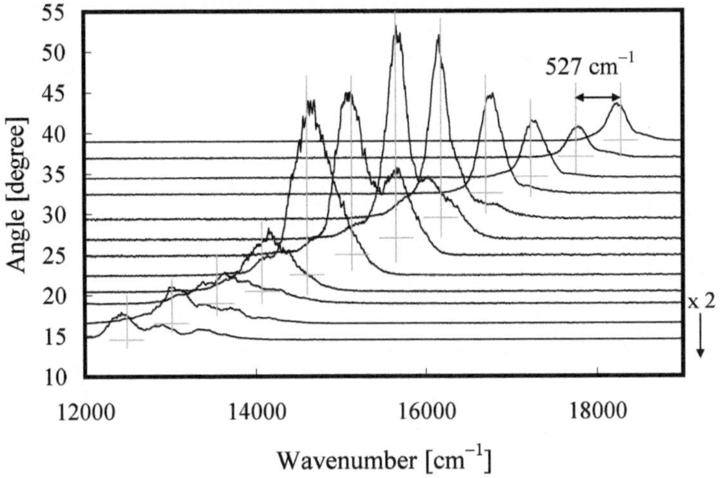

Fig. 7.26. The angular dependence of the multistep CARS spectra due to the originally Raman-forbidden $TO_4(X_5)$ phonon with frequency spacing $527\,\mathrm{cm}^{-1}$ in $KTaO_3$. The experimental condition is such that $\Delta\omega$=$770\,\mathrm{cm}^{-1}$ ($\omega_1 = 6760\,\mathrm{cm}^{-1}$ with $2.0\,\mu$J/ pulse and $\omega_2 = 5990\,\mathrm{cm}^{-1}$ with $3.0\,\mu$J/pulse) [121]

BZ edges have been derived from the mode assignments. All the phonon dispersion curves of $KTaO_3$ have been observed by using inelastic neutron scattering [123], and the result agrees well with that obtained from the assigned RS. When a combination of TO_4 and TO_2 modes at opposite BZ edges is resonantly pumped by two beams of $\omega_1 = 6770\,\mathrm{cm}^{-1}$ ($2.0\,\mu$J/ pulse) and $\omega_2 = 5900\,\mathrm{cm}^{-1}$ ($3.0\,\mu$J/pulse), a series of multistep CARS signals due to a single TO_4 phonon ($527\,\mathrm{cm}^{-1}$) at the BZ edge, which is originally Raman-inactive, is observed in the visible region as Fig. 7.26 shows.

This means symmetry-breaking for the signal of a single phonon Raman scattering. On the other hand, both conventional CARS and the symmetry-breaking single-phonon signals are observed to coexist beyond the critical pumping power. Figure 7.27 shows that the conventional CARS signal at $2\omega_1 - \omega_2$ increases as P_1^2, i.e., the square of the ω_1 incident power P_1 but no signal of the symmetry-breaking at $\omega_1 + TO_4$ is observed below the threshold power P_{th}. Beyond P_{th}, the first-order CARS signal intensity begins to saturate and the dynamical symmetry-breaking signal at $\omega_1 + TO_4$ increases rapidly. Two beams hit the crystal with an angle of several degrees so that the CARS signals and a symmetry-breaking single phonon signals depend on the observing degree measured from the $\boldsymbol{k}_1$ vector, as shown in Fig. 7.28. Note that the nth conventional CARS signal is observed in the direction $\boldsymbol{k}_1+n(\boldsymbol{k}_1 - \boldsymbol{k}_2)$, which is denoted by the closed (experimental) and open (calculated) squares,

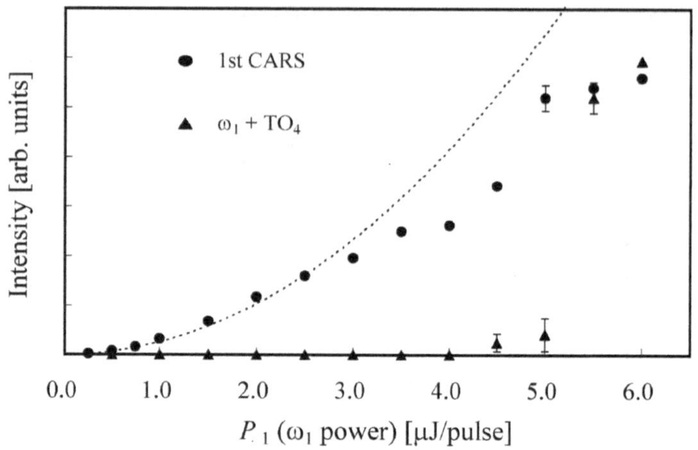

Fig. 7.27. Dependences of intensity of conventional first-order CARS signal and dynamical symmetry-breaking signal at $\omega_1 + TO_4$ on ω_1-pulse power of P_1 in $KTaO_3$. Note the quasithreshold power for the $\omega_1 + TO_4$ signal and the P_1^2-dependence for the conventional first-order CARS signal. A value of 4.0 μJ for the pulse corresponds to an excitation photon density of 1.0×10^{11} W/cm^2 [122]

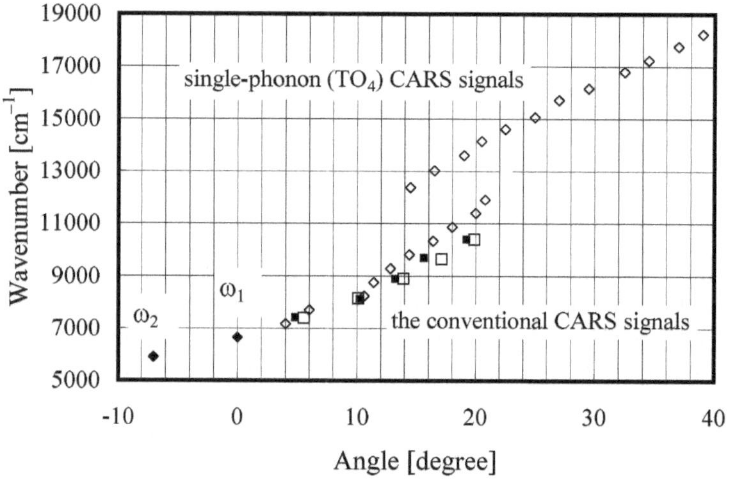

Fig. 7.28. Angle vs. frequency relation for observed multistep CARS signals of $KTaO_3$ shown in Fig. 7.26 under pumping of the $TO_4 + TO_2$ mode at the BZ edge. The open diamonds indicate the signals with the spacing of the TO_4 mode frequency at the BZ edge; the data points in the lower and upper branches correspond to the signals having frequencies $\omega_1 + m\mathrm{TO}_4$ ($m = 1, 2, \ldots, 10$) shown in Fig. 7.26. The closed and open squares indicate the observed and calculated conventional CARS signals, respectively, with the frequencies $\omega_1 + n\Delta\omega$ ($\Delta\omega = 749\,\mathrm{cm}^{-1}$, $n = 1, 2, \ldots, 5$) [122]

and that the dynamical symmetry-breaking signal (denoted by the open diamond) is observed with the constant neighboring angle $\Delta\boldsymbol{k}/2$ independent of the order n. Note here that $\boldsymbol{\Delta k}/2$ is a vector component parallel to the surface of $\boldsymbol{k}_1 - \boldsymbol{k}_2$. This phenomenon is explained as follows:

(1) When we pump resonantly, e.g., a combination of TO_4 (at the $(-X)$-point) and TO_2 (at the X-point) or vice versa, coherent standing waves of TO_4 and TO_2 modes at the X-point persist for incident powers I_1 and I_2 beyond threshold.
(2) Once the X-point phonons are condensed, the unit cell is doubled in the X-direction because the phases of vibrational modes in the neighboring unit cells differ by π.
(3) All phonon pairs are given the difference vector $\Delta\boldsymbol{k}$ of two incident beams so that the single mode of a phonon is given $\Delta\boldsymbol{k}/2$. Therefore the phonon grating has the wavevector of $\Delta\boldsymbol{k}/2$.
(4) When the nth signal ω_n with wavevector $\boldsymbol{k}_n$ is scattered dynamically by this phonon grating, the $(n+1)$th signal $\omega_n + \omega(TO_4)$ is observed in the direction $\boldsymbol{k}_n + \Delta\boldsymbol{k}/2$.

The existence of the threshold for this dynamical symmetry-breaking looks reasonable because the creation rate of the phonon pair should overcome the dissipation rate of these phonons. The dynamical symmetry-breaking is understood as one of nonlinear optical responses which is beyond the perturbational treatment of the electron–photon interaction and is characteristic of the crystal.

References

Chapter 1

1. C. Cohen-Tannoudji, J. Dupont-Roc, G. Grynberg: *Photons and Atoms, Introduction to Quantum Electrodynamics* (John Wiley, 1989)
2. C. Cohen-Tannoudji, J. Dupont-Roc, G. Grynberg: *Atom–Photon Interactions. Basic Processes and Applicatoiins* (John Wiley, 1992)
3. P.A. Dirac: Proc. Royal Soc. London Sec A **114**, 243 (1927)
4. W. Heitler: *The Quantum Theory of Radiation*, 3rd edition (Oxford University Press, 1954)
5. P.L. Knight, L. Allen: *Concept of Quantum Optics* (Pergamon, 1983)
6. R. Loudon: *The Quantum Theory of Light* (Oxford University Press, 1973)
7. A. Messiah: *Quantum Mchanics*, (John Wiley, 1958); Dover edition (Dover, 1999)
8. L.-A. Wu, H.J. Kimble, J.L. Hall, H. Wu: Phys. Rev. Lett. **57**, 2520 (1986)
9. S. Machida, Y. Yamamoto, Y. Itaya: Phys. Rev. Lett. **58**, 1000 (1987)
10. Y. Yamamoto, N. Imoto, S. Machida: Phys. Rev. A **33**, 3243 (1986)

Chapter 2

11. See, e.g., C. Cohen-Tannoudji, J. Dupont-Roc, G. Grynberg: *Photons and Atoms* (John Wiley, 1989)
12. A. Einstein: Phys. Z **18**, 121 (1917)
13. D. Kleppner: Physics Today **58**(2), 30 (2005)
14. W. Heitler: *The Quantum Theory of Radiation* (Clarendon Press, 1954)
15. R.G. Hulet, E.S. Hilfer, D. Kleppner: Phys. Rev. Lett. **55**, 2137 (1985)
16. D.J. Heinzen, J.J. Childs, J.E. Thomas, M.S. Feld: Phys. Rev. Lett. **58**, 1320 (1987)
17. T.W. Hänsch, A. L. Schawlow: Opt. Commun. **13**, 68 (1975)
18. For example, see S. Chu: Rev. Mod. Phys. **70**, 685 (1998)
19. For example, see W.D. Phillips: Rev. Mod. Phys. **70**, 7020 (1998)
20. For example, see C.N. Cohen-Tannoudji: Rev. Mod. Phys. **70**, 707 (1998)
21. M.H. Anderson, J.R. Ensher, M.R. Mathews, C.E. Wieman, E.A. Cornell: Science **269**, 198 (1995)

22. C.C. Bradley, C.A. Sackett, J.J. Tollett, R.G. Hulet: Phys. Rev. Lett. **75**, 1687 (1995)
23. K.B. Davis, M.-O. Mewes, M.R. Andrewa, N.J. van Druten, D.S. Durfee, D.M. Kurn, W. Ketterle: Phys. Rev. Lett. **75**, 3969 (1995)
24. M. Greiner, C.A. Regal, D.S. Jin: Nature **426**, 537 (2003)
25. S. Jochim, M. Bartenstein, A. Altmeyer, G. Hendel, S. Riedl, C. Chin, J. Hecker Denschlag, R. Grimm: Science **302**, 2101 (2003)
26. N.M. Zwierleim, C.A. Stan, C.H. Schunck, S.M.F. Raupack, S. Gupta, Z. Hadzibabic, W. Ketterle: Phys. Rev. Lett. **91**, 250401 (2003)
27. C.A. Regal, M. Greiner, D.S. Jin: Phys. Rev. Lett. **92**, 040403 (2004)

Chapter 3

28. R. Hanbury-Brown, R.Q. Twiss: Nature **177**, 27 (1956)
29. F.T. Arrecchi, E. Gatti, A. Sona: Phys. Lett. **20**, 27 (1966)
30. P.L. Kelly, W.H. Kleiner: Phys. Rev. **A136**, 316 (1964)
31. E. Jakeman, C.J. Oliver, E.R. Pike: J. Phys. **A1**, 406 (1968)

Chapter 4

32. R. Loudon: *The quantum Theory of Light*, 2nd edition (Oxford University Press, 1975)
33. L.S. Slater: *Confluent Hypergeometric Functions* (Cambridge University Press, 1960)
34. F.T. Arrecchi, G.S. Rodari, A. Sona: Phys. Lett. **25A**, 59 (1967)
35. M. Sargent III, M.O. Scully, W.E. Lamb Jr.: *Laser Physics* (Addison Wesley, 1974)
36. S. Sugano, Y. Tanabe, H. Kamimura: *Multiplets of Transition Metals in Crystal* (Academic Press, 1970)
37. R. Powell: *Physics of Solid-State Laser Materials* (AIP Press, 1998)
38. J.C. Walling, O.G. Peterson, H.P. Jenssen, R.C. Morris, E.W. O'Dell: IEEE J. Quantum Electron. **QE 16**, 1302 (1980)
39. P.F. Moulton: J. Opt. Soc. Am. B. **3**, 125 (1986)
40. S. Watanabe: Parity (in Japanese) **5**, 28 (1990)

Chapter 5

41. M.D. Perry, G. Mourou: Science **264**, 917 (1994)
42. G.A. Mourou, C.P.J. Barty, M.D. Perry: Physics Today, January 1998, pp. 22. This review article contains important references newer than [41].
43. D. Strickland, G. Mourou: Opt. Commun. **56**, 219 (1985)
44. P. Maine, D. Strickland, P. Bado, M. Pessot, G. Mourou: IEEE J. Quantum Electron. **24**, 398 (1988)
45. L.L. Mollenauer, W.J. Tomlinson: Opt. Lett. **8**, 186 (1983)
46. W.J. Tomlinson, R.H. Stolen, C.V. Shank: J. Opt. Soc. Am. **1**, 139 (1984)
47. B. Nikolaus, D. Grischkowsky: Appl. Phys. Lett. **43**, 228 (1983)
48. K. Yamane, T. Kito, R. Morita, M. Yamashita: in *Ultrafast Phenomena* VIV, ed. by T. Kobayashi *et al.* (Springer, 2004)

49. L.F. Mollenauer, R.H. Stolen, J.P. Gordon: Phys. Rev. Lett. **45**, 1095 (1980)
50. O.E. Martinez: IEEE J. Quantum Electron. **23**, 1385 (1987)
51. R.H. Dicke: Phys. Rev. **93**, 99 (1054)
52. N. Skribanowitz, I.P. Herman, J.C. Mac Gillivray, M.S. Feld: Phys. Rev. Lett. **30**, 309 (1973)
53. H.M. Gibbs, Q.H.F. Vrehen, H.M.J. Hikspoorso: Phys. Rev. Lett. **39**, 547 (1977); *Dissipative Systems in Quantum Optics*, p. 111 (Springer-Berlag, Berlin, 1982)
54. Q.E.F. Vrehen, M.F.H. Schuurmans: Phys. Rev. Lett. **42**, 224 (1979)
55. I.P. Herman, J.C. MacGilliuray, N. Skribanowitz, M.S. Feld: *Laser Spectroscopy* (Plenum Press, 1974)
56. E. Hanamura: Phys. Rev. **B37**, 1273 (1988)
57. E. Hanamura: Phys. Rev. **B38**, 1228 (1988)
58. T. Itoh, M. Furumiya, T. Ikehara: Solid State Commun. **73**, 271 (1990)
59. A. Nakamura, H. Yamada, T. Tokizaki: Phys. Rev. **B40**, 8585 (1989)

Chapter 6

60. P.D. Maker, R.W. Terhune, M. Nisenoff, C.M. Savage: Phys. Rev. Lett. **8**, 21 (1962)
61. J.-M. Halbout, S. Blit, W. Donaldson, T. Chung: IEEE J. Quantum Electron. **15**, 1176 (1979)
62. M.M. Fejer, G.A. Magel, D.H. Jundt, R.L. Byer: IEEE J. Quantum Electron. **28**, 2631 (1992) and references therein
63. R.L. Byer: J. Nonlinear Opt. Phys. Materials **6**, 549 (1997)
64. D. Madge, H. Mahr: Phys. Rev. Lett. **18**, 905 (1967)
65. R.L. Byer: in *Quantum Electronics*, ed. H. Rabin and C. L. Tang (Academic Press, 1975)
66. M.D. Levenson: IEEE J. Quantum Electron. **10**, 110 (1974)
67. R.K. Jain, R.C. Lind: J. Opt. Soc. Am. **73**, 647 (1983)
68. Y.R. Shen: *The Principles of Nonlinear Optics* (John Wiley, 1984)
69. M. Dagenais, W. Sharfin: J. Opt. Soc. Am. **B2**, 1179 (1985)
70. E. Hanamura: Phys. Rev. **B37**, 1273 (1988)
71. E. Hanamura: Phys. Rev. **B38**, 1228 (1988)
72. Y. Masumoto, M. Yamazaki, H. Sugawara: Appl. Phys. Lett. **53**, 1527 (1988)
73. M. Kuwata-Gonokami: J. Lumin. **38**, 247 (1987)
74. M. Inoue, Y. Toyozawa: J. Phys. Soc. Jpn. **20**, 363 (1965)
75. O. Akimoto, E. Hanamura: J. Phys. Soc. Jpn. **33**, 1537 (1972)
76. E. Hanamura: Solid State Commun. **12**, 951 (1972)
77. G.M. Gale, A. Mysyrowicz: Phys. Rev. Lett. **32**, 727 (1974)
78. E. Hanamura: J. Phys. Soc. Jpn. **39**, 1506 and 1516 (1975)
79. Vu Duy Phach, R. Lévy: Solid State Commun. **29**, 247 (1979)
80. E. Hanamura, N.T. Dan, Y. Tanabe: Phys. Rev. **B62**, 7033 (2000)
81. E. Hanamura, N.T. Dan, Y. Tanabe: J. Phys.: Condens. Matter **12**, L345 (2000)
82. A. Schülzgen, Y. Kawabe, E. Hanamura, A. Yamanaka, P.-A. Blonche, J. Lee, N.T. Dan, H. Sato, M. Naito, S. Uchida, Y. Tanabe, N. Peyghambarian: Phys. Rev. Lett. **86**, 3164 (2001)
83. T. Iizuka-Sakano, E. Hanamura, Y. Tanabe: J. Phys. Condens. Matter, **13**, 3031 (2001)

84. D. Fröhlich, S. Leute, V.V. Pavlov, R.V. Pisarev: Phys. Rev. Lett. **81**, 3239 (1998)
85. D. Fröhlich, S. Leute, V.V. Pavlov, R.V. Pisarev, K. Kohn: J. Appl. Phys. **85** 4762 (1999)
86. M. Fiebig, T. Lottermoset, D. Fröhlich, Goltsev, R.V. Pisarev: Nature **419**, 818 (2002)
87. K.L. Yakel, W.C. Koehler, E.F Bertaut, E.F. Forrat: Acta Crystallogr. **16**, 27 (1963)
88. E. Hanamura, Y. Tanabe: in *Magnetoelectric Interaction phenomena in crystals*, ed. by M. Fiebig *et al.* (Kluwer Academic, 2004) pp. 151–161
89. E. Hanamura, Y. Tanabe: J. Phys. Soc. Jpn **72**, 2959 (2003)
90. G.A. Smolenskii, I.E. Chupis: Soviet Phys. Usp. **25**, 475 (1983)
91. M. Muto, Y. Tanabe, T. Iizuka-Sakano, E. Hanamura: Phys. Rev. **B57**, 9586 (1998)
92. Y. Tanabe, M. Muto, M. Fiebig, E. Hanamura: Phys. Rev. **B58**, 8654 (1998)
93. M. Fiebig, D. Fröhlich, B.B. Krichevtsov, R.V. Pisarev: Phys. Rev. Lett. **73**, 2127 (1994).

Chapter 7

94. H.C. Kapteyn, M.M. Murnane, I.P. Christov: Physics Today, March 2005, pp. 39
95. R.W. Boyd: *Nonlinear Optics*, 2nd edition (Academic Press, 2003)
96. A. L'Huillier, P. Balcou: Phys. Rev. Lett. **70**, 774 (1993)
97. J.J. Macklin, J.D. Kmetec, C.L. Gordon III: Phys. Rev. Lett. **70**, 766 (1993)
98. J.L. Krause, K.J. Schafer, K.C. Kulander: Phys. Rev. Lett. **68**, 3535 (1992)
99. M. Lewenstein. P. Balcou, M.Yu. Ivanov, A. L'Huillier, P.B. Corkum: Phys. Rev. **A49**, 2117 (1994)
100. G.A. Mourou, C.P.J. Barty, M.D. Perry: Physics Today, January 1998, pp. 22
101. J. Zhou, J. Peatross, M.M. Murane, H. Kapteyn: Phys. Rev. Lett. **76**, 752 (1996)
102. A. Paul, R.A. Bartels, R. Tobey, H. Green. S. Weiman, I.P. Christov, M.M. Murnane, H.C. Kapteyn, S. Backus: Nature **421**, 51 (2003)
103. E.A. Gibson, A. Paul, N. Wagner, R. Tobey, D. Gaudiosi, S. Backus, I.P. Christov, A. Aquila, E.M. Gullikson, D.T. Attwood, M.M. Murnane, H.C. Kapteyn: Science **302**, 95 (2003)
104. E.A. Gibson, A. Paul, N. Wagner, R. Tobey, S. Backus, I.P. Christov, M.M. Murnane, H.C. Kapteyn: Phys. Rev. Lett. **92**, 033001 (2004)
105. M. Hentshel, R. Kienberger, C. Spielmann, G.A. Reider, N. Milesevic, T. Brabec, P. Corkum, U. Heinzmann, M. Dreschen, F. Krausz: Nature **414**, 509 (2001)
106. P.M. Paul, E.S. Toma, P. Bregen, G. Mullot, F. Augé, P. Balcou, H.G. Muller, P. Agostini: Science **292**, 1689 (2001)
107. P. Tzallas, D. Charalambidis, N.A. Papadogiannis, K. Witte, G.D. Trakiris: Nature **426**, 269 (2003)
108. A. Baltuška, T. Udem, M. Uiberacker, M. Hentschel, E. Goulielmakis, C. Gohle, R. Holzwarth, Y.S. Yakovlev, A. Scrinzi, T.W. Hänsch, F. Krausz: Nature **421**, 611 (2003)
109. S.E. Harris, A.V. Sokolov: Phys. Rev. Lett. **81**, 2894 (1998)

110. A. Nazarkin, G. Korn, M. Wittmann, T. Elsaesser: Phys. Rev. Lett. **83**, 2560 (1999)
111. E. Sali, P. Kinsler, G.H.C. New, K.J. Mendham, T. Halfmann, J.W.G. Tisch, J.P. Marangos: Phys. Rev. **A72**, 013813 (2005)
112. A. Nazarkin, G. Korn, T. Elsaesser: Opt. Commun. **203**, 403 (2002)
113. A. Nazarkin, G. Korn, M. Wittmann, T. Elsaesser: Phys. Rev. Lett. **83**, 2560 (1999)
114. A.V. Sokolov, D.R. Walker, D.D. Yavuz, G.Y. Yin, S.E. Harris: Phys. Rev. Lett. **85**, 562 (2000)
115. M.Y. Skverdin, D.R. Walker, D.D. Yavuz, G.Y. Yin, S.E. Harris: Phys. Rev. Lett. **94**, 033904 (2005)
116. M. Katsuragawa, K. Yokoyama, T. Onose, K. Misawa: Opt. Exp. **13**, 5628 (2005)
117. K. Hakuta, M. Suzuki, M. Katsuragwa, J.Z. Li: Phys. Rev. Lett. **79**, 209 (1997)
118. S.E. Harris: Physics Today July (1997) p. 36
119. A.M. Weiner, D.E. Leaird, G.P. Wiederrecht, K.A. Nelson: Science **247**, 1317 (1990)
120. H. Matsuki: private communication
121. E. Matsubara, K. Inoue, E. Hanamura: Phys. Rev. **B73**, 134101 (2005)
122. E. Matsubara, K. Inoue, E. Hanamura: J. Phys. Soc. Jpn. **75**, 024712 (2006)
123. C.H. Perry, R. Currat, H. Buhay, R.M. Migoni, W.G. Stirling, J.D. Axe: Phys. Rev. **B39**, 8666 (1989)

Index